T0201285

Hydrophilic Interaction Chromatography

A Guide for Practitioners

Edited by

BERNARD A. OLSEN
BRIAN W. PACK

A JOHN WILEY & SONS, INC., PUBLICATION

Cover design: John Wiley & Sons, Inc.
Cover illustration: Copyright Brian W. Pack

Published by John Wiley & Sons, Inc., Hoboken, New Jersey
Published simultaneously in Canada

For general information on our other products and services or for technical support, please
contact our Customer Care Department within the United States at (800) 762-2974, outside the
United States at (317) 572-3993 or fax (317) 572-4002.

Wiley also publishes its books in a variety of electronic formats. Some content that appears in
print may not be available in electronic formats. For more information about Wiley products,
visit our web site at www.wiley.com.

Library of Congress Cataloging-in-Publication Data:

Olsen, Bernard A., 1953– author.
  Hydrophilic interaction chromatography : a guide for practitioners / Bernard A. Olsen,
Brian W. Pack.
        pages cm
  Includes bibliographical references and index.
  ISBN 978-1-118-05417-8 (hardback)
  1. Hydrophilic interaction liquid chromatography.   I. Pack, Brian W., 1970– author.
II. Title.
  QD79.C454O47   2013
  543'.84–dc23

                                                                   2012027157

Printed in the United States of America

10 9 8 7 6 5 4 3 2 1

CONTENTS

CHAPTER 4 PHARMACEUTICAL APPLICATIONS OF HYDROPHILIC INTERACTION CHROMATOGRAPHY   **111**

Bernard A. Olsen, Donald S. Risley, V. Scott Sharp, Brian W. Pack, and Michelle L. Lytle

PREFACE

The popularity of hydrophilic interaction chromatography (HILIC) has grown rapidly in recent years. The HILIC mode can provide retention and separation of polar compounds that are difficult to analyze by reversed-phase high-performance liquid chromatography (RP-HPLC) or other means. HILIC has been utilized for a wide variety of applications including drugs and metabolites in biological fluids, biochemicals, pharmaceuticals (from drug discovery to quality control), foods, and environmental. Multiple HILIC stationary phases have been developed, and methods employing mass spectrometric detection or HILIC as part of two-dimensional separation systems are becoming more common.

Many researchers do not have an extensive background or experience with HILIC, particularly as compared with RP-HPLC. Several established references are available for information on RP-HPLC theory, mechanisms, and method development. Despite the recent growth in the use of HILIC, less information is available to guide potential practitioners in the understanding and development of robust HILIC separations. The lack of familiarity with HILIC can lead to trial-and-error method development and perhaps less-than-optimal results for a given application.

We sought to compile a book that would be a good reference for HILIC fundamentals as well as to provide a broad overview of popular areas of application. Our goal was to provide a resource for several important topics to those who want to explore HILIC as a separation mode. We believe that a basic understanding of retention mechanisms and the impact of stationary phase and mobile phase properties on separations can lead to more efficient and effective development of robust separation methods.

The first three chapters of the book are devoted to HILIC retention mechanisms, stationary phases, and general aspects of method development. These chapters provide a foundation for subsequent chapters dealing with different areas of application. The application chapters focus on specific areas of interest to workers in the respective fields being addressed. Unique separation challenges are presented for bioanalytical, environmental, pharmaceutical, and biochemical applications, as well as a thorough discussion of HILIC in two-dimensional chromatography. Illustrative examples of several HILIC methods and development approaches are highlighted, and references for further details are provided.

We are indebted to all the authors who contributed to the book. We believe they provided discussions of their subject area in a concise fashion with minimal redundancy with the other chapters. Research to understand the fundamentals of HILIC separations and further application of HILIC to analytical problems will certainly continue. Our goal is that this book will be a useful reference for current and future HILIC practitioners.

We are also grateful to Dr. Mark F. Vitha, Chemical Analysis Series Editor, for his valuable help throughout the preparation of the book.

November 7, 2013   BERNARD A. OLSEN
                                                          BRIAN W. PACK

CONTRIBUTORS

Alfonso Espada, Analytical Technologies Department, Centro de Investigacion Lilly S.A., Madrid, Spain

Aikaterini M. Gremilogianni, Laboratory of Analytical Chemistry, Department of Chemistry, University of Athens, Athens, Greece

Stephen R. Groskreutz, Department of Chemistry, Gustavus Adolphus College, St. Peter, MN

Yong Guo, School of Pharmacy, Fairleigh Dickinson University, Madison, NJ

Mohammed E.A. Ibrahim, Department of Chemistry, University of Alberta, Gunning/Lemieux Chemistry Centre, Edmonton, Alberta, Canada

Michael A. Koupparis, Laboratory of Analytical Chemistry, Department of Chemistry, University of Athens, Athens, Greece

Charles A. Lucy, Department of Chemistry, University of Alberta, Gunning/Lemieux Chemistry Centre, Edmonton, Alberta, Canada

Michelle L. Lytle, Analytical Sciences Research and Development, Lilly Research Laboratories, A Division of Eli Lilly and Company, Indianapolis, IN

David V. McCalley, Department of Applied Sciences, University of the West of England, Bristol, UK

Nikolaos C. Megoulas, Laboratory of Analytical Chemistry, Department of Chemistry, University of Athens, Athens, Greece

Hien P. Nguyen, Department of Chemistry and Biochemistry, The University of Texas at Arlington, Arlington, TX

Bernard A. Olsen, Olsen Pharmaceutical Consulting, LLC, West Lafayette, IN

Brian W. Pack, Analytical Sciences Research and Development, Lilly Research Laboratories, A Division of Eli Lilly and Company, Indianapolis, IN

Fred Rabel, ChromHELP, LLC, Woodbury, NJ

Donald S. Risley, Pharmaceutical Sciences Research and Development, Lilly Research Laboratories, A Division of Eli Lilly and Company, Indianapolis, IN

Kevin A. Schug, Department of Chemistry and Biochemistry, The University of Texas at Arlington, Arlington, TX

V. Scott Sharp, Pharmaceutical Sciences Research and Development, Lilly Research Laboratories, A Division of Eli Lilly and Company, Indianapolis, IN

Dwight R. Stoll, Department of Chemistry, Gustavus Adolphus College, St. Peter, MN

Mark Strege, Analytical Sciences Research and Development, Lilly Research Laboratories, A Division of Eli Lilly and Company, Indianapolis, IN

Heather D. Tippens, Department of Chemistry and Biochemistry, The University of Texas at Arlington, Arlington, TX

Xiande Wang, Johnson & Johnson, Raritan, NJ

CHAPTER

1

SEPARATION MECHANISMS IN HYDROPHILIC INTERACTION CHROMATOGRAPHY

DAVID V. MCCALLEY

Department of Applied Sciences, University of the West of England, Bristol, UK

1.1 INTRODUCTION

Hydrophilic interaction chromatography (HILIC) is a technique that has become increasingly popular for the separation of polar, hydrophilic, and ionizable compounds, which are difficult to separate by reversed-phase chromatography (RP) due to their poor retention when RP is used. HILIC typically uses a polar stationary phase such as bare silica or a polar bonded phase, together with an eluent that contains at least 2.5% water and >60% of an organic solvent such as acetonitrile (ACN). However, these values should not be regarded as definitive of the rather nebulous group of mobile and stationary phase conditions that are considered to constitute HILIC. Figure 1.1 shows the number of publications on HILIC between the years 1990 (when the term was first employed) and 2010 according to the Web of Knowledge [1] using the search terms "HILIC" or "hydrophilic interaction (liquid) chromatography." For the first 12 years or so, the number of publications remained between 1 and 15, but after this period, interest increased rapidly from 19 publications

Hydrophilic Interaction Chromatography: A Guide for Practitioners, First Edition.
Edited by Bernard A. Olsen and Brian W. Pack.
© 2013 John Wiley & Sons, Inc. Published 2013 by John Wiley & Sons, Inc.

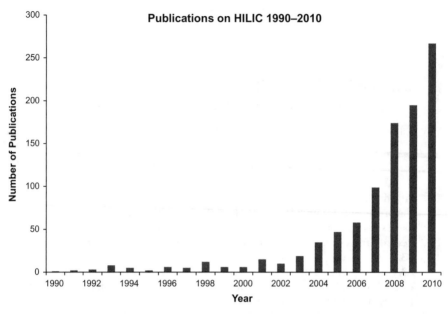

Figure 1.1. Yearly publications containing the terms "hydrophilic interaction chromatography" or "hydrophilic interaction liquid chromatography" or HILIC according to Thomson Web of Knowledge [1].

in 2003 to 267 in 2010. While HILIC has unique retention characteristics for hydrophilic compounds, this increase in interest also reflects the advantages of HILIC over RP methods in situations where either technique is applicable. These advantages result mostly from the high organic content of typical mobile phases and their resultant high volatility and low viscosity. A particular advantage is in coupling HILIC to mass spectrometry (MS) as mobile phases are more efficiently desolvated in interfaces such as electrospray, giving rise to better sensitivity than with RP methods. Thus, Grumbach and coworkers demonstrated sensitivity increases of 3–4 orders of magnitude when comparing the analysis of the drugs salbutamol and bamethan by HILIC on a bare silica column using a gradient analysis starting at 90% ACN with that on a C18 RP column using a gradient starting at 0% ACN [2]. Columns can be used at considerably lower pressures than in RP; the viscosity of 80–90% ACN mixtures with water as typically used in HILIC is only about half that of 20–30% ACN mixtures that might be used in RP separations [3]. Alternatively, longer columns can be used at pressures typically found in RP analysis, allowing high efficiencies to be obtained [4]. For example, when combining the low viscosity of HILIC with the efficiency gains shown by superficially porous (shell) particle columns, it is possible to generate column efficiencies in excess of 100,000 plates with reasonable analysis times, and using pressures that are well within the capabilities of conventional HPLC systems (pressure < < 400 bar). Low viscosity also results in increased solute diffusion in the mobile phase, giving rise to smaller van Deemter C terms and improved mass transfer, and the possibility of operating columns at high flow rates with reduced losses in efficiency for fast analysis [5]. Surprisingly good peak shapes can be obtained for some basic compounds. For example, efficiencies of around 100,000 plates/m with asymmetry factors (A_s) close to 1.0 were reported for basic drugs such as nortritpyline (pK_a ~10) using a 5-μm particle size bare silica HILIC phase. In comparison, such solutes often give rise to peak asymmetry in RP separations.

A separate advantage of HILIC is its compatibility with sample preparation methods using solid-phase extraction (SPE). Some such methods incorporate an elution step that uses a high concentration of an organic solvent, which gives rise to a potential injection solvent of the eluate that is stronger than typical RP mobile phases [2]. This mismatch in solvent strengths can give rise to peak broadening or splitting, necessitating evaporation of the SPE eluate and reconstitution in the mobile phase. SPE eluates with high organic solvent concentrations can be injected directly in HILIC, as they are weak solvents in this technique. The combination of different retention mechanisms in sample purification and analysis steps (HILIC/RP) can be advantageous in giving extra selectivity compared with an RP/RP procedure, where in some cases the SPE column may act merely as a sort of filter for the analytical column [6].

While HILIC is simple to implement in practice, some recent papers have concluded that the separation mechanism is a complex multiparametric process that may involve partition of solutes between a water layer held on the surface and the bulk mobile phase, adsorption via interactions such as hydrogen bonding and dipole–dipole forces, ionic interactions, and even nonpolar retention mechanisms (similar to RP interactions), depending on the stationary and mobile phases [7–9]. In this chapter, we will consider in some detail the various mechanisms that contribute to HILIC separations.

1.2 HISTORICAL BACKGROUND: RECOGNITION OF THE CONTRIBUTION OF PARTITION, ION EXCHANGE, AND RP INTERACTIONS TO THE RETENTION PROCESS

The term "hydrophilic interaction chromatography" was coined in 1990 by Alpert [10]. He carefully avoided the acronym HIC to avoid confusion with the technique of hydrophobic interaction chromatography, the latter being an adaptation of the RP technique where decreasing salt concentrations are used to progressively elute large biomolecules from the stationary phase. However, it is possible that HILIC dates back to the earliest days of liquid chromatography, when Martin and Synge separated amino acids on a silica column using water-saturated chloroform as the mobile phase. These authors explained the separation mechanism as being the partitioning of the solutes between a water layer held on the column surface and the chloroform [11].

The silica was considered to act merely as a mechanical support. It later became clear that use of a solvent that is immiscible with water, such as chloroform, is not an essential requirement. Lindon and Lawhead [12] discussed the separation of sugars such as fructose, glucose, sucrose, melibiose, and raffinose on a micro-Bondapak carbohydrate column (an aminosilica column, 10 μm particle size). The mobile phase was ACN–water (75:25, v/v); the authors showed that increasing the concentration of water reduced the retention times of the sugars. The authors noted that while the α- and β-anomers of sugars are readily separated by gas-liquid chromatography, they were not separated by this LC method, removing an unnecessary complication. However, no explanation for this lack of separation, or for the retention mechanism, was presented. It was shown later that aminopropyl silica in the presence of ACN–water greatly increased the mutarotation rate of the sugars compared with the effect of bare silica [13]. This effect is due to the basic environment generated in the column pores by the presence of the amino groups [14]. With a refractive index detector, it was shown that water was retained on the aminopropyl silica when pumping mobile phases of ACN–water and that the volume fraction of water in the liquid associated with the stationary phase was much higher than that in the corresponding eluent. The extent of water enrichment in the stationary liquid was found to be relatively high when the eluent contained a low water concentration. The separation of the sugars was explained as being due to their partition between the water-rich liquid in the stationary phase and the bulk mobile phase. Using a similar experimental procedure, other workers showed a reduced uptake of water on an aminosilica column when methanol–water was used as the mobile phase compared with ACN–water, as the competition between water and methanol for polar sites on the column was increased [15].

While the reports on sugar analysis were clearly classical HILIC separations in their use of a polar phase together with an ACN–water mobile phase containing a high concentration of organic solvent, a number of other early papers using bare silica columns demonstrated separations that contain at least some of the mechanisms that are now considered contributory to HILIC. Bidlingmeyer and coworkers [16] separated organic amines on a silica column using "reversed-phase eluents" consisting typically of ACN–water (60:40, v/v) containing ammonium phosphate buffer, pH 7.8. They showed that increasing the salt concentration decreased the retention of ionized basic compounds, indicating the contribution of ionic retention to the overall mechanism. It was demonstrated that over the range 70–30% ACN, if buffer strength and pH were held constant, retention decreased with increasing proportion of ACN as would be expected in a reversed-phase separation. Good peak symmetry was obtained for basic compounds on these bare silica columns. The authors concluded from a comparison with RP that the key to good peak shapes with these solutes was not the presence or absence of silanols but more probably the accessibility of these surface groups. Nevertheless, the concentrations of

ACN employed in this work were at the lowest end of the range generally used for HILIC separations, and it is questionable whether the important partition element of the HILIC mechanism was involved to any extent in such separations, as the mobile phase becomes more hydrophilic and thus competitive with the stationary phase. Other early work by Flanagan and Jane also showed the separation of basic drugs on bare silica columns, but this time using nonaqueous ionic eluents [17,18]. The nonaqueous, primarily methanolic eluents, contained additives such as perchloric acid or ammonium perchlorate of appropriate pH and ionic strength. The authors demonstrated that the retention of quaternary compounds increased with eluent pH, particularly in the pH range of 7–9, whereas the retention of bases decreased steadily at a high pH, where they were unprotonated. The observations were consistent with an ionic retention mechanism on ionized silanols. However, in the absence of water in the mobile phase, the conditions are clearly not consistent with those of HILIC. Euerby and coworkers [19] separated a variety of basic analytes on bare silica columns of varying metal content using buffered methanol and ACN mobile phases of again rather low organic concentration (typically 20–40%). Their experimental conditions were somewhat similar to those of Bidlingmeyer (organic solvent concentrations were lower than classic HILIC conditions), and their conclusions were also that ionic and hydrophobic mechanisms were the main contributors to retention.

Cox and Stout [20] studied the retention of a set of nitrogenous bases such as thiamine and morphine on some bare silica columns. Their work was inspired by the difficulties that were encountered in the separation of basic compounds using typical RP columns available at that time (mid-1980s), which often gave long and variable retention, poor separation efficiency, and excessive peak tailing. While their studies indicated that ion exchange was a major contributor to retention in these systems, they reported that the mechanism appeared to be more complex, incorporating more than a single retention process. Linear plots of retention factor versus the reciprocal of buffer cation concentration (see Section 1.3.3.4) were obtained with retention decreasing as the concentration of the buffer increased, indicative of an ion-exchange mechanism. These experiments were performed at low concentrations of methanol (15% or 30%). These are not typical HILIC conditions, and little, if any, contribution of a classical HILIC partitioning mechanism seems likely. However, all of the plots showed a positive intercept on the y-axis. For a pure ion-exchange mechanism, straight lines passing through the origin should be generated. The positive intercept of the plot was cited as evidence for a competing mechanism, which existed at an infinite buffer concentration (i.e., when the reciprocal of the buffer concentration is zero). The authors first considered changes in the ionization of the solutes with addition of the organic solvent that could have influenced the results, but discounted this hypothesis on the basis that morphine was a strong base and should be completely ionized. The authors therefore concluded that some nonionic interaction of the solute with silanol groups might occur. It seems that this contributory mechanism might in fact be of the same nature as that suggested in Bidlingmeyer's work; that is, it is hydrophobic in origin [16]. Most of this work was carried out with low concentrations of methanol, that is, remote from classical HILIC conditions. However, a plot of k derived from retention at an infinite buffer concentration for thiamine and morphine against methanol concentration from 15–75% v/v showed a U-shaped plot with retention maxima shown at 15% and 75% methanol, the maximum at 75% methanol in hindsight perhaps being indicative of the onset of a HILIC retention process.

While the paper of Alpert [10] was clearly not the first to demonstrate analysis using HILIC conditions, it was certainly a landmark publication because of the quality of the separations demonstrated for peptides, nucleic acids, and other polar compounds, and its careful discussion of the separation mechanism. Alpert showed that retention of peptides on hydrophilic columns, including a strong cation exchange material, PolySulfoethyl A, and a (largely) uncharged material, PolyHydroxyethyl A, increased dramatically when concentrations of ACN greater than 70% were used and that the order of their elution was from the least to the most hydrophilic, that is, the opposite from the order in RP separations. For the cation exchange material, electrostatic effects were superimposed on the HILIC mechanism. In agreement with the conclusions of previous workers [13,15], Alpert interpreted the earlier retention of sugars on amino columns as being not due to any electrostatic effects but caused by the hydrophilic nature of the basic column groups, demonstrating that the separation of carbohydrates could also be performed on the neutral PolyHydroxyl A phase, albeit giving elution in doublets corresponding to the α- and β-anomers. Clearly, this neutral phase could not generate the alkaline mobile phase environment required to speed up the mutarotation of sugars. However, the problem was overcome by addition of a small amount of amine to the mobile phase to speed up the mutarotation process. In analogy with the partition mechanism that had previously been suggested for the separation of sugars, Alpert proposed that the same mechanism could also explain the separation of other classes of polar solutes, such as peptides and amino acids. He also cited the relatively small differences obtained in the separations of peptides between uncharged and charged stationary phases as the organic content of the mobile phase was increased as further evidence that partition was the dominant mechanism. As the partition contribution to retention is increased, the proportional contribution of ion exchange to the total retention is reduced. It was, nevertheless, clearly shown on the cation exchange phase that ionic retention effects could be superimposed on HILIC retention and could give useful selectivity effects. Alpert noted distinct similarities in the separation of thymidilic acid oligomers between HILIC and classical partitioning systems, citing this result as being further indicative of a partition mechanism in the chromatographic technique. He speculated that some form of dipole–dipole interactions might be involved, although retention of sugars had been shown to correlate better with their hydration number than with their potential to form hydrogen bonds [15].

1.3   RECENT STUDIES ON THE CONTRIBUTORY
MECHANISMS TO HILIC RETENTION

1.3.1   Overview

As HILIC retention depends on the hydrophilicity of the solutes, attempts have been made to correlate this retention with physical descriptors of this property. Log P values represent the log of the partition coefficient when a solute is distributed between an aqueous phase and n-octanol, which in simple terms (using concentrations as an approximation for activities) can be written as

$$\log P = \log([C_o]/[C_w]),$$

where C_o is the concentration of the compound in octanol, and C_w is the concentration of the compound in water. Strictly, log P refers to the distribution of the nonionized form of ionogenic compounds. Alternatively, the distribution coefficient D is defined as the equilibrium concentration ratio of a given compound in both its ionized and nonionized forms between octanol and water. The use of log D instead of log P requires knowledge or estimation of the pK_a of the compound to calculate its ionization at a particular pH. Kadar and coworkers [21] studied the application of log D values produced by the ACD (Advanced Chemistry Development Inc.) calculation program to the prediction of a compound's suitability for HILIC analysis. The ACD program generates estimations of log D for mono- or polyprotic acids and bases over the pH range of 0–14 with increments of 0.1 pH units [22]. The authors tested the hypothesis that a relationship exists between the analyte's retention factor, k, and its log D at pH 3.0. The value of pH 3.0 was chosen due to the consideration that the majority of active pharmaceutical ingredients are basic amines that will be protonated under acidic conditions, and that these conditions are frequently used for their analysis. In this work the authors assumed that a partially immobilized layer of water existed on the phase and that the pH of this immobilized water layer was 3.0. The authors debated at some length whether aqueous pH data ($_w^w$pH) should be used rather than $_w^s$pH values (where the pH is measured in the aqueous–organic solution). There is a considerable difference in these quantities when large concentrations of ACN are utilized, as in typical HILIC separations. However, due to the paucity of data concerning pK_a values in aqueous–organic mixtures, the authors decided to use $_w^w$pH and $_w^w$pK_a data. A further consideration not mentioned by the authors is that if a water layer exists on the column surface in HILIC, it is possible that data measured in water are more appropriate. For this work, the authors selected 30 probe compounds representative of pharmaceutical compounds used in the therapeutic areas of anti-infectives, cancer, cardiovascular and metabolic disease, and the central nervous system. They determined the retention factor of each compound experimentally using a bare silica HILIC column and a mobile phase consisting of 85%, 90%, or 95%

ACN containing a total ammonium formate buffer concentration of 10 mM at pH 3.0. Linear regression analysis of log k versus log D produced correlation coefficients of 0.751, 0.696, and 0.689 for 85%, 90%, and 95% ACN concentrations, respectively, giving relationships (for example with 85% ACN) of the form

$$\log k = -0.132 \ (\log D) - 0.234.$$

These equations could then potentially be used to predict the value of k for a given calculated log D value. The authors interpreted the deviations of the correlation coefficients from unity as being due to secondary interactions in addition to the partitioning that was initially assumed in the hypothesis to be the only retention mechanism on the bare silica column. In particular, they considered that electrostatic interactions would occur with negatively charged silanols, giving increasing retention for charged basic compounds relative to that expected for a pure partition mechanism. Conversely, charged acidic compounds should experience repulsion and therefore give retention less than expected from a pure partitioning mechanism. Indeed, they showed that the predicted k from log D values of several compounds that contained at least one basic functional group that was fully protonated at the experimental pH was significantly underestimated compared with the experimentally measured retention. The authors concluded that there is a direct correlation between a compound's HILIC retention and its distribution ratio, although an accurate prediction of k could not be made due to these secondary interactions. They concluded that the work also supports Alpert's theory of a partition mechanism to describe HILIC separations. Bicker and coworkers [23], who studied the retention of nucleosides and nucleobases on a series of silica packings bonded with neutral trimethoxysilylpropylurea ligands, obtained variable results with prediction of retention based on log D values. They cautioned that these predictions should only be regarded as a simplistic concept for estimating the relative strength of HILIC-type interactions because the underlying molecular processes of retention and their correlations with solute polarity are not sufficiently understood as yet. They reported severe limitations of the predictions in the case of charged solutes, where other types of interaction than a partition mechanism come into effect. West and coworkers [24] acquired retention data for 76 model compounds using two zwitterionic phases and $_w^w$ pH 4.4 ammonium acetate buffer in 80% ACN, with an overall salt concentration of 20 mM. The coefficients of determination (r^2) of 0.70 and 0.87 for ZIC-HILIC and a Nucleodor phase, respectively, gave evidence according to the authors that hydrophilic partitioning was only one of the mechanisms involved in HILIC separations, and thus log D values could only give a rough estimate of retention. They argued that the relatively high salt concentrations used should have suppressed some ionic interactions of ionized stationary phase groups, possibly improving the correlation with log D. No particular groups of solutes (neutral, anionic, cationic, or zwitterionic) appeared to be responsible for the poor correlation, as all were scattered more or less uniformly about the regression line. However, it seemed that the fit was poorest for solutes with low retention, where the accuracy of the measurement could be a factor.

Some reports have shown the separation of the same mixture of solutes on a number of different stationary phases, in studies designed to contribute to elucidation of the separation mechanism. Guo and Gaiki examined the retention characteristics of four polar silica-based stationary phases (amide, amino, silica, and sulfobetaine—a zwitterionic phase containing quaternary amine and sulfonic acid groups) using small polar compounds as solutes. The solutes studied included salicylic acid and derivatives, some nucleosides and nucleic acid bases, selected because they are usually difficult to retain on RP columns [25]. Figure 1.2 shows the separation of the salicylic acid derivatives on the four columns using ACN–water (85:15 v/v) containing 20 mM ammonium acetate as the mobile phase. The retention and elution order clearly varied

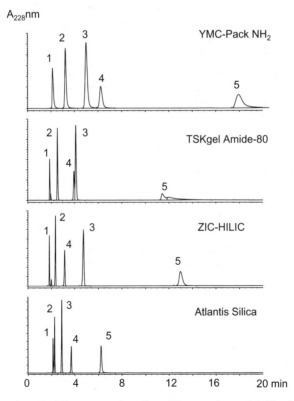

Figure 1.2. Separation of acidic compounds on four different columns. Mobile phase ACN–water (85:15, v/v) containing 20 mM ammonium acetate. Column temperature 30°C. Flow rate 1.5 mL/min. Ultraviolet (UV) detection. Compound identities: 1 = salicylamide, 2 = salicylic acid, 3 = 4-amino salicylic acid, 4 = acetylsalicylic acid, 5 = 3,4-dihydroxyphenylacetic acid. All columns 25 × 0.46 cm containing 5-μm particle size packing. Reprinted from Reference 25 with permission from Elsevier.

from column to column. The acids were most retained on the amino column. As this column contained positively charged groups with the mobile phase conditions used, the negatively charged acids could undergo ionic interactions, increasing their retention. The acids had weaker retention on the amide column, and aspirin and 4-aminosalicylic acid (peaks 4 and 3) were only partially resolved. In contrast, the resolution of these two solutes was greatly improved on the zwitterionic column, however with a reversed order of elution compared with the amino column. On the bare silica column, the peaks were also well resolved, but their elution order was more similar to that on the amino column. The authors considered that the different elution patterns of the acids on the four columns indicated that the polar stationary phases had significant differences in retention and selectivity. Similar differences in selectivity were noted for a mixture of nucleic acid bases and nucleosides on the four columns. Specific interactions between the solutes and surface functional groups were thought to be most likely to be responsible for these selectivity differences. Such interactions could not be considered under a pure partition model, nor would they be accounted for in predictions using log D values. The authors also investigated the contribution of ionic processes to the overall retention, examining the effect of different ammonium salts (ammonium acetate, formate, and bicarbonate) on the retention of the acid compounds. They showed some differences in retention of the solutes, which they attributed in part to different eluting strengths of the competing anions in ionic interactions with the positively charged column groups. They also investigated the effect of salt concentration by varying the concentration of ammonium acetate from 5 to 20 mM in a mobile phase of ACN–water (85:15, v/v). For salicylic acid and aspirin, they showed increases in retention on the amide, bare silica, and zwitterionic column of 20–40% as the buffer strength increased. The authors considered the possibility that an increase in the buffer strength could be reducing repulsive effects of the acids from negatively charged silanol groups on the silica-based phases. However, they observed smaller but significant (8–20%) increases in the retention time of cytosine on all four columns. As cytosine was not charged under the mobile phase conditions used, electrostatic effects could not contribute to retention increases for this solute. The authors concluded that in this case, the retention increase might be related to increased hydrophilic partitioning, instead of any specific interactions with the functional groups on the stationary phases. Higher salt concentrations should drive more solvated salt ions into the water-rich liquid layer on the column surface, resulting in an increase in volume or hydrophilicity of the liquid layer, leading to stronger retention of the solutes. The authors suggested that this experiment provided indirect evidence to support the retention mechanism of HILIC that had been proposed by Alpert [10]. Nevertheless, increases in the salt concentration produced considerable decreases in the retention of salicylic acid and aspirin on the amino column. The ion exchange interactions of the acids on this phase were reduced by increasing competition of the buffer ions. It was interesting to note that no such decreases in retention were observed on the zwitterionic phase. The authors speculated that electrostatic repulsion from the negatively charged sulfonic groups was balanced by the influence of the quaternary amine groups on this phase.

A comparison of the retention properties of HILIC phases using a rather different set of solutes was performed recently by McCalley [26]. Figure 1.3 shows the separation of a mixture of two neutral compounds (phenol and caffeine) two strong acids (*p*-xylene-2-sulfonic acid and naphthalene-2-sulfonic acid) and four basic compounds (nortriptyine, diphenhydramine, benzylamine, and procainamide) on five different silica-based HILIC phases of the same dimensions and particle size (5 μm). The mobile phase was 5 mM ammonium formate, pH 3.0, in either 85% ACN (Fig. 1.3a) or 95% ACN (Fig. 1.3b). The structure of the bonded groups and the physical characteristics of

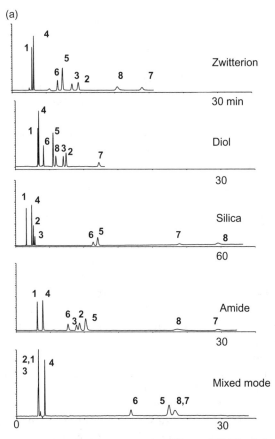

Figure 1.3. (a) Chromatograms of eight solutes on five different HILIC columns (all 25 × 0.46 cm, 5 μm particle size). Mobile phase ACN–water (95:5, v/v) containing 5 mM ammonium formate, pH 3.0, 1 mL/min. Peak identities: (1) phenol, (2) naphthalene-2-sulfonic acid, (3) *p*-xylenesulfonic acid, (4) caffeine, (5) nortriptyline, (6) diphenhydramine, (7) benzylamine, (8) procainamide.

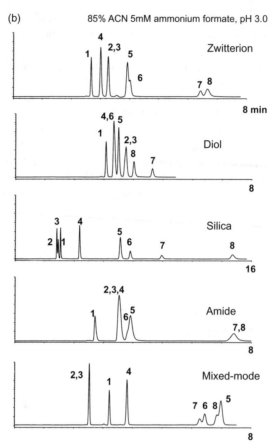

Figure 1.3. (b) Mobile phase ACN–water (85:15, v/v) containing 5 mM ammonium formate, pH 3. Reprinted from Reference 26 with permission from Elsevier.

these phases are given in Table 1.1. These were zwitterionic sulfobetaine, bare silica, diol, amide, and a mixed mode phase. The mixed mode phase was developed to exhibit both hydrophilic and reversed-phase characteristics [27], consisting of a long carbon chain with a diol grouping on the outlying carbon atoms. It was suggested that this phase has a dual operation mode. For example, the separation of cytosine and naphthalene could be achieved in the RP mode using ACN–ammonium acetate buffer pH 5 (52:48 v/v) with naphthalene eluting last, and in the HILIC mode using ACN–buffer (92:8, v/v) with naphthalene eluting first. It is immediately clear from this comparison that considerable differences exist in the selectivity of the various columns toward this group of solutes. For the basic solutes (peaks 5–8), the silica column is much more retentive than the other phases (note that the time axis is about double that for the other phases). This high retention is likely to result from ionic

Table 1.1. Specifications of the HILIC Columns Used

Column	Manufacturer	Bonded Group	Pore Diameter (Å)	Surface Area (m^2/g)	Void Volume (mL^b)
Zwitterion (Zilic)	Merck	$-CH_2N^+(CH_3)_2-CH_2-$ $CH_2-CH_2-SO_3^-$	200	140	2.6
Silica	Phenomenex	~SiOH	100	400	3.0
Diol (Luna HILIC)	Phenomenex	~cross-linked diol/ ethylene bridges	195	185	3.0
Amide	TSK	$-CONH_2$ nonionic carbamoyl	80	a	2.7
Acclaim mixed-mode HILIC-1	Dionex	$~Si(CH_3)_2C_9H_{19}CH$ $(OH)CH_2-OH$	120	300	3.3

[a]Value not disclosed by manufacturer.
[b]Value measured using toluene in 90:10 ACN: water (v/v) containing 5 mM ammonium formate, pH 3.0.

interactions with ionized silanol groups. Nortriptyline, diphenhydramine, benzylamine, and procainamide all have pK_a values > 9 and thus should be protonated under the conditions of the experiment. In contrast, the bare silica phase shows low retention of the ionized acidic solutes p-xylene sulfonic and naphthalene 2-sulfonic acids (peaks 2 and 3). While ionization of solutes in general should increase their hydrophilicity and thus their retention by partitioning into the aqueous layer, low retention in this case can be explained by repulsion of these charged solutes from ionized column silanols. This low retention of acids is also shown on the mixed-mode phase. However, the relative retention of these acids is greater than that of the ionized base diphenhydramine (peak 6) on the zwitterionic, diol, and amide phases. On the diol phase, the acids show highest retention of all solutes apart from benzylamine, although the average retention of all solutes on this phase is rather low. The particular variety of zwitterionic, diol, and amide phases used in this study had a polymeric bonded phase layer [26], and it is possible that this shields the silanols from interaction with ionized solutes. A comparison of the separation of the probes when using either 85% or 95% ACN in the mobile phase, while maintaining the buffer concentration constant, shows some differences in selectivity. For example, the retention of benzylamine (peak 7) is increased relative to that of procainamide (8) on the amide column at the higher concentration of ACN, and the order of elution of the basic compounds (peaks 5, 6, 7, 8) on the mixed mode phase was also changed. These differences could indicate changes in specific interactions between the solutes and the column groups as the mobile phase is changed.

The differences in selectivity that occur between different columns when used with the same mobile phase as shown in these studies confirm that the mechanism of separation in HILIC is complex and that the stationary phase gives a considerable contribution to retention. Thus, the stationary phase cannot be considered merely as an inert support for a layer of water into which solutes selectively partition. Ion-exchange processes have long been recognized as contributory to the overall mechanism. The same conclusions were reached recently by Bicker and coworkers [23], who cited three major retention mechanisms on bare silica, or columns bonded with the neutral ligand trimethoxysilylpropyl urea: (1) HILIC-type partitioning, (2) HILIC-type weak adsorption such as hydrogen bonding between solutes and the bonded ligands or the underlying silanols (which could be influenced by the experimental conditions), and (3) strong electrostatic forces for ionized solutes, which could be attractive or repulsive. They summarized that multi- or mixed mechanism separations seemed to be common under HILIC conditions, which are associated with useful selectivity effects.

These various contributors to retention in HILIC will be considered in more detail in the following sections.

1.3.2 Contribution of Adsorption and Partition to HILIC Separations

Studies on the retention of sugars showed that water from the mobile phase is retained on the surface of HILIC columns [13,15] and thus that the concentration of water is higher in the stationary phase than in the mobile phase, providing evidence for a partition mechanism. A more recent study [28] used the retention of benzene and toluene, which are unretained void volume markers in HILIC, to indicate more exactly the presence of a water layer on the stationary phase surface. Figure 1.4 shows the decrease in retention time of these solutes as the water concentration in an ACN–water mobile phase

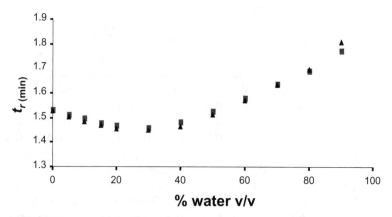

Figure 1.4. Retention time (t_r, min) of benzene (squares) and toluene (triangles) as a function of water content of aqueous ACN mobile phase. Flow rate 1.0 mL/min. Detection UV at 254 nm. Injection volume 5 μL. Column bare silica shell (2.7 μm particles, 15 × 0.46 cm). Reprinted from Reference 28 with permission from Elsevier.

was increased from 0% to about 30% v/v. Due to the limited solubility of these hydrophobic compounds in water, it can be assumed that they partition almost entirely into the bulk mobile phase and cannot penetrate the water layer. The difference between the retention volume of the probe in pure ACN and in a given mobile phase can be used to estimate the proportion of the pore volume occupied by water. However, the method breaks down at water concentrations of >30% v/v. The increasing retention of the probes at these higher water concentrations can be attributed to an RP-type retention mechanism on column siloxane groups, similar to that proposed originally by Bidlingmeyer [16]. The presence of a water layer has also been shown recently by the studies of Tallarek and coworkers [29] that involve molecular simulation dynamics using high-speed computers, modeling cylindrical silica pores that have a diameter of 3 nm. The simulations were performed with water/ACN mixtures of molar ratios 1/3, 1/1, and 3/1, which corresponded to approximate volumetric ratios of 10/90, 25/75, and 50/50 v/v. The results indicated that the water/ACN ratios in the pores were considerably higher at 1.5, 3.2, and 7.0 for the respective mixtures. The *relative* water fraction in the pores thus increased with decreasing water content of the bulk mobile phase. The simulations suggested a layer close to the surface (<0.45 nm) where water hydrogen bonds preferentially to silanol groups, with only scarce silanol–ACN bonds. The water molecules in this region appeared to be nearly immobilized to the silanol groups. Outside this immediate region close to the surface of the stationary phase, water–water hydrogen bonding was preferred, although some ACN–water hydrogen bonds were indicated.

Irgum and coworkers reported the use of ^{2}H nuclear magnetic resonance (NMR) spectroscopy for probing the state of water in a number of different HILIC phases including bare silica particles of pore size 60–100 Å, and in silica bonded with polymeric sulfobetaine zwitterionic functionalities [30]. They distinguished three types of water that could be present in polymer systems: free water that resembles ordinary bulk water, freezable bound water that has a slightly shifted transition temperature between the solid and liquid state compared with bulk water, and water that is bound within the polymeric network such that it does not freeze in the expected temperature range for bulk water. ^{2}H NMR was chosen rather than ^{1}H NMR because of the problems experienced when combining high-field NMR spectrometers with porous samples. Deuterated water enriched to 20–30 mL % was used for the study, which increased the freezing point of water by a small amount (0.8–1.2°C). The use of ^{2}H NMR enabled signals from liquid and frozen water to be clearly distinguished from each other. On freezing, the NMR line width increases significantly, spreading out over nearly a 300-kHz broad frequency range. Because the peak becomes so broad, the signal from frozen water effectively disappears from the NMR spectrum, allowing only the signal from water in the liquid state to be observed. For neat silica, the pore size had an influence on the depression of the freezing point of water, and thus on the percent of nonfreezable water, that is, water that was strongly associated with the stationary phase. This amount decreased as the pore size of the silica increased. A 3-µm 100 Å silica was shown to contain 14% of nonfreezable water. The polymeric zwitterionic stationary phases were shown to contain higher amounts of water compared with the neat silica. A difficulty with these measurements is that they were apparently carried out under purely aqueous conditions, that is, in the absence of organic solvents like acetonitrile, and thus different from normal conditions used in HILIC.

While the existence of a water layer is a prerequisite for the partitioning model in HILIC, its existence does not preclude the occurrence of an alternative adsorption mechanism. Indeed, the different selectivities for some solutes on HILIC phases containing different polar bonded ligands could be interpreted as being caused by differential adsorption on these groups, as suggested in the previous discussion, even if the polar column groups are deactivated by the presence of water. It is possible that adsorption is a more likely contributor to the mechanism when low concentrations of water are present in the mobile phase. Hemström and Irgum [31] considered the relative contributions of adsorption and partition on the basis of fitting retention data to the relevant equations that describe these two mechanisms. Retention in adsorption chromatography can be described by the Snyder–Soczewinski equation:

$$\log k = \log k_B - n \log X_B \qquad (1.1)$$

where X_B is the mole fraction of the strong solvent B (in this case water) in the mobile phase, k_B is the retention factor with pure B as the eluent, n is the number of B solvent molecules displaced by the solute. Alternatively, for a partition-like mechanism, the empirical equation

$$\log k = \log k_w - S\varphi \qquad (1.2)$$

describes retention approximately, where φ is the volume fraction of the strong solvent B in the mobile phase, and k_w is the hypothetical retention factor when the mobile phase contains no B solvent (i.e., solely the weak solvent). Thus, a plot of log k versus log (mole fraction water) should yield a straight line for an adsorption mechanism, whereas a plot of log k versus (volume fraction water) should yield a straight line for a partition mechanism. The authors considered data from a number of studies (e.g., References 25, 32, and 33) to determine the relative linearity of these plots, although for the log–log plots they used the more approximate volume fraction instead of the mole fraction of water. Clearly, these quantities are not collinear. The original authors of Reference 32 had shown that for the separation of sugars on amino-ethylenediamino and diethylenetriamino silicas, the mechanism appeared to be constant for all phases (the selectivity was the same, although the absolute retention increased as the number of amino groups on the stationary phase increased). They argued, therefore, that the bonded groups on the column were not directly involved in the separation, and served only to trap water; that is, this was strong evidence for a partition mechanism. However, Hemström and Irgum, in interpretation of the earlier data [32], showed considerably better linearity in the log–log plots than the log–lin plots, consistent with an adsorption process. Nevertheless, their attempts to model other data showed equally poor fits using either type of plot. Overall, it was not possible to reach any firm conclusions from these studies as to whether partition or adsorption was likely to be the dominant mechanism.

Lindner and coworkers [9] constructed similar plots for the retention of cytosine and cytidine on some novel oxidized and nonoxidized 2-mercaptoethanol and thioglycerol phases, and some commercial diol phases. They found considerable intercolumn differences in the log–log and log–lin plots over the examined range (5–40% v/v water in the eluent). Similar trends were also found for other nucleobases and nucleosides examined. For example, with the mercaptoethanol phase, the log–lin plot was shown to give a better linear relationship, while for the more polar phases only the log–log model delivered adequately linear relationships. In this latter case, linear regression analysis of the log–log plots generated r^2 values for the oxidized phases of between 0.996 and 0.999. The authors suggested that the nonlinear behavior of the nonoxidized mercaptoethanol in the log–log but not in the log–lin plots indicated that partitioning was the prevailing mechanism under the specified elution conditions for the particular solutes examined. Furthermore, they showed that at low water concentrations, there was a transition to curvature in the log–lin plots for this column, which corresponded to a linear behavior in this range for the log–log plots. They interpreted this behavior as being indicative of adsorptive interactions that come into play when the water content of the mobile phase is low and the ACN content is high. However, this clear trend was not shown for the nonoxidized thioglycerol phase. With an increase in phase polarity produced by oxidation of these two phases, the authors surmised that adsorptive interactions become more relevant or even dominant for retention, as shown by the greater linearity of the log–log plots. They also noted that the change in selectivity of nucleoside separations when changing from the nonoxidized to the oxidized forms of the phases could not be explained satisfactorily on the basis of a pure partitioning process that is commonly invoked to describe retention in the HILIC mode. While a complete transition from a partitioning-dominated mechanism to an adsorption-driven mechanism was considered unlikely in the oxidized variants of these new phases, these data do indicate the potential impact of the nature of the stationary phase on the retention mechanism in HILIC. The conclusion of the authors was that a mixed-mode process for these new packings was operating that involved contributions of both partition and adsorption.

Li and coworkers [34] examined the retention of four zwitterionic tetracyclines on an amino bonded HILIC column, using buffered aqueous solutions in the range 10–50% v/v (90–50% ACN). They found r^2 values of the log–log plots of between 0.9953 and 0.9987 (the log of the volume fraction of water in the mobile phase was again used, rather than the mole fraction), whereas the log–lin plots were less linear, giving values from 0.9649 to 0.9978. However, the authors noted that r^2 for the log–lin plots improved to 0.9777–0.9915 when the percentage of water was in the reduced range of 20–50%. They concluded that this result indicated that the relative contributions of partition and adsorption changed depending on the mobile phase composition.

A further study examined the relative linearity of log–log and log–lin plots for the same five columns and five of the eight solutes shown in Figure 1.3 [26]. The results are shown in Figure 1.5. This mixture of probes contained basic compounds that are also retained by ion-exchange as well as HILIC processes.

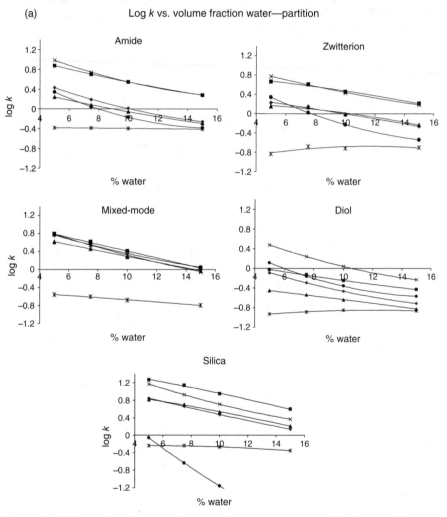

Figure 1.5. (a) Plots of log k versus volume fraction of water in the mobile phase and (b) plots of log k versus log mole fraction of water in the mobile phase. Solute identities: diamonds = nortriptyline, squares = procainamide, triangles = diphenhydramine, crosses = benzylamine, stars = caffeine, circles = p-xylenesulfonic acid. For other conditions see Figure 1.2. Reprinted from Reference 26 with permission from Elsevier.

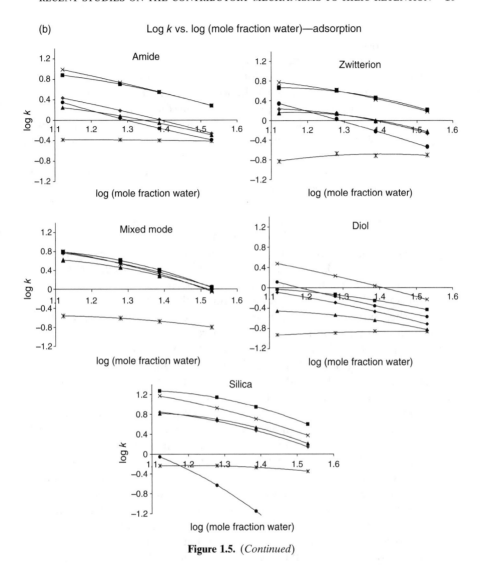

Figure 1.5. (*Continued*)

The overall buffer concentration was maintained constant at 5 mM and the $_w^w$pH of the aqueous component held fixed at 3.0 in an effort to maintain ionic interactions constant. However, the change in ACN concentration actually brought about a small change in the $_w^s$pH (the pH measured in the mobile phase, with the meter calibrated in aqueous buffers) from $_w^s$pH 6.1 in 95% ACN to $_w^s$pH 5.2 in 85% ACN, and it is conceivable that the contribution of ionic interactions could therefore change. Nevertheless, retention of these solutes was shown to be rather insensitive to small pH changes in this region in a previous study [5]. It is also debatable whether $_w^s$pH or $_w^w$pH is more appropriate to use in such a study, considering that the solutes are held in a water layer close to the stationary phase. The plots show typical HILIC behavior, in that increasing concentration of water in the mobile phase results in decreased solute retention. A comparison of Figure 1.5a (log–lin plots) and Figure 1.5b (log–log plots) shows that these alternative plots were again rather inconclusive. For the bare silica and mixed mode phases, the log–lin plots seem to be more linear, indicating a partition mechanism, whereas for the diol and amide phases, the log–log plots appear to be more linear, indicating a partitioning mechanism. A complication of these plots is the possibility of the changing contribution of other mechanisms as the concentration of organic solvent changes (e.g., hydrophobic retention is possible at low concentrations of ACN; see Section 1.3.4). Furthermore, there are approximations involved with the equations themselves. We do not believe, therefore, that plots of this kind can give conclusive evidence of the predominance of either the partition or adsorption mechanism. Nevertheless, such plots can still be useful in indicating selectivity changes that occur as a function of organic solvent concentration. For example, Figure 1.5a,b indicate the increased relative retention of p-xylenesulfonic acid at low percent water on both the zwitterionic and amide phases.

It seems very likely that the predominant mechanism could change dependent on the stationary and mobile phase conditions. Irgum and coworkers [35] compared the performance of 22 commercial HILIC phases with a set of probe compounds designed to reflect the different possible contributions to the HILIC mechanism, evaluating the results using principal components analysis. They concluded that unmodified silica columns relied mainly on adsorption and oriented hydrogen bonding for selectivity. It was interesting in this respect that silica hydride phases, as prepared by Pesek and coworkers, appeared to exhibit similar behavior to conventional type A and type B silica phases [36]. Pesek and coworkers preferred the term "aqueous normal phase" (ANP) to describe separations with these phases. They argued that silica hydride has a lower hydrophilicity than type B silica, due to a lower concentration of silanol groups, and thus the hydration layer on the surface should be much thinner than for typical HILIC columns. If this is the case, then adsorption might have been expected to be the dominant mechanism for such columns. Irgum proposed that columns with highly hydrophilic polymeric interactive layers such as the zwitterionic column ZIC–HILIC generally showed a selectivity pattern that could be attributed to partitioning. Neutral and amino columns were stated to occupy an intermediate position between silica and zwitterionic columns.

1.3.3 Further Studies on the Contribution of Ionic Retention in HILIC

1.3.3.1 Introduction

As discussed previously, ionic retention has been recognized as contributory to retention on bare silica and other HILIC columns for many years. The contribution of ionic retention to total retention should increase as the concentration of water in the mobile phase is increased, because the relative contribution of the HILIC mechanism (i.e., partition or adsorption) to the total retention should decrease. Ion exchange groups have been deliberately introduced into phases since the earliest days of HILIC. For instance, amino groups used in columns for the separation of sugars can also be used as anion exchangers and the PolySulfoethyl phase used by Alpert [10] gives retention of cations superimposed on HILIC retention effects in the separation of peptides. In Irgum's study of 22 different stationary phases [35], cation exchange was shown to be a very strong contributor to the selectivity of separations on bare silica columns. However, the group of columns studied included older type B silica phases that might be expected to show strong interactions of this type, due to the presence of acidic silanols, which are more readily ionized. The study was also carried out at neutral pH in ammonium acetate solutions, conditions under which silanols might be expected to be at least partially ionized. Zwitterionic columns such as the sulfobetaine phase ZIC–HILIC were also shown to exhibit cation exchange properties. Electrostatic interactions, however, were reported to be much weaker than for the (particular) underivatized silica columns examined in the study.

1.3.3.2 Mobile Phase Considerations for the Separation of Ionogenic Compounds

For the separation of ionic compounds, buffer solutions are preferred to stabilize the solute charge. Stabilization of the charge on column groups is also an important factor for ionized solutes. However, the charge on these groups might conceivably affect the formation of a water layer on the column and thus influence the separation even of uncharged compounds. Olsen [37], in a study of the separation of some pyrimidines, purines, and amides on silica and amino columns, concluded that mobile phases should contain a buffer of acid for pH control in order to achieve similar and reproducible results among columns from different sources. As HILIC is a particularly advantageous separation technique to use in conjunction with MS (see Section 1.1), volatile buffer solutions are often favored. Simple aqueous solutions of organic acids such as formic and acetic acids are also recommended by some column manufacturers for HILIC although sometimes at higher concentration (e.g., 0.2%) than used typically in RP separations. Figure 1.6d shows the analysis of 100 mg/L solutions of three neutral compounds (3-phenylpropanol, caffeine, and phenol), whereas Figure 1.6a shows three ionized compounds (2-naphthalene sulfonic acid, nortriptyline, and propranolol) at the same concentration, using an Atlantis silica column with acetonitrile–water (85:15, v/v) containing an overall concentration of 0.2% formic acid. While the neutral compounds gave excellent peak shape, the charged compounds gave broad fronting peaks and column efficiencies only about one tenth that for the neutral compounds. Reduction of the concentration of the ionized solutes, from 100 to 10 and 1 mg/L (Fig. 1.6b,c), gave improved peak shapes showing that the poor peak

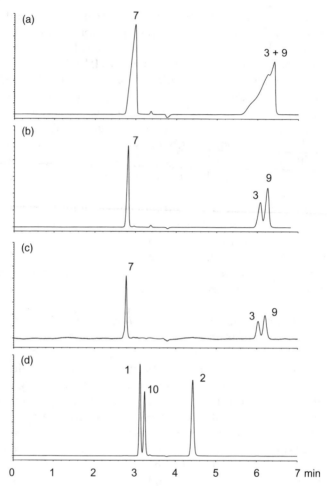

Figure 1.6. Analysis of selected compounds on Atlantis silica. Peak identities 1 = phenol, 2 = caffeine, 3 = nortriptyline, 7 = 2-naphthalenesulfonic acid, 9 = propranolol, 10 = 3-phenylpropanol. Solute concentrations (a) and (d) 100 mg/L; (b) 10 mg/L; (c) 1 mg/L. Mobile phase ACN–water 85:15, overall 0.2% formic acid. Reprinted from Reference 5 with permission from Elsevier.

shapes are due to some overloading effect. At 1 mg/L, efficiencies for ionized compounds approached those for neutral compounds, although some peak fronting was still shown (asymmetry factor $A_s = 0.8$–0.9). Eighty-five percent ACN causes formic acid to become such a weak acid that the ionic strength of this mobile phase is very low, which could cause overloading. In contrast, ammonium formate should be completely ionized even in high concentrations of ACN. Figure 1.3 has already shown the excellent peak shapes that can be obtained for these ionized acidic and basic compounds, when using mobile phases at similar pH containing ammonium formate. Clearly, analysis of charged compounds is impractical in HILIC using solely formic acid as an additive. Problems of overloading with formic acid have also been reported in RP separations, although these are not so serious, as the ionization of the weak acid is not so greatly affected by the lower concentrations of ACN typically used in such separations [38,39]. Equally poor results in HILIC were also obtained with 0.2% acetic acid (results not shown), which is a weaker acid than formic acid.

Ammonium formate or ammonium acetate at low pH is often used as a buffer in HILIC separations; the latter is also quite frequently used [35] without further pH adjustment (pH of aqueous solutions is typically about 6.8). At neutral pH, ammonium acetate is not a buffer. However,no problems with reproducibility of retention times have been noted using this solution[35], at least not at the low solute concentrations used.

1.3.3.3 Ionization State of the Column as a Function of pH

Some studies, particularly with bare silica columns, have attempted to investigate the ionization of column groups by studying the retention of acids and bases as a function of the mobile phase pH. Figure 1.7 demonstrates the variation in the retention of a quaternary ammonium compound (benzyltriethylammonium chloride [BTEAC], always completely ionized under the conditions of the experiment) as a function of mobile phase pH on a bare silica column (Atlantis, Waters Associates). The mobile phase was 85% ACN containing 15 mM ammonia adjusted to various pH over the range $_w^w$pH 3.0 ($_w^s$pH 5.2) to $_w^w$pH 10.2 ($_w^s$pH 9.0). Figure 1.7 shows that retention increases only gradually as the pH is increased from $_w^w$pH 3.0 to $_w^w$pH 8.0 followed by a marked increase from $_w^w$pH 8.0 to $_w^w$pH 10.0, indicating a large increase in silanol ionization and thus increased ionic retention of this cationic species. Retention of the eight

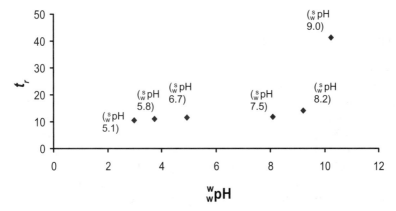

Figure 1.7. Retention of benzyltriethylammonium chloride (BTEAC) as a function of mobile phase pH on an Atlantis silica column (Waters). Mobile phase 85% ACN—0.1 M NH₃; pH adjusted with formic acid. Reprinted from Reference 5 with permission from Elsevier.

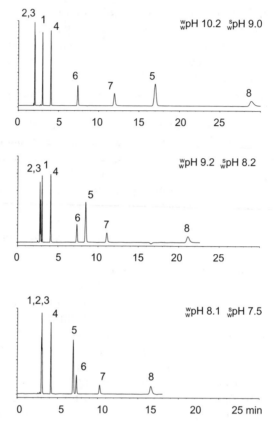

Figure 1.8. Analysis of test solutes on Atlantis silica. Mobile phase ACN—0.1 M HCOONH$_4$ $_w^w$pH 8.2 to 10.2 (85:15, v/v). Peak identities as for Figure 1.3. Reprinted from Reference 5 with permission from Elsevier.

test compounds used in Figure 1.3 was also investigated over the higher pH range of $_w^w$pH 8.1 to $_w^w$pH 10.2, as shown in Figure 1.8. Useful selectivity differences were demonstrated as the pH was varied. As pH increases, silanol ionization should increase, thus increasing the retention of cations. This effect was indeed shown by the strong base nortriptyline ($_w^w$pK_a 10.2), whose retention follows the same pattern as BTEAC. However, the effect of decreasing solute ionization at higher pH was superimposed, particularly for the weaker bases (but not for the quaternary compound, which remains ionized), on that of increasing column ionization. Decreasing solute ionization reduces ionic retention, and retention might also be decreased by reduced solubility of the uncharged compound in the water layer associated with the silica. This effect is particularly shown by the weakest base, diphenhydramine ($_w^w$pK_a 9.0), which elutes well before nortriptyline at $_w^w$pH 10.2 but after nortriptyline at $_w^w$pH 8.1 and below. However, ionic retention of cations occurs on this column even at

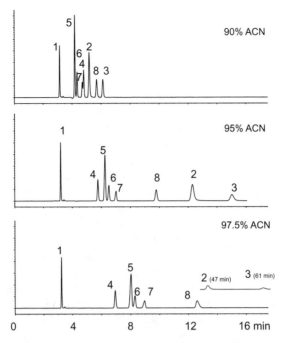

Figure 1.9. Analysis of test solutes on Atlantis silica. Mobile phase ACN–water containing overall 0.1% TFA. Peak identities as for Figure 1.3. Reprinted from Reference 5 with permission from Elsevier.

the lower pH studied ($_w^w$pH 3.0, see Section 1.3.3.4). The acids eluted below the column void volume at $_w^w$pH 10.2, presumably due to exclusion caused by the ionized silanols. The low retention of acids on a similar bare silica column at $_w^w$pH 3.0 has already been noted (see Fig. 1.3). In the same study, trifluoroacetic acid (TFA, overall concentration 0.1%) was used to determine whether the retention of the acids could be improved at a lower mobile phase pH than is obtainable with formate buffers. Figure 1.9 shows that this was indeed possible, and moreover a reversal in the order of elution of acids and bases on the Atlantis column was obtained with TFA, with the acids now eluting *after* the bases. Note that some variation in the order of elution of peaks was obtained by changing the ACN concentration from 90% to 95%. Increasing the ACN concentration further to 97.5% ACN gave retention for the acid *p*-XSA in excess of 1 h, whereas it eluted close to the void volume in ACN–ammonium formate buffer (compare results with Fig. 1.3). The change in elution pattern is rather surprising because the $_w^w$pH of 0.1% aqueous TFA is ~2.1, not vastly different from the $_w^w$pH 3.0 of aqueous ammonium formate as used previously. However, there are much larger differences in $_w^s$pH of these mobile phases, which were 1.35 for 85% ACN containing 0.1% TFA compared with $_w^s$pH 5.2 for 85% ACN–ammonium formate pH 3.0. The true thermodynamic pH ($_s^s$pH) can be calculated from the pH measured after the addition of solvent, with the meter calibrated in aqueous buffers ($_w^s\text{pH}$) according to the relationship:

$$_s^s\text{pH} = {_w^s}\text{pH} - \delta \qquad (1.3)$$

The δ term incorporates both the Gibbs energy for the transference of 1 mol of protons from the standard state in water to the standard state in the hydro-organic solvent at a given temperature, and the residual liquid junction potential (the difference between the liquid junction potential established during calibration in aqueous solutions, and that established in the hydro-organic mixture). Large negative δ values have been measured in aqueous–ACN mixtures with high ACN concentrations [40]. For example, $\delta \sim -1.1$ in 85% ACN and ~ -1.6 in 90% ACN. Thus, the thermodynamic $_s^s\text{pH}$ (calculated from Eq. 1.3), which is directly related to quantities such as the ionized fraction of the analyte, is ~6.3 in the ammonium formate buffer but ~2.5 for 0.1% TFA, both in 85% ACN. Clearly, formic acid becomes a very weak acid in high ACN concentrations, whereas the much stronger acid TFA is relatively unaffected. The differences in the $_s^s\text{pH}$ of these mobile phases could explain the difference in the elution pattern of the solutes; silanol ionization could be almost completely suppressed at the low pH of the TFA solution, thus preventing stationary phase exclusion of acidic solutes. Suppression of ionic repulsion could facilitate HILIC retention of the acids (solubility of the acids in the stationary phase water layer). At the same time, the retention of the basic compounds by ionic processes would be reduced. However, the situation is complex: while the average pK_a of silanols is considered to be ~7 in purely aqueous solutions, it is unknown in the presence of such high concentrations of ACN. Silanols are weak acids, and as such their pK_a might increase in 85% ACN, counteracting the higher effective mobile phase pH compared with that in purely aqueous solution. However, as already mentioned, a major complication in these deliberations is that the presence of a water layer on the phase surface may indicate that $_w^w\text{pH}$ and $_w^w pK_a$ are more appropriate, when considering either the solutes or the silanol groups. It may be that while there is a population of silanols that become ionized over the range of $_w^w\text{pH}$ 8 to $_w^w\text{pH}$ 10, there may be a further population of silanols whose ionization is suppressed only at the low pH of TFA.

1.3.3.4   Quantitation of Ionic Retention Effects on Different Columns

While ionic retention has been shown in many studies to contribute to the retention of ionized solutes in HILIC, particularly on bare silica phases, quantitation of these effects would be of interest such that the relative magnitude of the contribution could be gauged for different stationary phases. This ionic contribution could arise from residual silanols on silica-based phases (relatively few HILIC columns are based on an organic polymer matrix) as well as from ionogenic ligands deliberately bonded to the phase. Ionic retention can be studied by examining retention as a function of the mobile phase buffer concentration. Cox and Stout's studies of the retention mechanisms for basic compounds on bare silica columns under "pseudo-reversed-phase conditions" have already been mentioned [20]. The ion-exchange contribution to the retention of bases on silica can be expressed as

$$BH^+ + SiO^-M^+ \rightarrow SiO^-BH^+ + M^+ \tag{1.4}$$

where B is the base and M^+ represents the mobile phase counterions.

The ion-exchange equilibrium constant is:

$$K_{ix} = ([SiO^-BH^+][M^+])/([BH^+][SiO\text{-}M^+]). \tag{1.5}$$

The pH of the buffer controls the concentration of BH^+ through its ionization constant K_a

$$BH^+ \rightarrow B + H^+ \tag{1.6}$$

$$K_a = [B][H^+]/[BH^+] \tag{1.7}$$

Assuming that only the charged form BH^+ interacts with ionic sites on the stationary phase, the distribution coefficient between the stationary phase and mobile phase D_{ix} can be written as

$$D_{ix} = [SiO^-BH^+]/([BH^+]+[B]). \tag{1.8}$$

Rearranging Eq. 7 gives:

$$[B] = [BH^+]K_a/[H^+]. \tag{1.9}$$

Rearranging Eq. 5 gives

$$[BH^+] = ([SiO^-BH^+][M^+])/(K_{ix}[SiO^-M^+]). \tag{1.10}$$

Substituting Eq. 9 in Eq. 8 yields

$$D_{ix} = \frac{[SiO^-BH^+]}{[BH^+]+[BH^+]K_a/[H^+]} = \frac{[SiO^-BH^+]}{BH^+(1+K_a/[H^+])}. \tag{1.11}$$

Substituting Eq. 10 in Eq. 11 gives

$$D_{ix} = \frac{[SiO^-BH^+]}{\dfrac{[SiO^-BH^+][M^+](1+K_a/[H^+])}{K_{ix}[SiO^-M^+]}}. \tag{1.12}$$

This simplifies to

$$D_{ix} = \frac{K_{ix}[SiO^-M^+]}{[M^+]} \cdot \frac{1}{1+K_a/[H^+]}. \tag{1.13}$$

Thus, the distribution coefficient, and the retention factor, which is directly proportional to the distribution coefficient through the phase ratio, varies with the inverse of the counterion concentration in the mobile phase. A plot of the retention factor against the inverse of [M⁺] should be a straight line passing through the origin (assuming that no other retention mechanism exists), with the slope proportional to the ion-exchange equilibrium constant and the number of ionized sites (e.g., silanols on the silica surface). Alternatively, the presence of other retention mechanisms would be indicated by an intercept on the k-axis, which corresponds to an infinite competing ion concentration.

Figure 1.10 shows plots of k versus $1/[M^+]$ for each of the columns (used also in Fig. 1.3; specifications in Table 1.1) with 90% ACN containing overall buffer

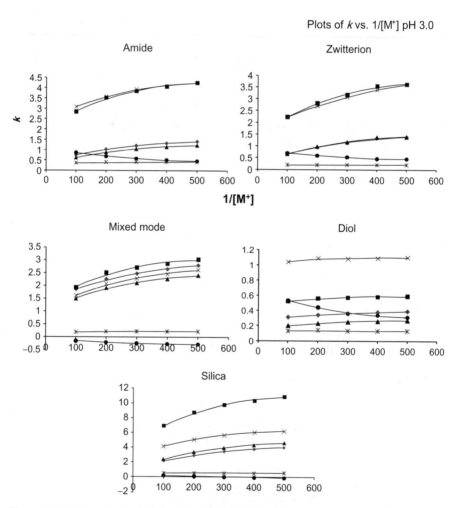

Figure 1.10. Plots of retention factor versus 1/[counter ion concentration] for five different HILIC columns. Solute identities: diamonds = nortriptyline, squares = procainamide, triangles = diphenhydramine, crosses = benzylamine, stars = caffeine, circles = p-xylenesulfonic acid. Mobile phase ACN–water (90:10, v/v) containing ammonium formate (concentration varied) ^w_w pH 3.0. Reprinted from Reference 26 with permission from Elsevier.

Table 1.2. Percentage Contribution of Ion Exchange to k at Two Different Levels of Counterion Concentration for Four Basic Solutes

Column	Procainamide 2 mM	Procainamide 10 mM	Benzylamine 2 mM	Benzylamine 10 mM	Diphenhydramine 2 mM	Diphenhydramine 10 mM	Nortriptyline 2 mM	Nortriptyline 10 mM
Silica	55	28	50	25	70	43	70	42
Diol	18	7	7	3	39	17	29	11
Zwitterionic	58	32	58	32	78	55	78	55
Mixed mode	54	27	56	27	55	27	49	22
Amide	49	24	41	19	73	46	72	47

concentration of 2–10 mM ^w_wpH 3.0 as mobile phase, and using the 4 basic solutes procainamide, benzylamine, diphenhydramine, and nortriptyline together with the neutral compound caffeine. The bases all have $pK_a > 9.0$ and thus should be completely protonated under the conditions of the experiment [26]. The plots were curves rather than straight lines, indicating that the buffer concentration has some additional effect on the separation mechanism other than merely competing with solute cations for stationary phase ionic retention sites. The points were fitted to a second-order polynomial expression, and the equation of the curves (which produced excellent empirical fits) was extrapolated to yield k for each solute at an infinite buffer concentration, which corresponds to the retention due to mechanisms other than ion exchange (i.e., the HILIC partition or adsorption mechanism). From the experimentally measured values of k for a particular solute and buffer concentration, the percent contribution of ion exchange could be determined, with results shown in Table 1.2. The percent contribution of ion exchange is much greater at a low counterion concentration, as expected due to reduced competition from the buffer ions. Thus, for the bare silica column, 55% of the retention of procainamide is estimated to be due to ionic processes at a 2-mM counterion concentration compared with only 28% at a 10-mM concentration. Overall, ionic retention is of considerable importance for the bare silica column, accounting for 50–70% of the total retention for the four basic probes studied at a 2-mM counterion concentration. Somewhat surprisingly, the contribution of ion exchange at a 2-mM counterion concentration was high for all columns, except the diol column. For the zwitterionic phase, retention of solute cations on the sulfonic acid functionality of the phase could explain the high retention due to ionic processes. For the amide and mixed-mode columns, the ionic retention must be due to ionized silanols on the underlying silica of these phases. It is possible even that an acidic silica (i.e., type A silica) might deliberately be used for the preparation of some bonded HILIC columns to increase their retention properties, although this comment is speculative, as few details of column preparation are revealed by commercial manufacturers. In contrast to the other columns, the diol column gave a contribution of ionic retention of only 3–39% for the four basic test compounds at a 2-mM counterion concentration, indicating possibly that the cross-linked stationary phase layer used in the particular type of column used in the study might give some shielding of the ionized silanols on the phase surface, or that the phase is bonded on a silica of low acidity.

Despite these considerations of the importance of ionic retention processes, it is clear that the retention on the silica column by nonionic processes is somewhat greater than the retention on the bonded-phase columns (k at an infinite buffer concentration is higher for the silica column as evidenced by the y-axis scales in Fig. 1.10). It is conceivable that there is a more extensive water layer on the silica column due to the greater concentration of polar silanol groups compared with the bonded silica phases; the surface area of the bare silica packing was also considerably higher than that of the other phases, as shown in Table 1.1. However, it seems likely that direct measurements of retention due to hydrophilic processes [35] will give a more accurate assessment of the situation.

Table 1.2 shows also that on all columns apart from the mixed-mode phase, the hydrophilic bases procainamide and benzylamine have a smaller proportion of their retention attributable to ion exchange than the hydrophobic bases diphenhydramine and nortriptyline. For example, on the zwitterion column, the percent contribution of ion exchange to retention was 58% for both benzylamine and procainamide using a 2-mM buffer concentration, but 78% for diphenhydramine and nortriptyline. For hydrophobic bases, there is likely to be smaller retention by conventional HILIC processes (e.g., solubility in a stationary phase water layer); thus, the contribution of ionic retention to overall retention on a given column will be greater. For the mixed-mode phase containing a long hydrocarbon chain, the hydrophilic and hydrophobic bases give rather similar percent contribution of ion exchange to total retention (range = 49–56%). It is likely that some hydrophobic retention of diphenhydramine and nortriptyline contributes to the overall retention of these compounds, making the ion-exchange contribution somewhat lower.

The question remains as to the cause of the curvature of the plots for the bases in Figure 1.10. The intercept on the k versus $1/[M^+]$ plots is indicative of a secondary retention mechanism (as discussed earlier), which could be partitioning into a water layer held on the column surface. If the secondary mechanism was simple and did not change with counterion concentration, then a straight line with a positive intercept should still result. It is possible that increasing the salt concentration might affect solute retention, for example, by increasing the thickness of a layer of water associated with the surface, in accord with the suggestions of Guo and Gaiki [25]. Thus, increasing salt concentration could be expected to increase the retention for all compounds.

It is interesting also to consider further the influence of the underlying base material of these bonded silica HILIC phases. Kumar et al. [6] studied the retention properties of catecholamines on a number of different phases. Catecholamines are biological amines released mainly from the adrenal gland in response to stress. The analysis of the compounds dopamine, epinephrine, and norepinephrine in biological fluids is important in hospital laboratories, as elevated levels can be indicative of tumors of the adrenal gland or neural tumors. Good separations of these compounds were shown on a silica-based zwitterionic sulfobetaine phase (ZIC–HILIC). A phase with the same bonded zwitterionic groups is also available based on a polymer matrix (ZIC–pHILIC). The catecholamines were analyzed using a mobile phase consisting of ammonium formate buffer, pH 3.0, at various concentrations over the range of 3.7–25 mM in 75% ACN, and plots of k versus the inverse of the buffer cation concentration on each column are shown in Figure 1.11. The catecholamines

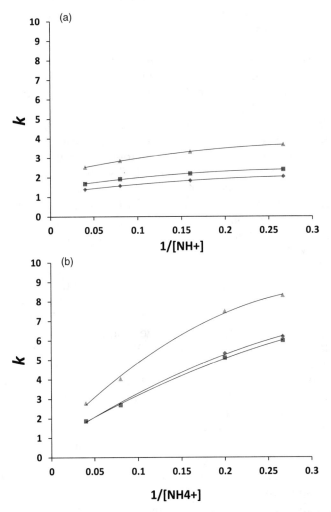

Figure 1.11. (a) Plots of k versus the inverse of buffer cation concentration using a ZIC–HILIC column. Mobile phases ammonium formate, pH 3.0 (3.7–25 mM) in 75% ACN. (b) As for (b) but using a ZIC–pHILIC column. Peak identities: diamonds = dopamine, squares = epinephrine, triangles = norepinephrine. Reprinted from Reference 6 with permission from Elsevier.

are strongly basic and should be protonated under the mobile phase conditions used. Figure 1.11 shows that retention decreases with increasing buffer strength on both columns, again giving curved plots, as had been shown for the basic solutes in Figure 1.10. Fitting the data to a second-order polynomial expression gave coefficients of determination $r^2 = 0.9984, 0.9978$, and 0.9991 for dopamine, norepinephrine, and epinephrine, respectively, using the polymer column, and similar fits were obtained for the silica-based column. However, it is clear that the slopes of the plots on the polymer column are much steeper than that for the silica column. Extrapolating the plots to an infinite buffer concentration allowed the HILIC contribution to retention to be determined. This contribution was estimated to be 8–14% for the three catecholamines on the polymer column using the lowest buffer concentration (3.75 mM), and 25–44% at the highest buffer concentration (25 mM). In comparison, for the silica-based column, the HILIC contribution was estimated to be 58–60% at a buffer concentration of 3.75 mM, and ~85% at 25 mM concentration. Thus, the HILIC contribution to retention of the polymer-based column was much smaller than that of the silica-based column, resulting in a much larger proportion of its retention being due to ionic processes. It seems likely that the ionic retention of the bases on the polymer column arises from interactions with the terminal sulfonic acid groups on the phase, although ionic contributions from the base polymer material itself cannot be completely discounted. Residues of catalyst material used in the production of the polymer can give rise to charged sites on polymers [41]. It is possible that the more hydrophobic matrix of the polymer column does not allow the formation of a water layer that is as extensive as that on a polar silica column. Thus, the contribution of HILIC retention to the overall process diminishes, leaving the influence of ion exchange to be more significant. However, much caution is necessary in such interpretations, as the exact details of column preparation are proprietary, and the silica and polymer columns are likely to differ in many other ways. A final practical consideration was that considerably lower efficiencies were obtained on the polymer column for the catecholamines (only about half those on the silica phase). This factor, which is also typical of the performance of RP columns, probably explains the predominance of silica-based columns in HILIC separations.

Of some further interest is the effect of the proportion of the ionic component of retention on efficiency. In RP separations, a mixed-mechanism process involving also ion exchange is generally considered detrimental to column efficiency and thus separation performance [42]. Figure 1.12 shows plots of column efficiency versus ammonium formate buffer concentration (2–10 mM at pH 3) for the hydrophilic base procainamide and the hydrophobic base nortriptyline on the five different columns of Table 1.1. The overall efficiencies shown were somewhat variable, with that of the amide column being lower than that of the other columns, which may be attributable to the polymeric nature of the bonded phase on this silica-based column. Nevertheless, there is little evidence on any of the columns of a detrimental effect of ionic retention on column efficiency for the solutes studied. This result may not hold, however, for other compounds. Efficiencies for the bare silica and mixed-mode (Dionex)

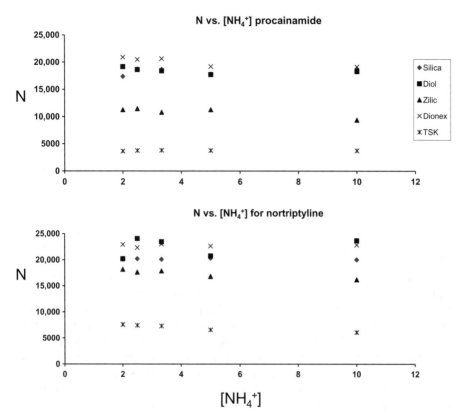

Figure 1.12. Plots of column efficiency against buffer cation concentration for five different HILIC phases (specifications detailed in Table 1.1). Mobile phase ACN–water (90:10, v/v) containing ammonium formate ^w_wpH 3.0, various concentrations. Reprinted from Reference 26 with permission from Elsevier.

column were high, while these columns have a large proportion of retention due to ionic processes at low buffer concentrations (see Table 1.2). Efficiency was maintained even at a low buffer concentration, where the contribution of ionic processes to retention was highest. This result also indicates that low buffer concentrations can be used successfully for MS applications, where sensitivity can be compromised in concentrated buffers [43]. However, the effect of solute structure on efficiencies for other compounds is not well understood. For example, the catecholamines gave poor peak shapes on bare silica columns, but excellent results on the zwitterionic phase [6]. Thus, the effect of ion-exchange processes on column efficiency may be solute dependent.

1.3.4   RP Retention on Bare Silica

Evidence for reversed-phase retention on bare silica columns at relatively low concentrations of organic solvent has existed at least since the publication of Bidlingmeyer [16]. A separation of the same test solutes used in Figure 1.3 was

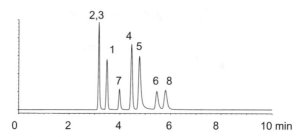

Figure 1.13. Analysis of test compounds using Atlantis silica. Mobile phase 10% ACN in 1.5 mM HCOONH₄, pH 3.0. Peak identities are the same as for Figure 1.3. Reprinted from Reference 5 with permission from Elsevier.

demonstrated on a bare silica column [5] using 10% ACN containing 1.5 mM ammonium formate, $_w^w$pH 3.0 (see Fig. 1.13). Reasonable peak shapes were still obtained, although some tailing was shown (e.g., the asymmetry factor for peak 5, nortriptyline was 2.1, compared with 1.0 for the silica column under true HILIC conditions in Fig. 1.3). The order of elution of the peaks differed from that obtained using the same column under HILIC conditions, showing lower retention of the hydrophilic base benzylamine (peak 7) and greater retention of caffeine (peak 4). Decreasing the concentration of ACN over the range of 10–2.5% ACN increased the retention of all solutes as would be expected for a RP-type mechanism, but in contradiction to changing its concentration when in the HILIC range (ACN concentration > 60%) [5]. These results may reflect RP retention on siloxane bonds of the bare silica, as suggested previously [16].

Lindner and coworkers [9] examined the separation properties of novel and commercial polar stationary phases in both the HILIC and the RP-LC mode. The novel phases included oxidized 2-mercaptoethanol and oxidized 1-thioglycerol packings, as well as three commercial diol phases. These phases were all devoid of ionic stationary phase groups, although the authors noted that ionic interactions arising from the base silica could be present. The effect of the ACN fraction in the hydro-organic eluent was studied in the range of 5–95% v/v for solutes such as adenosine and uridine. Both solutes experienced markedly increased retention at higher concentrations of ACN (>60% v/v), in line with the supposition that a HILIC retention mechanism was operating at these levels. On the other hand, at low concentrations of ACN (<20% v/v), a significant RP type of retention was observed for the purine base adenosine, to a lesser extent also for guanosine, generating U-shaped curves of retention factor versus ACN concentration. This effect was not observed for the pyrimidine bases uridine and cytidine or the nucleobases cytosine and uracil. The effect was most pronounced with the nonoxidized and therefore less polar phases.

More recently, Sandra and coworkers have coined the term "per aqueous liquid chromatography" (PALC) to describe separations obtained on polar columns using eluents that are 100% aqueous or contain only low concentrations of organic solvents. These applications are recommended as examples of "green chemistry" and overcome one of the main drawbacks of classical HILIC separations—the environmental cost of using toxic solvents [44]. While not strictly using HILIC conditions, these separations are of interest here because the retention mechanism may be contributory to that of HILIC, at least in situations where the organic content of the mobile phase is relatively low. Separations of some neurotransmitters on a bare silica column have been reported using this technique [44]. However, in later papers, Gritti and Guiochon (together with the original authors) showed that the surface of silica columns when operated with PALC mobile phases was seriously heterogeneous, with up to five different adsorption sites, including a small number of very strong sites [45,46]. They concluded from theoretical and practical considerations that better sensitivity, higher efficiency, and better resolution could be obtained in the conventional HILIC mode where the adsorption mechanism was found to be much more homogeneous. A problem with the PALC mode was found to be serious overloading of the few strong column sites even with very small amounts of some solutes, giving a poor peak shape. PALC separations with moderate solute k gave the worst column efficiencies, and only addition of ACN, resulting in very small solute k, and apparent blocking of these strong sites by this solvent, gave reasonable column efficiency. In a more recent study, Sandra and coworkers showed some separations on polyethyleneglycol (PEG) or diol columns with either 100% water or water containing 1% ACN or ethanol at 60°C giving efficiencies of up to 76,000 plates/m for solutes such as caffeine, acetophenone, aniline, phenol, toluene, and benzene. However, the authors reported some phase instability of the PEG column under PALC conditions [47].

1.3.5   Electrostatic Repulsion Hydrophilic Interaction Chromatography (ERLIC): A New Separation Mode in HILIC

In 2008, Alpert proposed a new variant of HILIC that has the potential to make significant contributions to the separation of biomolecules [48]. He noted that an elution gradient is often used to ensure that all solutes in a mixture elute in the same time frame. This is true particularly for the separation of biologically important molecules such as peptides, where the individual components of the mixture can have widely different retention times. However, an alternative strategy is to superimpose a second mode of chromatography on the primary separation mechanism that selectively reduces the retention of solutes that are usually the most strongly retained. The new mode uses coulombic effects superimposed on the usual HILIC separation mechanism and was termed ERLIC or electrostatic repulsion hydrophilic interaction chromatography. It uses an ion-exchange column of charge of the same sign as that on the solutes. In this paper, Alpert re-iterated the model for the retention mechanism of HILIC being mostly partitioning between the dynamic mobile phase and a slow moving layer of water with which the polar stationary phase is hydrated.

Alpert noted that gradients for HILIC involve increasing the polarity of the mobile phase, typically by decreasing the concentration of organic solvent. However, it is also possible to use increasing salt concentrations in a mobile phase containing 60–70% organic solvent. If a cation exchange column is used to separate acidic amino acids, the solutes will elute prior to the void volume of the column, as electrostatic repulsion prevents access of these solutes (which have the same charge as the column groups) to the full pore volume of the stationary phase. However, if the mobile phase contains >60% organic solvent, then acidic amino acids show almost the same retention on cation exchange columns as given by neutral columns, as had been shown by Alpert's original experiments using PolySulfoethylA and PolyHydroxyethyl phases [10]. The rationale given for this result was that hydrophilic interactions are independent of electrostatic effects. If sufficient organic solvent is used in the mobile phase, then hydrophilic interaction dominates solute retention. Another example discussed was that phosphate groups decrease the retention of basic histone proteins on a cation exchange column (presumably because the increased negative charge on the solute produces repulsion from the negatively charged column sites) in the absence of an organic solvent [49]. However, under HILIC conditions with the mobile phase containing 70% ACN, the phosphate groups lead to a net increase in retention of the protein. The hydrophilic interaction conferred by the phosphate groups acting to increase retention is stronger than the electrostatic repulsion, which decreases retention.

Alpert argued that basic solutes are usually the most retained in HILIC, followed by phosphorylated solutes [10]. Thus, gradients are necessary to separate samples that contain very basic peptides or strongly phosphorylated compounds such as adenosine triphosphate (ATP). However, if an anion exchange column was used in the HILIC mode, gradients should be unnecessary. An example of the application of this type of ERLIC separation is shown in Figure 1.14, which demonstrates the simultaneous separation of basic and acidic peptides using an isocratic method. Usually, basic peptides have much stronger retention than acidic peptides on a neutral HILIC column like PolyHydroxyethyl A. However, use of a PolyWAX LP column, a weak anion exchange material, gives repulsion of the positively charged basic compounds, reducing their retention to values similar to that of acidic peptides (whose retention is increased). Another possible application is the use of an anion exchange column at low pH, under which conditions tryptic peptides from protein digests are mostly uncharged at the carboxyl end, leaving peptides with a net positive charge. These peptides are repelled from the positively charged stationary phase, leaving peptides with phosphate groups or glycopeptides with sialic acid residues that retain negative charge under these conditions to be retained selectivity. Note that if a classical anion exchange column is used, the presence of a single phosphate still produces low retention of the peptide, due to repulsion of the positive ends of the peptide.

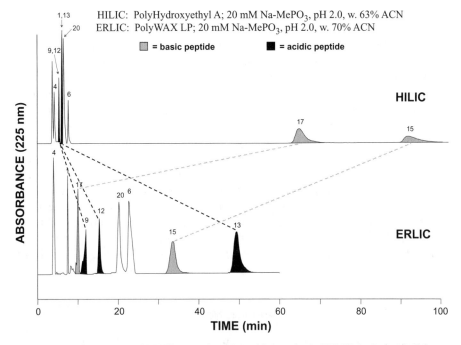

Figure 1.14. HILIC versus ERLIC separation of peptide standards. HILIC mode (top). Column: PolyHydroxyethyl A. Mobile phase: 20 mM Na-MePO₄, pH 2.0, with 63% ACN. Flow rate: 1.0 mL/min. ERLIC mode (bottom). Column: PolyWAX LP. Mobile phase: 20 mM Na-MePO₄, pH 2.0, with 70% ACN. Flow rate: 1.0 mL/min. Reprinted with permission from *Anal. Chem.* 2008; **80**: 62–76. Copyright (2008) American Chemical Society.

However, with the superimposed HILIC mechanism, retention of such compounds can be achieved.

Salt concentration is a critically important parameter in ERLIC separations in determining selectivity. Increasing levels of salt shield solutes from all electrostatic effects, both attractive and repulsive, and at high salt concentrations, the selectivity converges on that of HILIC. Using an anion exchange column in the ERLIC mode, retention of acidic peptides was shown to decrease as expected for acidic peptides (which undergo coulombic attraction with the column groups) but to increase for basic peptides (which undergo coulombic repulsion) with increasing salt concentrations. At the highest salt concentrations studied (120 mM NaMePO₄, pH 2.0) with 65% ACN, basic peptides once again became the most retained.

Alpert showed a number of other applications of ERLIC, including the separation of acidic, basic, and neutral amino acids, and the separation of nucleotides without recourse to gradients. Clearly, this new separation mechanism has much potential, particularly for the separation of molecules of biological significance.

1.4 CONCLUSIONS

Interest in HILIC separations has increased rapidly, in particular over the last 5 years. For the separation of polar, hydrophilic, or ionized compounds, HILIC shows many advantages over RP-LC. A better understanding of the mechanism of these separations is emerging, although the technique is not nearly so well understood as RP-LC. Contributory mechanisms are likely to be partition, adsorption, ionic interactions, and even hydrophobic retention depending on the experimental conditions. Although for samples to which it is applicable, HILIC has many advantages over RP-LC, the limitations of HILIC should also not be overlooked. These include problems with the solubility of some solutes, particularly in preparative separations, the longer time required for column equilibration than in RP, and the lack of applicability to the large number of solutes that are insufficiently polar. The lack of understanding of the HILIC method is also a barrier to the development of new analytical methods. Nevertheless, it seems that HILIC is a technique that is now firmly established as a complementary approach to RP analysis.

REFERENCES

1. Web of Knowledge, Thomson-Reuters, 2011. http://wok.mimas.ac.uk/
2. Grumbach ES, Wagrowski-Diehl DM, Mazzeo JR, Alden B, Iraneta PC. Hydrophilic interaction chromatography using silica columns for the retention of polar analytes and enhanced ESI-MS sensitivity. *LC-GC N. Am* 2004; **22**: 1010–1023.
3. Colin H, Diez-Masa JC, Guiochon G, Czajkowska T, Miedziak I. Role of temperature in reversed-phase high performance liquid chromatography using pyrocarbon-containing adsorbents. *J. Chromatogr.* 1978; **167**: 41–65.
4. McCalley DV. Evaluation of the properties of a superficially porous silica stationary phase in hydrophilic interaction chromatography. *J. Chromatogr. A* 2008; **1193**: 85–91.
5. McCalley DV. Is hydrophilic interaction chromatography with silica columns a viable alternative to reversed-phase liquid chromatography for the analysis of ionisable compounds? *J. Chromatogr. A* 2007; **1171**: 46–55.
6. Kumar A, Hart JP, McCalley DV. Determination of catecholamines in urine using hydrophilic interaction chromatography with electrochemical detection. *J. Chromatogr. A* 2011; **1218**: 3854–3861.
7. Bicker W, Wu JY, Lämmerhofer M, Lindner W. Hydrophilic interaction chromatography in nonaqueous elution mode for separation of hydrophilic analytes on silica-based packings with noncharged polar bondings. *J. Sep. Sci.* 2008; **31**: 2971–2987.
8. Lämmerhofer M, Richter M, Wu J, Nogueira R, Bicker W. Mixed-mode ion-exchangers and their comparative chromatographic characterization in reversed-phase and hydrophilic interaction chromatography elution modes. *J. Sep. Sci.* 2008; **31**: 2572–2588.
9. Wu J, Bicker W, Lindner W. Separation properties of novel and commercial polar stationary phases in hydrophilic interaction and reversed-phase liquid chromatography mode. *J. Sep. Sci.* 2008; **31**: 1492–1503.

10. Alpert AJ. Hydrophilic interaction chromatography for the separation of peptides, nucleic acids and other polar compounds. *J. Chromatogr.* 1990; **499**: 177–196.

11. Martin AJP, Synge RLM, Biochem J. A new form of chromatogram employing two liquid phases: A theory of chromatography. 2. Application to the micro-determination of the higher monoamino-acids in proteins. *Biochem. J.* 1941; **35**: 1358–1368.

12. Linden JC, Lawhead CL. Liquid chromatography of saccharides. *J. Chromatogr.* 1975; **105**: 125–133.

13. Verhaar LATh, Kuster BFM. Contribution to the elucidation of the mechanism of sugar retention on amine-modified silica in liquid chromatography. *J. Chromatogr.* 1982; **234**: 57–64.

14. Neue UD. *HPLC Columns: Theory, Technology and Practice.* New York: Wiley-VCH; 1997.

15. Nikolov ZL, Reilly PJ. Retention of carbohydrates on silica and amine-bonded stationary phases-application of the hydration model. *J. Chromatogr.* 1985; **325**: 287–293.

16. Bidlingmeyer BA, Del Rios JK, Korpi J. Separation of organic amine compounds on silica-gel with reversed-phase eluents. *Anal. Chem.* 1982; **52**: 442–447.

17. Jane I. Separation of a wide range of drugs of abuse by high pressure liquid chromatography. *J. Chromatogr.* 1975; **111**: 227–233.

18. Flanagan RJ, Jane I. High-performance liquid-chromatographic analysis of basic drugs on silica columns using non-aqueous ionic eluents.1. Factors influencing retention, peak shape and detector response. *J. Chromatogr.* 1985; **323**: 173–189.

19. McKeown AP, Euerby MR, Lomax H, Johnson CM, Ritchie H, Woodruff M. The use of silica for liquid chromatographic/mass spectrometric analysis of basic analytes. *J. Sep. Sci.* 2001; **24**: 835–842.

20. Cox GB, Stout RW. Study of the retention mechanisms for basic compounds on silica under pseudo-reversed-phase conditions. *J. Chromatogr.* 1987; **384**: 315–336.

21. Kadar EP, Wujcik CE, Wolford DP, Kavetskaia O. Rapid determination of the applicability of hydrophilic interaction chromatography utilizing ACD Labs Log D Suite: A bioanalytical application. *J. Chromatogr. B* 2008; **863**: 1–8.

22. ACD log D Suite version 9.0, Reference Manual, Advanced Chemistry Development Inc., 2005.

23. Bicker W, Wu J, Yeman H, Albert K, Lindner W. Retention and selectivity effects caused by bonding of a polar urea-type ligand to silica: A study on mixed-mode retention mechanisms and the pivotal role of solute-silanol interactions in the hydrophilic interaction chromatography elution mode. *J. Chromatogr. A* 2011; **1218**: 882–895.

24. Chirita RI, West C, Zubrzycki S, Finaru SL, Elfakir C. Investigations on the chromatographic behaviour of zwitterionic stationary phases used in hydrophilic interaction chromatography. *J. Chromatogr. A* 2011; **1218**: 5939–5963.

25. Guo Y, Gaiki S. Retention behavior of small polar compounds on polar stationary phases in hydrophilic interaction chromatography. *J. Chromatogr. A* 2005; **1074**: 71–80.

26. McCalley DV. Study of the selectivity, retention mechanisms and performance of alternative silica-based stationary phases for separation of ionised solutes in hydrophilic interaction chromatography. *J. Chromatogr. A* 2010; **1217**: 3408–3417.

27. Liu X, Pohl C. New hydrophilic interaction/reversed-phase mixed-mode stationary phase and its application for analysis of nonionic ethoxylated surfactants. *J. Chromatogr. A* 2008; **1191**: 83–89.

28. McCalley DV, Neue UD. Estimation of the extent of the water-rich layer associated with the silica surface in hydrophilic interaction chromatography. *J. Chromatogr. A* 2008; **1192**: 225–229.

29. Melnikov SM, Hoeltzel A, Seidel-Morgenstern A, Tallarek U. Composition, structure, and mobility of water-acetonitrile mixtures in a silica nanopore studied by molecular dynamics simulations. *Anal. Chem.* 2011; **83**: 2569–2575.

30. Wikberg E, Sparrman T, Viklund C, Johnsson T, Irgum K. A (2)H nuclear magnetic resonance study of the state of water in neat silica and zwitterionic stationary phases and its influence on the chromatographic retention characteristics in hydrophilic interaction high-performance liquid chromatography. *J. Chromatogr. A* 2011; **1218**: 6630–6638.

31. Hemström P, Irgum K. Hydrophilic interaction chromatography. *J. Sep. Sci.* 2006; **29**: 1784–1821.

32. Orth P, Engelhardt H. Separation of sugars on chemically modified silica-gel. *Chromatographia* 1982; **15**: 91–96.

33. Guo Y, Huang A. A HILIC method for the analysis of tromethamine as the counter ion in an investigational pharmaceutical salt. *J. Pharm. Biomed. Anal.* 2003; **31**: 1191–1201.

34. Li R, Zhang Y, Lee CC, Liu LM, Huang YP. Hydrophilic interaction chromatography separation mechanisms of tetracyclines on amino-bonded silica column. *J. Sep. Sci.* 2011; **34**: 1508–1516.

35. Dinh NP, Jonsson T, Irgum K. Probing the interaction mode in hydrophilic interaction chromatography. *J. Chromatogr. A* 2011; **1218**: 5880–5891.

36. Young JE, Matyska MT, Pesek JJ. Liquid chromatography/mass spectrometry compatible approaches for the quantitation of folic acid in fortified juices and cereals using aqueous normal phase conditions. *J. Chromatogr. A* 2011; **1218**: 2121–2126.

37. Olsen BA. Hydrophilic interaction chromatography using amino and silica columns for the determination of polar pharmaceuticals and impurities. *J. Chromatogr. A* 2001; **913**: 113–122.

38. McCalley DV. Overload for ionized solutes in reversed-phase high-performance liquid chromatography. *Anal. Chem.* 2006; **78**: 2532–2538.

39. McCalley DV. Rationalization of retention and overloading behavior of basic compounds in reversed-phase HPLC using low ionic strength buffers suitable for mass spectrometric detection. *Anal. Chem.* 2003; **75**: 3404–3410.

40. Gagliardi LG, Castells CB, Rafols C, Rosés M, Bosch E. Delta conversion parameter between pH scales ((s)(w)pH and (s)(s)pH) in acetonitrile/water mixtures at various compositions and temperatures. *Anal. Chem.* 2007; **79**: 3180–3187.

41. Buckenmaier SMC, McCalley DV, Euerby MR. Overloading study of bases using polymeric RP-HPLC columns as an aid to rationalization of overloading on silica-ODS phases. *Anal. Chem.* 2002; **74**: 4672–4681.

42. McCalley DV. The challenges of the analysis of basic compounds by high performance liquid chromatography: Some possible approaches for improved separations. *J. Chromatogr. A* 2010; **1217**: 858–880.

43. Temesi D, Law B. Factors to consider in the development of generic bioanalytical high-performance liquid chromatographic-mass spectrometric methods to support drug discovery. *J. Chromatogr. B* 2000; **748**: 21–30.

44. Pereira AD, David F, Vanhoenacker G, Sandra P. The acetonitrile shortage: Is reversed HILIC with water an alternative for the analysis of highly polar ionizable solutes? *J. Sep. Sci.* 2009; **32**: 2001–2007.

45. Gritti F, Pereira AD, Sandra P, Guiochon G. Comparison of the adsorption mechanisms of pyridine in hydrophilic interaction chromatography and in reversed-phase aqueous liquid chromatography. *J. Chromatogr. A* 2009; **1216**: 8496–8504.

46. Gritti F, Pereira AD, Sandra P, Guiochon G. Efficiency of the same neat silica column in hydrophilic interaction chromatography and per aqueous liquid chromatography. *J. Chromatogr. A* 2010; **1217**: 683–688.

47. Pereira AD, Higashi N, Mitsui K, Kanda H, David F, Sandra P. Evaluation of diol and polyethylene glycol columns for the analysis of ionisable solutes by different chromatographic modes. Poster presented at the 36th International Symposium on High Performance Liquid Phase Separations and Related Techniques (HPLC 2011), Budapest, Hungary, June 2011.

48. Alpert AJ. Electrostatic repulsion hydrophilic interaction chromatography for isocratic separation of charged solutes and selective isolation of phosphopeptides. *Anal. Chem.* 2008; **80**: 62–76.

49. Lindner H, Sarg B, Helliger W. Application of hydrophilic interaction liquid chromatography to the separation of phosphorylated H1 histones. *J. Chromatogr. A* 1997; **782**: 55–62.

CHAPTER

2

STATIONARY PHASES FOR HILIC

MOHAMMED E.A. IBRAHIM and CHARLES A. LUCY

Department of Chemistry, University of Alberta, Gunning/
Lemieux Chemistry Centre, Edmonton, Alberta, Canada

2.1 INTRODUCTION

As noted in Chapter 1, literature and research on hydrophilic interaction liquid chromatography (HILIC) has increased dramatically in recent years. This has been accompanied by a correspondingly rapid increase in stationary phases developed for HILIC. The term "HILIC" was first coined by Alpert for the separation of polar analytes such as peptides and carbohydrates [1]. Usually a mixture of water and a high percentage of an organic modifier, in most cases acetonitrile (ACN), is used with a polar stationary phase. This polar phase encourages the formation of a water layer on its surface. According to Alpert's theory [1], partitioning of analytes between the formed water layer and the mobile phase constitutes the major mechanism of retention in HILIC as shown in Figure 2.1.

Thus, all HILIC stationary phases should be hydrophilic to encourage the formation of a stagnant water layer into which the analytes partition. In general, the retentivity increases with the polarity of the stationary phase, that is, the more hydrophilic the functional groups on the stationary phase, the

Hydrophilic Interaction Chromatography: A Guide for Practitioners, First Edition.
Edited by Bernard A. Olsen and Brian W. Pack.
© 2013 John Wiley & Sons, Inc. Published 2013 by John Wiley & Sons, Inc.

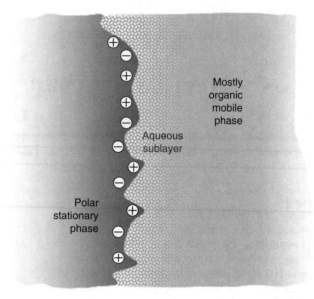

Figure 2.1. Formation of the water-rich layer on the surface of stationary phase under HILIC conditions. Reproduced from Reference 2 with permission from Sielc Technologies.

longer the retention of polar analytes due to the formation of a richer water layer. However, as discussed in Chapter 1, numerous secondary interactions (e.g., electrostatic attraction and repulsion) also affect retention and selectivity in HILIC. Section 2.2 discusses the chemistry of the different classes of HILIC stationary phases. Section 2.3 provides a comparison of various commercial HILIC columns.

2.2 HILIC STATIONARY PHASES

In this section, all classes of stationary phases that have been used in HILIC mode are discussed in terms of chemistry, available trade names, and representative applications.

2.2.1 Underivatized Silica

Although the term HILIC was coined by Alpert in 1990 using derivatized silica phases [1], the same approach had been applied by Jane in 1975 using an underivatized silica phase [3]. Bare silica columns (without any modifications) have been widely used in HILIC [4–13], particularly in liquid chromatography-mass spectrometry (LC-MS) methods. In general, HILIC is attractive for LC-MS as the high organic content in the mobile phase lessens the problem of ion suppression, in particular with electrospray ionization (ESI) [14]. Underivatized silica HILIC phases are further attractive due to the

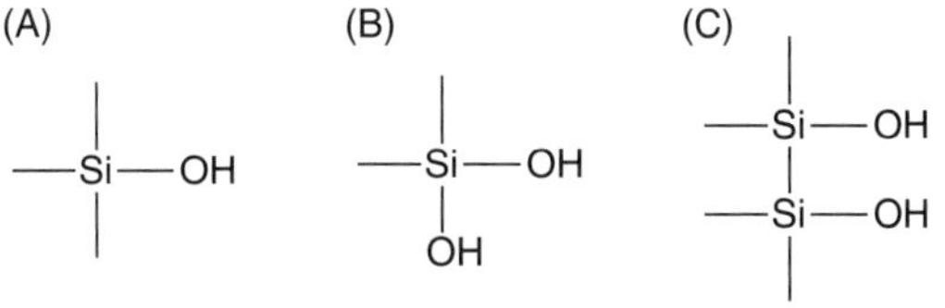

Figure 2.2. Different types of silanols on the surface of silica.

absence of ligands, which could otherwise leach from the column and appear as extra peaks in the mass spectrum [15].

Silanol groups (–SiOH) are the key chemical feature of hydrated silica surfaces [16]. Their surface concentration is 8 μmoles/m^2. These silanol groups can react with silanes to form the bonded phases discussed in Section 2.2.2. The silanol groups on the silica surface may be free, geminal, or associated, as shown in Figure 2.2A–C [17–21]. Treatment of the silica affects the distribution and type of the silanol groups. High-temperature treatment of silica converts geminal and associated silanols into free silanols, which are more acidic than geminal and associated silanols. The high acidity of free silanols contributes to the peak tailing and low efficiency (N) values for basic analytes. However, treatment at temperatures higher than 800°C removes all active silanols and leaves hydrophobic siloxane bridges only, rendering the silica more hydrophobic [22].

The acidity of silanols is also affected by the purity of the silica itself. The presence of contaminant metals such as Al^{3+} and Fe^{3+} increases the acidity of silica by withdrawing electrons from the oxygen atoms of the silanol groups. According to the degree of purity, silica can be classified into type A or type B. Type A silica is less pure and more acidic than type B silica and was widely used prior to 1990. Currently, type A silica is reserved for primitive applications including sample preparation and preparative chromatography. On the other hand, type B silica is prepared carefully in a metal-free environment to prevent any contamination. Thus, Type-B silica is less acidic and has a lower tendency to generate tailed peaks with basic solutes as compared with type A silica. The term type C silica does not relate to the silica purity but rather indicates the phase is produced through hydrosilylation [23], where surface silanols (Si–OH) are replaced by silicon hydride (Si–H). Cogent Silica-C™ from Microsolv is a good example of a Type-C silica column (Table 2.1). Separation of phenylalanine and phenyglycine has been achieved on a 4-μm Cogent Silica-C™ (100 Å) under HILIC conditions [24].

2.2.1.1 Totally Porous Silica Particles

Silica phases may be totally porous, superficially porous, or monolithic. Generally, silica is characterized by high mechanical strength so silica can withstand high pressures, as compared with polymeric phases, producing uniform peaks and higher N values. Totally porous silica particles (TPP) are widely used due to their greater column capacity, which enables injection of

Table 2.1. Characteristics of Commercial HILIC Phases

Column No.	Brand Name	Manufacturer	Support	Functionality	Particle Size (µm)	Pore Size (Å)	Surface Area (m²/g)	Column Length (mm)	Column Diameter (mm)
Underivatized silica phases									
14	Atlantis HILIC Si	Waters	Silica	Underivatized	3,5	100	330	30–250	1.0–4.6
	Betasil Silica	Thermo Electron Corporation	Silica	Underivatized	3,5	100	—	50–250	1.0–4.6
	Hypersil Silica	Thermo Electron Corporation	Silica	Underivatized	3–30	120	—	30–300	1.0–4.6
13	Kromasil Silica	Akzo Nobel	Silica	Underivatized	3.5–16	60–300	—	50–250	2.1–50
	Chromolith Si	Merck	Silica	Underivatized	N/A	130	300	100	4.6
15	Onyx Si	Phenomenex	Silica	Underivatized	N/A	130	300	100	4.6
	Purospher STAR Si	Merck	Silica	Underivatized	5	120	330	125–250	4.0–4.6
16	LiChrospher Si 100	Merck	Silica	Underivatized	5	100	400	50–300	2.1–4.6
17	LiChrospher Si 60	Merck	Silica	Underivatized	5	60	700	50–300	2.1–4.6
	Spheri-5 Silica	PerkinElmer Brownlee	Silica	Underivatized	5	80	180	250	4.6
18	Spherisorb Silica	Waters	Silica	Underivatized	3–10	80	220	20–250	1.0–4.6
	Cogent silica-C	Microsolv	Silica	Underivatized (Type C)	4	100	350	100	4.6
	HALO HILIC	Advanced materials technology	Silica	Underivatized (Type B)	2.7	90	150	50–150	2.1–4.6

	Manufacturer	Base	Bonded phase	Particle size	Pore	Surface area	Length	Diameter
BEH phases								
Acquity UPLC BEH HILIC	Waters	Silica	BEH	1.7–10	130–300	185	50–150	1.0–3.0
Amide silica phases								
7–8[a] TSKgel Amide-80	Tosoh Bioscience	Silica	Amide	3–10	80	450	50–300	1.0–21.5
GlycoSep N	ProZyme	Silica	Amide	5	—	—	250	4.6
Diol silica phases								
10 LiChrospher 100 Diol	Merck	Silica	2,3-dihydroxypropyl	5	100	350	50–300	2.1–4.6
LiChrosorb 100 Diol	Merck	Silica	2,3-dihydroxypropyl	5,10	100	300	50–300	2.1–4.6
Inertsil Diol	GL Sciences	Silica	2,3-dihydroxypropyl	3,5	100	450	33–250	1.0–4.0
YMC-pack Diol NP	YMC	Silica	2,3-dihydroxypropyl	5	60–300	100–330	50–250	2.0–4.6
11 Luna HILIC	Phenomenex	Silica	Cross-linked diol	3,5	200	185	30–100	2.0–21.2
Cyanopropyl silica phases								
LiChrospher 100 CN	Merck	Silica	3-cyanopropyl	5,10	100	350	125–250	4.0
Altima HP Cyano	Grace Alltech	Silica	3-cyanopropyl	3,5	190	200	50–250	2.1–4.6
Spherisorb CN	Waters	Silica	3-cyanopropyl	3,5,10	80	220	20–250	1.0–4.6

(*Continued*)

47

Table 2.1. (Continued)

Column No.	Brand Name	Manufacturer	Support	Functionality	Particle Size (μm)	Pore Size (Å)	Surface Area (m²/g)	Column Length (mm)	Column Diameter (mm)
Cyclodextrin-based silica phases									
	Nucleodex β-OH	Macherey-Nagel	Silica	B-cyclodextrin	5	100	—	200	4.0
	Cyclobond I 2000	ASTEC	Silica	B-cyclodextrin	5,10	100	—	50–250	2.1–10.0
Zwitterionic phases									
1–3[b]	ZIC-HILIC	Merck	Silica	Sulfoalkylbetaine	3,5,5	100,200	135,180	100–250	2.1–4.6
4	ZIC-pHILIC	Merck	Polymer	Sulfoalkylbetaine	5	—	—	50–150	2.1–4.6
5	Nucleodur HILIC	Macherey-Nagel	Silica	Sulfoalkylbetaine	1.8–5	110	340	30–250	2.0–4.6
	Obelisc N	SiELC	Silica	Unspecified	5,10	100	—	10–250	1.0–4.6
6	PC HILIC	Shiseido	Silica	Phosphoryl-choline	5	100	450	100–250	2.0–4.6
Aminopropyl silica phases									
19	LiChrospher 100 NH₂	Merck	Silica	3-Aminopropyl	5	100	350	125–300	3.2–4.6
20	Purospher STAR NH₂	Merck	Silica	3-Aminopropyl	5	120	330	125–250	4.0–4.6
	Luna NH₂	Phenomenex	Silica	3-Aminopropyl	3–10	100	400	250	4.6
	Hypersil APS-2 (Amino)	Thermo scientific	Silica	3-Aminopropyl	3–10	120	170	30–250	2.1–4.6
	Spherisorb NH₂	Waters	Silica	3-Aminopropyl	3–10	80	220	20–250	1.0–4.6
	Zorbax NH₂	Agilent	Silica	3-Aminopropyl	5,7	70	300	50–250	4.6–21.2
21	TSKgel NH₂-100	Tosoh Bioscience	Silica	Aminoalkyl	3	100	450	50–150	2.0–4.6

Latex coated silica phases									
	AS9-SC silica	Home made	Silica monolith	Quaternary ammonium salts	N/A	130	300	100	4.6
Poly(succinimide) silica phases									
12	Polysulfo-ethyl A	PolyLC	Silica	Poly(2-sulfoethyl aspartamide)	3-12	200	188	35-250	1.0-50.8
	PolyCAT A	PolyLC	Silica	Poly(aspartic acid)	3-12	200	188	35-250	1.0-50.8
9	Polyhydroxy-ethyl A	PolyLC	Silica	Poly (2-hydroxyethyl aspartamide)	3-12	200	188	35-250	1.0-50.8
Amino phases									
	Styros AminoHILIC	Orachrom Inc	Poly(styrene-divinyl benzene)	Amino	N/A	1000–2000	—	33-250	1.0-20
Sulfonated S-DVB phases									
	Agilent Hi-Plex H	Agilent	Styrene-divinyl benzene	Sulfonic acid	8	—	—	300	6.5-7.7

[a]Column no. 7 for TSKgel Amide-80 (5 μm, 100 × 4.6 mm inner diameter [ID]); Column no. 8 for TSKgel Amide-80 (3 μm, 50 × 4.6 mm ID).
[b]Column no. 1 for ZIC-HILIC (5 μm, 100 × 4.6 mm ID, 200 Å); Column no. 2 for ZIC-HILIC (3.5 μm, 150 × 4.6 mm ID, 200 Å); Column no. 3 for ZIC-HILIC (3.5 μm, 150 × 4.6 mm ID, 100 Å).

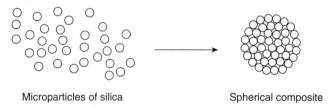

Microparticles of silica   Spherical composite

Figure 2.3. Assembly of small silica particles into spherical aggregates. Reprinted from Reference 16 with permission from Wiley.

larger sample mass, and due to their availability in a wider variety of dimension options. The most common particle diameters are in the range of 1.5–5.0 μm. TPP are prepared either by sol–gel synthesis or aggregation (assembly) of smaller particles.

The sol–gel procedure involves the emulsification of a silica solution (*sol*) in an immiscible nonpolar solvent. Droplets of this emulsified sol are converted into spherical beads of silica hydrogel. These beads are then dried and classified into a narrower particle size range. Controlling the pH, temperature, and concentration of the silica sol enables production of silica particles with the desired particle and pore sizes [16].

Aggregation of smaller particles (Fig. 2.3) is an alternative approach for preparation of TPP. In this approach, a silica sol of a definite particle size is dispersed into a polar liquid followed by addition of a polymerizable material such as melamine. The polymerizable material initiates the coacervation of the silica particles to form spherical aggregates of uniform size. These aggregates are then sintered at high temperature to strengthen the network of the silica sol particles. Generally, the size of the silica sol particle used to prepare the aggregate particle dictates the size of the resultant pores [16].

2.2.1.2   *Superficially Porous (Core Shell) Silica Particles*

Superficially porous particles (SPP) consist of a solid core (1.5–5 μm) coated with a porous outer silica shell (0.25–0.5 μm). These phases are characterized by their high N values [25,26]. The surface areas of SPP are about three-fourths that of totally porous particles, but substantially greater than pellicular particles. The HALO HILIC phase (Table 2.1) is a representative example of the core-shell silica particles [27]. The HALO HILIC phase consists of a 1.7-μm solid core particle with a 0.5-μm type B porous silica layer fused to the surface. Gritti et al. recently demonstrated the van Deemter behavior of a 150 × 4.6 mm HILIC column packed with 2.7-μm HALO particles under HILIC conditions [28].

2.2.1.3   *Monolithic Silica*

A monolithic column consists of a single piece of porous material, rather than a column packed with discrete particles. Monolithic columns are characterized by the presence of large macropores (1–3 μm) for through-flow of mobile phase and relatively small mesopores (10–25 nm) to provide the surface area for retention [29–31]. Monoliths may be either polymeric or inorganic, with the latter being predominantly silica. Polymeric monoliths usually swell or shrink in the presence of organic solvents such as used in HILIC, leading to poor mass transfer and much lower efficiencies [32]. The highly porous structure of the monolith offers high permeability, which allows fast separation of analytes at very high flow rates with minimal backpressures [33–36]. However, monoliths have a low phase ratio, that is, lower sample capacity and hence lower retentivity than particulate columns [30,33,37,38]. Although many research studies have been made on silica monoliths, HILIC applications using silica monoliths are limited. Some applications are separation of inorganic ions for example, Li^+, Na^+, K^+, and Cl^- and some drugs including naproxen and warfarin on a Chromolith Si column (Table 2.1) [39].

2.2.1.4   Ethylene Bridged Hybrids (BEH)

Although silica-based packings are characterized by high chromatographic efficiency and excellent mechanical stability, silica-based bonded phases are chemically unstable at pH values lower than 2 (due to hydrolysis of the bonded phase) or higher than 8 (due to dissolution of the silica itself) [40–42]. This results in loss of column efficiency, an increase in column backpressure, and bed collapse of the silica packing material [43]. Additionally, the high acidity of free silanols causes peak tailing, particularly with basic analytes.

One of the ways to overcome these problems was the invention of BEH. These hybrid phases (Fig. 2.4) vary in particle size from 1.7 to 10 μm and have been derivatized to form a variety of bonded phases including C_8, C_{18}, phenyl, and HILIC. Reversed-phase (RP) BEH columns are usually synthesized by the co-condensation of 1,2-bis(triethoxysilyl) ethane (BTEE) with tetraethoxysilane (TEOS) [44] as shown in Figure 2.4.

These hybrid materials are spherical and mechanically strong so they are frequently utilized in ultra-high pressure LC (UHPLC) [44]. Column stability at extreme pH values (up to pH 10) can be attributed to the increased hydrolytic stability of the ethyl-bridged groups within the particle matrix. Additionally, the reduced acidity of the bridged silanols in these hybrid phases suppresses peak tailing for basic analytes. Neue and coworkers studied the HILIC behavior of a 1.7-μm underivatized BEH phase and compared it with other underivatized silica phases [45]. They concluded that the retention mechanism of the BEH phase includes partitioning, adsorption, and secondary interactions that are quite similar to other underivatized silica. Factors affecting retention such as pH, organic modifier, and % ACN were studied for the 1.7 μm BEH phase. Smaller particle size enhanced the efficiency and produced narrower peaks; hence, higher sensitivity for the BEH phase in the ESI-MS mode was

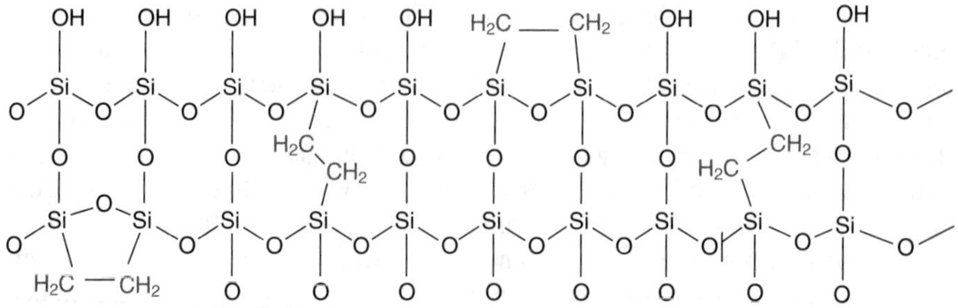

Figure 2.4. Schematic of Waters BEH phase. Reprinted from Reference 45 with permission from Wiley.

observed compared with other RP silica-based phases [45]. Stable performance has been demonstrated for over 2000 injections [45].

2.2.2   Derivatized Silica

Chemical modification of the silica surface yields a variety of silica-based polar derivatized phases. These attached polar groups are able to induce formation of a water-enriched layer (due to their hydrophilic nature) into which polar analytes can partition [46]. In this chapter, these silica modified polar phases are classified according to the net charge on their surfaces, that is, neutral, zwitterionic, positively charged, and negatively charged derivatized silica phases.

2.2.2.1   Neutral Derivatized Silica

Amide Silica. The TSKgel Amide-80 (Tables 2.1 and 2.2) from Tosoh is a good example of amide silica-based columns and has been available since 1985. It is available in 3-, 5-, or 10-μm particles. The surface functionality consists of nonionic carbamoyl groups bonded to the silica backbone through a short alkyl chain. Unlike amino phases, the amide group is not basic. Hence, retention of unionized analytes should be unaffected by the pH of the mobile phase. Moreover, the absence of amino groups prevents the formation of Schiff's bases with sugars and other carbonyl derivatives [15]. This phase was used for multidimensional mapping of oligosaccharides along with octadecyl silica (ODS) and diethylaminoethyl (DEAE) phases [47–49]. After this work, Yoshida used the same column for separation of peptides where the amide silica-based column showed good recovery and stability after 500 injections [50,51].

During the last decade, HILIC applications of TSKgel Amide-80 have increased dramatically. Recently, a TSKgel Amide-80 column has been used for the simultaneous analysis of α-amanitin, β-amanitin, and phalloidin in toxic mushrooms by LC-time-of-flight MS [52]. Additionally, applications of

Table 2.2. Chemical Structure of Selected Bonded Silica Phases

Phase Type	Phase Name	Chemical Structure[a]
Neutral derivatized silica	Amide silica	
	Diol silica	
	Cross-linked diol	
	Cyanopropyl silica	
Zwitterionic derivatized silica	Sulfoalkylbetaine silica	
Positively charged silica	Aminopropyl silica	

[a]Chemical structures of the other phases are included in different figures.

the TSKgel Amide-80 column include the separation of melamine and cyanuric acid that have been used for adulteration of milk, with excellent recovery and resolution [53], and the fast separation of both inactive and active ingredients in mannitol injections, with a high degree of robustness and accuracy [54].

Diol Silica. Diol phases (Table 2.2) were some of the first bonded silica phases to be developed. The diol phase was developed primarily to overcome the problems of adsorption caused by the free silanols on bare silica phases [15]. Diol phases are prepared by reaction of silica with glycidoxypropyltrimethoxy silane, followed by acid-catalyzed ring-opening hydrolysis of the oxirane group to form the diol hydrophilic neutral phase. Diol silica phases contain hydrophilic hydroxyl groups, and silanols can be blocked by a silylating reagent to overcome the adsorption of analytes on the surface. Hence, diol phases are considered as one of the best phases for HILIC due to the presence of the hydrophilic hydroxyl groups and absence of adsorptive properties of free silanols. The overall polarity of diol phases is quite similar to that of bare silica [55].

Although the first HILIC applications of diol silica phases were for separations of proteins, nucleic acids, and polysaccharides [56], diol phases are now commonly used for the separation of small-sized polyols. Diol silica phases were evaluated against amino-bonded silica phases [57]. Diol silica showed no irreversible retention of reducing sugars. Diol phases are the best for separation of carbohydrates due to the absence of amino groups and thus also the absence of Schiff's base formation [58]. Residual silanol activity can influence retention of some analytes on diol phases used for HILIC [59]. For instance, on an Inertsil Diol, 5-μm phase (Table 2.1) retention of glycine changed with the addition of trifluoroacetic acid (TFA), whereas retention of urea and sucrose remained constant. More recently, different anomeric forms of monosaccharides have been resolved on a diol silica column, which allows monitoring of the rate of the mutarotation [60].

A HILIC/RPLC mixed-mode phase, prepared by attaching an alkyl linker to a silica column, was released by Dionex under the trade name Acclaim Mixed Mode HILIC-1. This linker consists of an alkyl chain to provide RP retention (hydrophobic interaction) and a glycol terminal group, which contributes to diol-type HILIC properties. This mixed-mode phase was used for analysis of nonionic ethoxylated surfactants in ACN-rich eluents [61].

Diol columns may slowly release the bonded phase under acidic conditions. One of the approaches for increasing its stability against hydrolysis is the synthesis of cross-linked diol phases (Table 2.2). Luna HILIC 200 (Table 2.1) is a good example of these cross-linked phases, which shows high hydrolytic stability, stronger hydrophobic interactions, and better peak shape than noncross-linked diol phases [62]. Furthermore, the Luna (cross-linked diol) column showed a dual HILIC/RPLC retention mode depending on the percentage of the organic modifier in the mobile phase [62].

Cyanopropyl Silica. Although cyanopropyl silica phases (Table 2.2) can be used in both normal phase and RP chromatography, only a few HILIC applications have been reported [63,64]. Due to lower hydrogen bond donor capabilities compared with silanols, cyanopropyl silica phases are less retentive in normal phase chromatography than silica and other normal phase packings [22]. One of the major disadvantages of cyanopropyl silica phases is their mechanical instability, that is, collapse of these particles in solvents of intermediate polarity. This instability is mechanical (not chemical) in nature [22]. In solvents of intermediate polarity, the adhesion of particles to each other decreases, which causes the particle bed to collapse. On the other hand, the adhesion of these particles is strong in either nonpolar or polar solvents, which prevents the collapse of the bed. The limited number of applications of cyanopropyl phases in HILIC conditions can be attributed to the previously mentioned mechanical instability. Some hydrophilic analytes (e.g., uracil, cytosine, and dihydroxyacetone) actually eluted *faster* than dead time markers on a LiChrospher CN (Table 2.1), verifying the low potential of cyanopropyl silica as a HILIC stationary phase [65].

Cyclodextrin-Based Silica. Cyclodextrins (CDs) are formed through enzymatic hydrolysis of starch. Chemically, CDs consist of sugar units bound together in the form of a ring; hence, they can be considered cyclic oligosaccharides. Figure 2.5 shows the common types of cyclodextrins: α-CD, β-CD, and γ-CD that consist of 6, 7, and 8 D-glucopyranoside units, respectively, which are 1–4 linked together. CDs can be topologically represented as toroids, the rims of which are covered with the hydroxyl groups of the sugar units. This arrangement makes the interior of the CD hydrophobic and thus able to host other hydrophobic molecules. In contrast, the exterior is sufficiently hydrophilic to act as a HILIC stationary phase.

CDs exhibit chiral recognition characteristics because they consist of optically active sugars. CDs have been used as normal phase stationary phases for the separations of sugars, sugar alcohols, flavones, and aromatic alcohols [66,67]. Due to the hydrophilic nature of the hydroxyl groups of the sugars in CDs, they have been used for separation of plant extracts [68] and amino acids [69] under HILIC conditions. As the number of sugar units of the separated oligosaccharides increases, the retention increases due to stronger interactions with the sugar hydroxyl groups located on the exterior of CD rather than penetration inside the cavity of CD [70]. CD columns show greater retention for amino acids than the TSKgel Amide-80 phase, and more reproducibility and stability than aminopropyl silica-based phases [69]. Detailed descriptions of pharmaceutical applications of chiral HILIC stationary phases are discussed in Chapter 4.

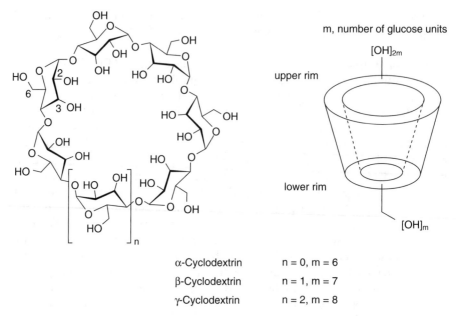

α-Cyclodextrin	n = 0, m = 6
β-Cyclodextrin	n = 1, m = 7
γ-Cyclodextrin	n = 2, m = 8

Figure 2.5. Chemical structure of cyclodextrins (CDs). Reprinted from Reference 16 with permission from Wiley.

2.2.2.2 Zwitterionic Derivatized Silica

Sulfoalkylbetaine Silica. Irgum and coworkers introduced a group of zwitterionic silica-based stationary phases for HILIC [71,72]. These zwitterionic phases are synthesized via grafting an active layer containing sulfoalkylbetaine groups (Table 2.2) onto wide pore silica (ZIC-HILIC) or a polymeric support (ZIC-pHILIC). Sulfoalkylbetaine phases are zwitterionic in nature due to the presence of basic quaternary groups and acidic sulfonic groups as shown in Table 2.2. While these phases contain both positive and negative charges, they have poor ion-exchange characteristics. Their net charge is approximately zero since the oppositely charged groups exist in a molar ratio of 1:1. The poor ion exchange characteristics of these phases may be attributed to their low surface areas [15] and shielding of free silanols by the oppositely charged functionalities [73]. As zwitterions are strong osmolytes [74], that is, they encourage the binding of water to their surfaces, such phases are suitable for HILIC.

However, sulfoalkylbetaine silica phases carry a very small negative charge arising from the distal sulfonic acid groups [15]. This excessive negative charge is pH independent [75,76]. Indeed, Guo et al. [73] found that the retention on sulfoalkylbetaine silica columns is the least affected by the pH of the HILIC columns studied.

Although these phases were initially designed for the separation of inorganic anions and cations [71,72], many HILIC applications have been reported including separations of nucleobases [77], peptides [78–80], metabolites [81,82], ions [83], and other polar analytes [84,85]. Separation of inorganic ions and zwitterionic solutes was achieved on a zwitterionic micellar coated stationary phase using pure water as the mobile phase [86].

Obelisc R and Obelisc N Columns. Obelisc is a trade name of a group of zwitterionic stationary phases that are produced by SiELC. Figure 2.6 shows a schematic diagram of these phases. The manufacturer suggested that they are the first available columns with liquid separation cell technology, that is, a new chemical modification of silica pores into liquid separation cells with their own characteristics. These columns are distinguished by three main characteristics [2]: (1) the high density of cationic and anionic charges on the surface makes Obelisc columns suitable for preparative chromatography; (2) the ionic strength inside the cells is higher than that of the mobile phase, leading to higher mass transfer rates and hence higher efficiency; and (3) both anionic and cationic charges are involved in electrostatic interactions with analytes.

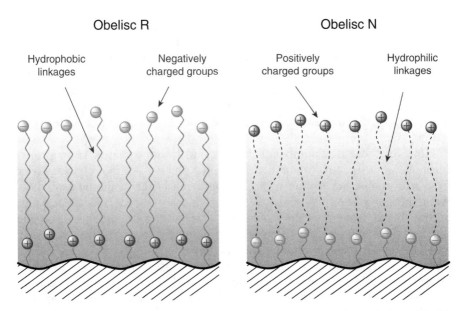

Figure 2.6. Schematic diagram of Obelisc R and Obelisc N. Reproduced from Reference 2 with permission from Sielc Technologies.

Obelisc R has RP character while Obelisc N has normal phase character. As shown in Figure 2.6, Obelisc R and N phases differ in the position of the charged groups on surface. However, because Obelisc R has an RP character and hence cannot be used in the HILIC mode, we will focus on the Obelisc N phase. Obelisc N has anionic groups close to the silica surface separated from cationic groups by a hydrophilic spacer. Packings are available as 5 and 10 μm with a 100-Å pore size as shown in Table 2.1. Unfortunately, the exact chemical structure of these zwitterionic phases is not specified by the manufacturer. Obelisc N columns are characterized by high polarity due to the presence of charged groups and hydrophilic chains on their surfaces. Hence, Obelisc N can be utilized in ion-exchange chromatography due to charged groups on the surface, and in HILIC due to the water layer formed on its surface. The stability of Obelisc N is limited to a pH range of 2.5–4.5 and a temperature range of 20–45°C [10].

However, a few HILIC applications on the Obelisc N phase have been reported. In 2011, the retention behavior of dexrazoxane (a model bisdioxo-piperazine drug) and its three polar metabolites was studied on Obelisc N and other HILIC phases [10]. Obelisc N showed a significantly lower hydrophobic selectivity than other mixed-mode stationary phases and showed a comparable behavior to hydrophilic amino phases [87].

2.2.2.3   *Positively Charged Derivatized Silica*

Aminopropyl. Aminopropyl silica phases (Table 2.2) are among the oldest amine-based phases. These phases have been introduced for either normal phase LC or HILIC purposes. Aminopropyl phases have been used extensively under HILIC conditions for separation of carbohydrates [57,88–90], amino acids, proteins [91], and some antibiotics [92]. These phases have become more popular than bare silica in carbohydrate separations as they promote fast mutarotation, which prevents formation of double peaks due to anomer resolution [15].

However, there are a number of challenges with amino phases. Aminopropyl silica is more reactive than other HILIC phases. Amino phases suffer from irreversible adsorption problems, particularly for acidic analytes [93]. Aminopropyl silica phases also exhibit significant bleed (i.e., detachment of the ligand from the silica skeleton) compared with other hydrophilic bonded silica phases [94]. Slow pH equilibration can be observed [92]. Finally, Schiff's base formation with aldehydes leads to problems during separation of some sugars [37]. Stationary phases containing secondary or tertiary amine groups, for example, YMC-Pack Polyamine II, cannot form Schiff bases with reducing sugars, resulting in an improved column lifetime [95].

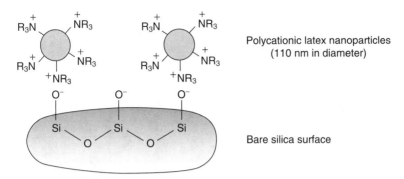

Polycationic latex nanoparticles
(110 nm in diameter)

Bare silica surface

Figure 2.7. Structure of the agglomerated latex coated silica monolith. Reprinted from Reference 38 with permission from Wiley.

Latex Coated Silica. Recently, an agglomerated silica monolithic column was prepared by electrostatically attaching polycationic latex particles onto a silica monolith (Fig. 2.7) by simply flushing a suspension of a latex possessing hydrophilic quaternary amines through a silica monolith [34,36]. This agglomerated phase was tested for separation of polar analytes, for example, benzoates, nucleotides, and amino acids, under HILIC conditions [36]. The high permeability offered by the monolith structure allowed fast (<15 s) separation of naphthalene, uracil, and cytosine with similar selectivity to other HILIC phases [36]. The positive charge on the agglomerated phase exhibited electrostatic repulsion hydrophilic liquid interaction chromatographic (ERLIC, see Section 1.3.5 in Chapter 1) for amino acids.

2.2.2.4 Negatively Charged Derivatized Silica

Poly(succinimide) based-silica represents the majority of stationary phases in this class. Figure 2.8 shows the synthesis of the poly(succinimide)-based phases. First, the aminopropyl silica skeleton reacts with poly(succinimide) where a fraction of the succinimide rings are opened and linked to the aminopropyl backbone through multiple amide linkages. Second, the unopened succinimide rings are reacted with different nucleophiles to form the various poly(succinimide)-based phases shown in Figure 2.8.

Alpert proposed the synthesis of poly(aspartic acid) silica through hydrolysis of the succinimide rings to yield a weak cation exchanger [96]. These columns were stable and durable for the separation of proteins. Poly(aspartic acid) silica was used for separation of anions and/or cations due to its

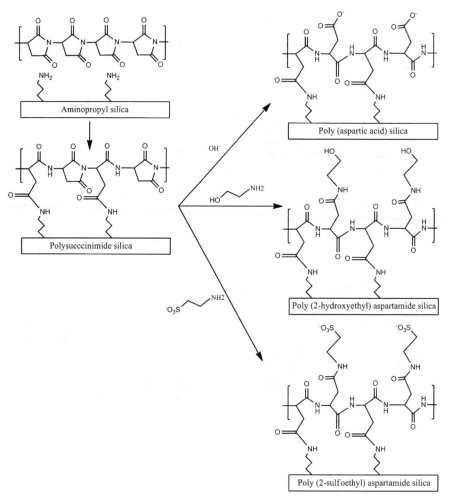

Figure 2.8. Synthesis of poly(succinimide)-based silica stationary phases. Adapted from References 15 and 96.

zwitterionic nature, which is attributed to the presence of both protonated aminopropyl and dissociated carboxylic groups [97].

These poly(succininmide) silica-based phases are manufactured by PolyLC and are available under the brand names of PolyCAT A™ for poly(aspartic acid) silica, PolySulfoethyl A™ for poly (2-sulfoethyl) aspartamide silica and PolyHydroxyethyl A™ for poly (2-hydroxyethyl) aspartamide silica (Table 2.1).

PolyHydroxyethyl A has been used for separations of carbohydrates, phosphorylated and nonphosphorylated amino acids, peptides, glycopeptides, oligonucleotides, glycosides, metabolites, and small polar analytes [98–101]. However, poly(2-hydroxyethyl aspartamide) columns suffer from a lower efficiency than other HILIC phases [99], a limited stability [102] and column bleed [103].

PolySulfoethyl A has also exhibited bleeding troubles, resulting in several interfering peaks during a two-dimensional LC-tandem mass spectrometry (MS/MS) study [103]. The level of interfering peaks was negligible for new columns but became significant after 1 month of use [103].

2.2.3 Nonsilica Phases

2.2.3.1 Amino Phases

The Styros™ AminoHILIC phase (Table 2.1) manufactured by Orachrom Inc. is an amino phase on a monolithic polymer support [104]. Cross-linked poly(styrene-divinyl benzene) was functionalized with surface amino groups. The highly cross-linked nature of the matrix minimizes shrinking and/or swelling (<0.2%). The high permeability of the monolith structure and the 4000-psi maximum operating pressure enable fast separation of polar analytes. In contrast to silica-based phases, the polymeric nature of the column support makes it compatible with buffers of extreme pH values. A mixture of nucleotides including AMP, ADP, ATP, and GMP were separated in 2 min using a flow rate of 4 mL/min [104]. Separation of nucleobases like cytosine and uracil and benzoic acid derivatives were achieved on the same column under HILIC conditions. However, the efficiencies obtained by the AminoHILIC were lower than the efficiencies obtained by a latex coated silica monolith discussed in Section 2.2.2.3 (e.g., H = 250 μm vs. 59 μm for cytosine) [36]. The lower efficiency of the AminoHILIC column might be attributable to the poorer mass transfer characteristics of polymeric monoliths.

2.2.3.2 Sulfonated S-DVB Phases

These resins consist of styrene-divinyl benzene (S-DVB) functionalized with negatively charged sulfonates, and have been used mainly as cation exchangers [105,106]. These phases have been used for aqueous normal phase chromatography [107,108] for separation of oligosaccharides, and for the HILIC separations of propylene glycol, glycerol, and dextrose [109]. HILIC retention increases with the sulfonation capacity, and with % ACN in the 70–95% range [109]. The Ca^{2+} form of S-DVB gave much more retention relative to the H^+ form [109].

2.3 COMMERCIAL HILIC PHASES

Table 2.1 summarizes important characteristics of some selected commercial HILIC phases. The table classifies HILIC phases according to their chemical nature. The following sections compare these HILIC phases in terms of efficiency, retention, and selectivity.

2.3.1   Efficiency Comparison

Two rich sources of information about the efficiency of commercial columns are the review of Ikegami et al. [37] and the recent study by Kawachi et al. [110]. Ikegami et al. provide an interesting and comprehensive review of efficiency of HILIC phases relative to the type of solute, type of stationary phase, and format of stationary phase (particulate vs. monolithic) [37]. Many of the efficiencies and plate heights quoted in this section were measured from published chromatograms by Ikegami et al. Kawachi et al. compared retention, selectivity, and efficiency of numerous model solutes on 14 commercial columns. Selected data from [110] are presented in Table 2.3.

In general, efficiency in HILIC follows similar broadening behavior to other forms of liquid chromatography. For instance, Figure 2.9 shows van Deemter curves for 3- and 5-μm HILIC packings. The Tosoh TSKgel Amide-80 column packed with 5 μm yields an efficiency of 20,000 plates for mannitol on a 250-mm column [111]. Reducing the particle size enhances the mass transfer, resulting in lower plate heights and thus higher efficiency (Fig. 2.9). Similar improvements in efficiency with decreasing particle size were observed by Kawachi et al. [110] for amide and zwitterionic HILIC phases (Table 2.3).

Table 2.3 also shows that high efficiency can be achieved in HILIC using monolithic (Chromolith Si) and core-shell particles (HALO). However, the minimum reduced plate height (H/d_p ~4) achieved on HALO columns is greater than the theoretical minimum (1.5), possibly due to the slow mass transfer of analyte from the water-rich adsorbed phase to the ACN-rich mobile phase [28]. HILIC peaks are generally more symmetrical than those in *per* aqueous liquid chromatography (PALC), due to the more homogeneous sorption energetics in HILIC [28]. Nonetheless, a significant peak asymmetry (1.37–1.9) was observed for the HALO column under HILIC conditions [110]. Efficiencies are generally poorer if the retention mechanism is dominated by ion exchange (Fig. 2.10) [110].

Decreasing separation efficiency with increasing HILIC retention has been observed [37], although more recently this has been attributed to the effect of increased ionic exchange retention [110]. Figure 2.10 shows that under conditions where ion-exchange effects are reduced (i.e., choosing analytes that are less ionized under experimental conditions), the plate height for a given column remains constant, independent of retention.

A number of studies compare the model analyte separations achieved on numerous HILIC columns [73,110,112,113]. For example, Figure 2.11 allows comparison of the efficiency of the Atlantis HILIC silica and three other HILIC phases (all 5 μm, 250 × 4.6 mm inner diameter [ID]) for nucleobases and nucleosides [73]. The Atlantis HILIC silica showed the lowest retention but produced the highest efficiency (N ~25,000 plates, H ~10 μm). On the other hand, the YMC-Pack amino phase provided the lowest efficiency (N = 5000–17,000 plates, H = 15–50 μm), while the TSKgel Amide-80 and ZIC-HILIC phases showed intermediate efficiency (N = 15,000–20,000 plates, H = 12–16 μm, and

Table 2.3. Comparison of HILIC Column Efficiencies[a]

Column	Uridine			Adenosine			Theophylline			Theobromine		
	k	H (μm)	Asym	k	H (μm)	Asym	k	H (μm)	Asym	k	H (μm)	Asym
Underivatized silica phases												
Chromolith Si[b]	0.31	12	1.00	0.73	13	2.02	0.26	14	1.09	0.31	13	1.14
HALO HILIC (2.7 μm)	0.64	8	1.47	1.59	6	1.37	0.50	8	1.36	0.64	7	1.3
Amide silica phases												
Amide-80 (5 μm)	3.30	26	1.37	3.80	28	1.38	0.76	36	1.41	1.06	32	1.37
Amide-80 (3 μm)	4.58	9	0.99	5.26	11	1.07	1.08	9	1.07	1.43	10	1.15
X-Bridge Amide (3.5 μm)	2.55	12	1.42	2.81	11	1.23	0.52	20	3.26	0.71	24	1.37
Diol silica phases												
LiChrospher Diol (5 μm)	1.50	17	0.98	2.50	17	0.97	0.55	12	1.14	0.57	15	1.28
Cyclodextrin-based silica phases												
Cyclobond I (5 μm)[c]	0.70	18	1.73	1.36	25	1.84	0.43	12	1.14	0.44	11	1.1
Zwitterionic phases												
ZIC-HILIC (5 μm)	2.11	25	1.34	1.55	24	1.22	0.30	22	1.36	0.36	21	1.29
ZIC-HILIC (3.5 μm)	2.10	12	1.26	1.51	12	1.32	0.28	12	1.44	0.34	12	1.47
Nucleodur HILIC (3.5 μm)	2.20	14	0.88	2.33	14	0.97	0.52	16	1.00	0.52	16	1.00

(Continued)

63

Table 2.3. (Continued)

Column	Uridine			Adenosine			Theophylline			Theobromine		
	k	H (μm)	Asym	k	H (μm)	Asym	k	H (μm)	Asym	k	H (μm)	Asym
Aminopropyl silica phases												
NH$_2$-MS (5 μm)	2.44	12	1.04	2.13	11	1.09	0.80	12	1.15	0.43	12	1.18
Poly(succinimide) silica phases												
PolySulfoethyl (3 μm)[d]	1.58	62	1.11	1.15	44	2.56	0.23	114	1.91	0.23	138	2.17
PolyHydroxyethyl (3 μm)[d]	3.92	61	0.99	3.75	50	0.99	0.66	126	1.81	0.75	92	1.31
Amino phase												
Cosmosil HILIC (5 μm)[e]	1.60	12	1.11	2.20	12	1.16	0.55	16	1.37	0.49	16	1.7
Cosmosil Sugar-D (5 μm)[f]	1.58	17	1.12	1.88	15	0.98	0.59	22	1.11	0.31	15	1.44

[a]Conditions: dimensions, 150 mm × 4.6 mm ID unless otherwise noted; particle size, as noted in column 1; injection, 4 μL of 1 mg/mL analyte dissolved in mobile phase; dead time marker, 1 mg/mL toluene; flow rate, 0.5 mL/min for 4.6 mm ID columns and 0.1 mL/min for 2 mm columns; eluent, 90:10 v/v ACN–ammonium acetate buffer (pH 4.7, 20 mM in the aqueous portion); temperature, 30°C.

[b]Dimensions: 100 mm × 4.6 mm ID.

[c]Dimensions: 250 mm × 4.6 mm ID.

[d]Dimensions: 100 mm × 2.1 mm ID.

[e]Triazole functionality.

[f]Secondary/tertiary amine functionality.

Source: Reproduced from Reference 110 with permission from Elsevier.

64

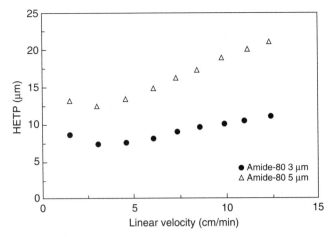

Figure 2.9. van Deemter plots of (●) TSKgel Amide-80 (150 mm × 4.6 mm ID, 3 μm), (△) TSKgel Amide-80 (250 mm × 4.6 mm ID, 5 μm); mobile phase, ACN/water = 75/25; temperature, 40°C; detection, RI; injection volume: 10 μL; sample: mannitol. Reprinted from Reference 37 with permission from Elsevier.

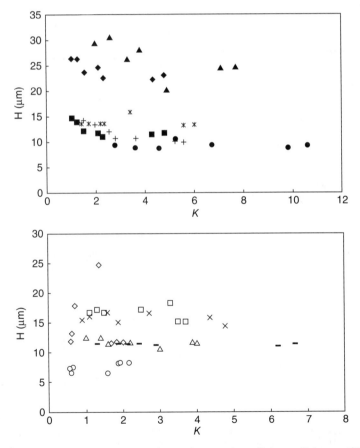

Figure 2.10. Relationship between retention and separation efficiency. Columns: ZIC-HILIC 5 μm (◆), ZIC-HILIC 3.5 μm (■), Amide-80 5 μm (▲), Amide-80 3 μm (●), Nucleodur (*), XBridge-Amide (+), Cyclobond I (◇), LiChrospher Diol (□), COSMOSIL HILIC (△), Sugar-D (×), MS-NH₂(–), and HALO HILIC (○). Reprinted from Reference 110 with permission from Elsevier.

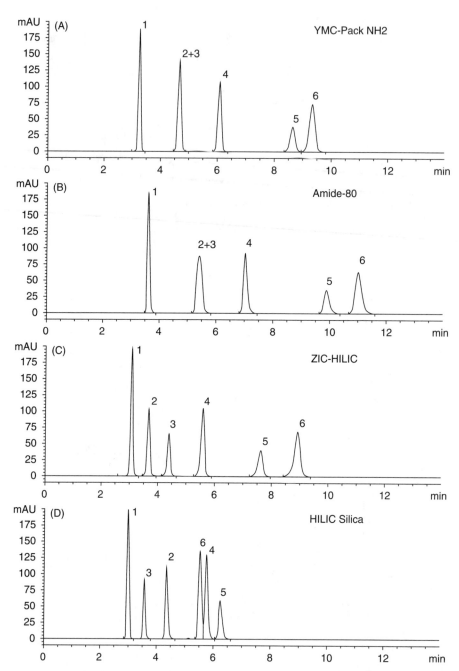

Figure 2.11. Separation of nucleic acid bases and nucleosides on: (A) YMC-Pack NH₂, (B) TSKgel Amide-80, (C) ZIC-HILIC, and (D) Atlantis HILIC Silica columns. Dimensions: 5 μm, 250 × 4.6 mm ID. Mobile phase: ACN/water (85/15, v/v) containing 10 mM ammonium acetate. Column temperature: 30°C. Flow rate: 1.5 mL/min. UV detection at 248 nm. Compounds: (1) uracil, (2) adenosine, (3) uridine, (4) cytosine, (5) cytidine, and (6) guanosine. Reprinted from Reference 73 with permission from Elsevier.

N = 12,000–22,000, plates, H = 11–21 μm, respectively). Figure 1.2 (Chapter 1) shows similar behavior for model carboxylic acids analytes.

Peak tailing and low efficiencies are commonly observed for amines on silica-based RPLC phases due to silanol activity. The degree of tailing depends on the nature and activity of the silanols [114,115]. This effect has not been studied in as much detail for HILIC [37]. However, as shown in Figure 2.12, silica phases from different manufacturers provide different retention and efficiency [93]. Peak tailing and hence lower efficiency was observed for pyrimidines (upper trace) on Nucleosil and Zorbax SIL. However, no significant tailing was observed for purines on any of the silica columns (lower trace Fig. 2.12).

Numerous studies report more limited comparisons of HILIC columns. Some interesting observations from these studies regarding efficiency are discussed below, in the order of column type discussed in Section 2.2 and Table 2.1.

The effect of retention on plate height has been studied on Betasil HILIC phase (5 μm, 50 × 4.6 mm ID) versus two RP phases [14]. Betasil gave an intermediate efficiency between the two RP phases (optimum flow rate ~1–2 mL/min), yielding a plate height of 20 μm for fluconazole.

As shown in Table 2.3, amide columns generally provide good efficiency. Similarly, Figure 1.2 (Chapter 1) shows efficiencies of 15,000 to 20,000 plates on an amide 80 phase (5 μm, 250 × 4.6 mm ID, H = 12–16 μm) for k below 5 [37]. However, N-acetylneuraminic acid and glucuronic acid exhibited efficiencies of only 3000 plates (H = 83 μm) and 2400 plates (H = 104 μm), respectively [116], on a TSKgel Amide-80 (5 μm, 250 × 4.6 mm ID). Similarly, strongly retained ($k = 7$–17) tetramers and pentamers of proanthocyanidins exhibited significant broadening (N ~1000) on the TSKgel Amide-80 phase [37,117].

Lichrosorb-DIOL (10 μm, 250 × 4.6 mm ID) is a commercial diol phase manufactured by Merck (Table 2.1). This diol phase provided N of 5300 and 3400 plates for sucrose (H = 47 μm) and lactose (H = 74 μm), respectively [57], which was comparable to an equivalent Lichrosorb NH_2 (amino silica phase). The galactose and lactose peaks were broad on the diol phase due to anomerization [57]. Thus, aminopropyl phases are favored for carbohydrates. Addition of an amine to the mobile phase or separation at higher temperatures reduces the anomerization broadening on a diol phase. For example, addition of 0.1% ethyldiisopropylamine increased the efficiency for lactose from 600 to 1300 plates. Plate height on an Inertsil Diol (5 μm, 150 × 4.6 mm ID, Table 2.1) decreased with column temperature for glycine and sucrose, while urea displayed a U-shaped behavior [59].

A cyano silica column (5 μm, 150 × 3.0 mm ID) yielded higher efficiency (N of 9500–17,000 plates) and greater retention than bare silica and amino silica-based phases for the analysis of denaturants in an alcohol formulation [118].

Separations of native oligosaccharides on a Cyclobond I 2000 (β-CD, 5 μm, 250 × 4.6 mm ID, Table 2.1) column yielded only 2300–7800 plates (H ~60 μm)

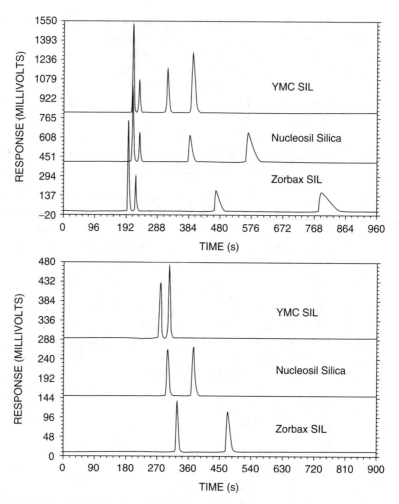

Figure 2.12. Pyrimidines (upper) and purines (lower) on silica columns. Conditions: mobile phase 5 mM phosphoric acid in ACN: water (75:25) for pyrimidines, (70:30) for purines; 275 nm detection; pyrimidines, ~0.05 mg/mL (in order of elution): 5-fluorouracil, uracil, 5-fluorocytosine, cytosine; purines, ~0.02 mg/mL (in order of elution): acyclovir, guanine. Reprinted from Reference 93 with permission from Elsevier.

[70], presumably due to use of a high mobile phase velocity [37]. The van Deemter plot for Cyclobond I shown in Figure 2.13 shows that a fast linear velocity (2 mm/s) was preferred for highly retained analytes such as sucrose and lactose, but a lower linear velocity (0.4 mm/s) was preferred for weakly retained analytes such as fructose ($k = 1.25$) [66].

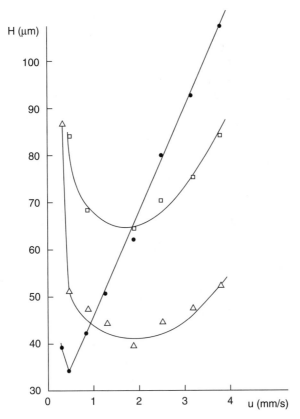

Figure 2.13. van Deemter plot for Cyclobond I HILIC Column (5 μm, 250 mm × 4.6 mm ID); mobile phase of ACN:water (85:15, v/v); compounds: (●) fructose; (△) sucrose; (□) lactose. Reprinted from References 37 and 66 with permission from Elsevier.

As illustrated in Figure 1.2 (Chapter 1) and Figure 2.11, Guo et al. compared four HILIC phases in terms of retention and efficiency [73]. The ZIC-HILIC (5 μm, 250 × 4.6 mm ID, Table 2.1) provided N of 12,000–22,000 plates (H = 11–21 μm) for carboxylic acids, nucleosides, and nucleobases [37,73]. Similarly 15-cm columns of 5 μm ZIC-HILIC yield 10,000–13,000 plates in both the 2.1-mm and 4.6-mm ID formats for well-behaved analytes (e.g., ascorbic and dehydroascorbic acid) [37,119], whereas lower efficiencies were observed for morphine and its glucuronide metabolite (1400–3500 plates) [37,119] and oligopeptides including neurotensin, Gly-His-Lys, and bradykinin (1800–3000 plates on a 100-mm column) [37,119].

The efficiencies observed in Figure 2.11 on the YMC-Pack NH_2 column (5 μm, 250 × 4.6 mm ID) were comparable to those obtained with other HILIC columns [37,73]. However, as noted previously, significantly lower efficiencies are observed in HILIC when ion exchange contributes significantly to retention. For instance, separations of benzoic acid derivatives on these same columns (Fig. 1.2, Chapter 1) yield distinctly different efficiencies (N ~5000 plates) on a YMC-Pack NH$_2$ column. Furthermore, significant differences in efficiency may exist between amino columns of the same dimensions but different manufacturers even for basic analytes (Fig. 2.14) [120]. Fronted peaks were observed on both YMC-Pack NH$_2$ and Nucleosil NH$_2$ phases, while higher efficiency (N = 4600–7300 plates, H = 21–33 μm) was observed with the Zorbax NH$_2$ phase [120].

Hypersil ASP-2 (Amino) (3 μm, 50 × 4.6 mm ID) was used for separation of tetracycline antibiotics with efficiencies of 3800 plates (H = 13 μm) for oxytetracycline [37,92].

Separation of amino acids on PolySulfoethyl A (5 μm, 200 × 4.6 mm ID) 2.0 mL/min [1] yielded N of 4000 plates (H = 50 μm) [37]. The flow rate (2.0 mL/

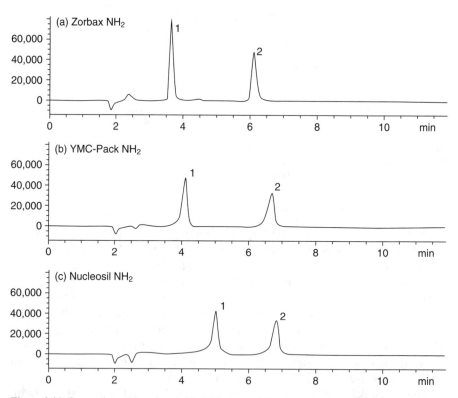

Figure 2.14. Separation of 2-amino-2-ethyl-1,3-propanediol (AEPD, peak 1) and tromethamine (peak 2) on: (a) Zorbax NH$_2$, (b) YMC-Pack NH$_2$, and (c) Nucleosil NH$_2$ columns (All, 5 μm, 150 × 4.6 mm ID); mobile phase: ACN/water (80:20, v/v): 25°C. Flow rate: 1 mL/min; injection 50 μL. Reprinted from References 37 and 120 with permission from Elsevier.

min) may have been too fast to achieve optimal efficiency. The efficiency of cyclopeptides on the same PolySulfoethyl A phase strongly depended on the mobile phase: 60% ACN displaying very poor efficiencies while 90% ACN yielded 9000 plates as estimated by Ikegami et al. [1,37]. Increasing % ACN further to 95% decreased efficiency to 5000 plates and increased the retention factor [1]. Alpert suggested that this kind of phases should be used for analytes of a small retention factor to provide a high separation efficiency [1].

Contrary to Table 2.3, Tolstikov et al. observed that the PolyHydroxyethyl A (5 μm, 150 × 1.0 mm ID, H = 15.8 μm for peak 12, maltoheptaose) is more efficient than the TSKgel Amide-80 phase (5 μm, 250 × 2.0 mm ID, H = 23.8 μm for peak 12, maltoheptaose) for amino acids and carbohydrates [99]. The two columns provided different elution order of analytes as well. The Poly-Hydroxyethyl A column (3 μm, 200 × 4.6 mm ID) yielded 2300–2900 plates (H = 69–87 μm) for dipeptide standards [1]. Increasing the buffer concentration or pH resulted in lower efficiency and lower retention. Separation of atosiban diastereomers on a PolyHydroxyl A column (5 μm, 200 × 4.6 mm ID) gave N of 5000–6000 plates (H = 33–40 μm) with lower resolution than an amino-silica phase [100].

Separation of inorganic ions, for example, Li^+, Na^+, Cl^-, and K^+, and drugs, for example, naproxen and warfarin, on Chromolith Si (100 × 4.6 mm ID) provided N of 1000–4800 plates (H = 21–100 μm) as estimated by Ikegami et al. [37,39]. These efficiencies were much lower than observed on RP silica-based monoliths. Monolithic silica capillaries (85 mm × 75 μm ID) yielded 9300–17,500 plates under HILIC conditions for separations of alkaloids [121]. An AS9-SC coated Chromolith silica monolith (100 × 4.6 mm ID, Fig. 2.7) was used to separate a variety of polar analytes under HILIC conditions [36], and provided H of 59–67 μm for model nucleobases, 71–110 μm (N = 900–1400 plates) for nucleotides, and 30–45 μm (N = 2200–3300 plates) for carboxylates [36]. The high column permeability enabled separations to be performed in less than 15 s. The Styros polymeric AminoHILIC phase (50 × 4.6 mm ID) yielded lower efficiencies (H = 154–400 μm and 67–167 μm, for nucleotides and benzoates, respectively) [36], possibly attributable to the poor mass transfer properties of its polymeric support.

2.3.2 Retention and Selectivity Comparisons

Several factors affect retention and selectivity in HILIC, including the type of organic modifier, % organic modifier, buffer strength, pH of the mobile phase, and stationary phase chemistry. In this section, we will focus on the effect of stationary phase chemistry on the selectivity in HILIC mode. Selectivity and retention of the HILIC phases have been studied and reviewed extensively during the last 10 years [65,73,112,122]. In 2005, Guo et al. demonstrated that different stationary phases showed different degrees of retention and selectivity for polar analytes [73]. Four stationary phases, silica, amino, amide, and zwitterionic columns (all of the same dimensions of 250 × 4.6 mm, 5 μm), were compared in terms of retention and selectivity using salicylic acid derivatives, nucleobases, and nucleosides as model analytes [73]. As shown in Figure 1.2 (Chapter 1), the acids are strongly retained on the YMC-Pack NH_2 phase due to the strong ion-exchange interaction between the negatively charged acids and the amino groups of the stationary phase. However, the TSKgel Amide-80 showed weaker retention, resulting in aspirin and 4-aminosalicylic acid being only partially resolved. HILIC silica showed the least retention of the acids. On the other hand, TSKgel Amide-80 provided the strongest retention for nucleobases and nucleosides but did not baseline resolve adenosine and uridine (Fig. 2.11). ZIC-HILIC and HILIC silica columns showed good resolution of adenosine and uridine. HILIC silica retained nucleosides the least of the studied phases but also showed completely different retention order from the other columns (Fig. 2.11) [73]. Thus, it is clear that the different elution patterns indicate that the chemistry of the stationary phases has a significant effect on retention and selectivity.

In 2010, McCalley investigated the retention and selectivity of strongly acidic and basic analytes on different HILIC stationary phases [122]. Five HILIC phases, ZIC-HILIC, Onyx Silica, Luna HILIC (diol), TSKgel Amide-80, and Acclaim mixed-mode HILIC (all 5 μm, 250 × 4.6 mm ID), were compared in terms of retention and selectivity. Figure 1.3a,b (Chapter 1) shows the different elution patterns observed on the various HILIC columns. The silica phase showed the greatest retention of basic analytes, which might be attributed to: (1) the high surface area of the silica phase (400 m^2/g) compared with the other phases; (2) strong ionic interaction between the protonated bases and ionized silanols; and (3) the hydrophilic nature of the silica surface [122]. On the other hand, the silica phase showed weak retention of the acidic analytes due to the repulsive forces between the negatively charged analytes and the ionized silanols [122]. The zwitterionic, amide, and particularly the diol phase provided reasonable retention of the acids, which could be explained by the use of low acidity silica and/or screening of ionized silanols by the bonded phase [122]. Finally, as has been discussed in Chapter 1, McCalley concluded that the difference in selectivity between different stationary phases, even when the same eluent is used, suggests that the stationary phase is not working simply as an inert support for the adsorbed water layer, but rather it has a considerable and characteristic contribution to retention [122].

Guo et al. compared the selectivity and retention of hydroxyl stationary phases including diol, cross-linked diol, polyhydroxy, and polyvinyl alcohol phases using nucleobases and nucleosides [123]. Figure 2.15 shows these phases have similar selectivity, with the cross-linked diol and polyhydroxy phases resolving cytidine and guanosine [112,123]. The selectivity of the silica phase (Fig. 2.11) is somewhat different from the HILIC phases included in Figure 2.15, attributable to specific interactions of the analyte with the silica surface [112,123]. Figure 2.16 compares the selectivity of three cationic phases (amino, imidazole, and triazole) using the same previously mentioned analytes

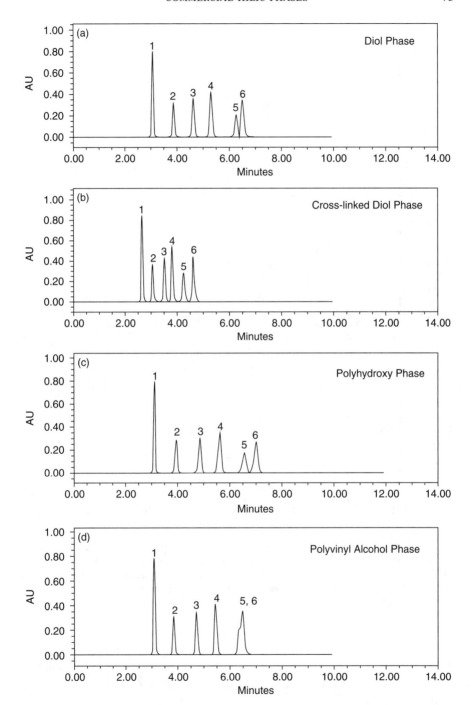

Figure 2.15. Separation of (1) uracil, (2) adenosine, (3) uridine, (4) cytosine, (5) cytidine, and (6) guanosine on (a) diol, (b) cross-linked diol, (c) polyhydroxy, and (d) polyvinyl alcohol phase. Column dimension: All, 250 × 4.6 mm ID and 5-μm particle size. Mobile phase: ACN/water (85/15, v/v) containing 10 mM ammonium acetate. Column temperature: 30°C. Flow rate: 1.5 mL/min. Reprinted from Reference 112 with permission from Elsevier.

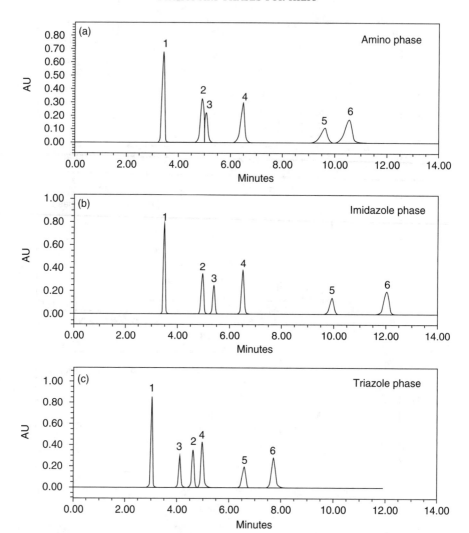

Figure 2.16. Separation of (1) uracil, (2) adenosine, (3) uridine, (4) cytosine, (5) cytidine, and (6) guanosine on (a) amino, (b) imidazole, and (c) triazole phase. Column dimension: All, 250 × 4.6 mm ID and 5-μm particle size. Mobile phase: ACN/water (85/15, v/v) containing 10 mM ammonium acetate. Column temperature: 30°C. Flow rate: 1.5 mL/min. Reprinted from Reference 112 with permission from Elsevier.

[112,123]. The imidazole and triazole phases display different selectivity. Although the amino phase did not resolve adenosine and uridine, the imidazole and triazole phases resolve them from each other but in a different elution order (Fig. 2.16).

Lammerhofer et al. compared the selectivity of six different HILIC phases as shown in Figure 2.17. The amide phase gave similar selectivity to the amino

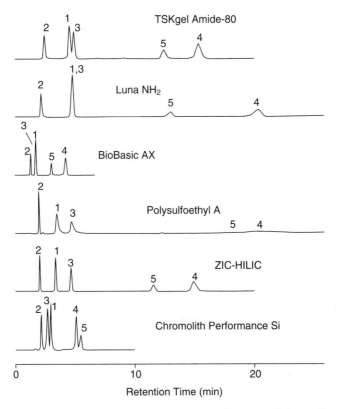

Figure 2.17. Separation of nucleosides on various HILIC phases. Column dimension: All, 100 mm × 4 mm ID, 5-μm particle size and 100 mm × 4.6 mm ID for Chromolith Performance Si with a macropore diameter of 2 μm and a mesopore diameter of 13 nm. Column temperature: 25°C. Mobile phase: ACN/5 mM ammonium acetate buffer (90/10, v/v), apparent pH ~8. The flow rate was adjusted to the same linear velocity (1.7 mm/s). Solutes: (1) adenosine, (2) thymidine, (3) uridine, (4) guanosine, and (5) cytidine Reprinted from References 87 and 112 with permission from Wiley and Elsevier.

and sulfobetaine phases; however, the silica phase provided a different elution order relative from the other phases [87].

Recently, Irgum and his coworkers probed the interactions taking place on 22 different HILIC stationary phases [65]. Using carefully selected pairs of analytes, interactions including hydrophilic, hydrophobic, electrostatic, hydrogen bonding, dipole–dipole, and π–π interaction were characterized. Then using principal component analysis, the 22 different HILIC columns were classified in terms of selectivity, as shown in Figure 2.18 [65]. The first two principal components explained more than 70% of the total variance. In Figure 2.18 the gray loading vectors are defined by the analyte pairs (triangles) whose relative retention characterizes a single type of interaction. For instance, the relative retention of benzoate versus cytosine (BA/CYT)

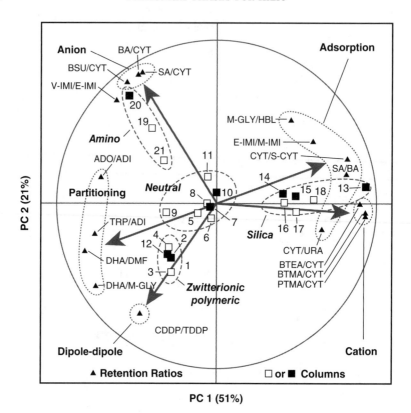

Figure 2.18. Score and loading biplot of the two first components of the model where the observations were the columns and the variables were the retention ratios of all 15 pairs of test analytes. For column numbering, see Table 2.1. Reprinted from Reference 65 with permission from Elsevier.

reflects selectivity based on anion exchange. Moving clockwise from the anion exchange loading vector are the loading vectors reflecting adsorption, cation exchange, dipole–dipole, and partitioning. Specific columns (as identified in Table 2.1) are plotted as squares in Figure 2.18. The higher the score of a column (i.e., the further out from the center along a loading vector), the more a column possesses that particular character. For example, Purosphere STAR NH$_2$ (square 20) has more anion exchange character than the TSKgel NH$_2$-100 (21).

In Figure 2.18 the HILIC columns cluster into four main groups: neutral columns such as amide (7,8) and diol (10, 11) phases; amino columns (19–21); silica columns (14–17) including monolithic silica (13) and type C silica (18); and zwitterionic phases (1–4). The numbers in parentheses indicate the phase number in Table 2.1. A few columns show selectivity that is surprising based on the manufacturer's stated functionality. For instance, both the Nucleodur

HILIC (5) and Shiseido PC HILIC (6) columns are purported to be zwitter-ionic phases, but both behaved as neutral columns. Also the sulfoethyl func-tionality of PolySulfoethyl A (12) would be expected to yield a column with strong cation exchange character, whereas it groups with the zwitterionic columns in Figure 2.18. Irgum and coworkers [65] speculate that this zwitter-ionic character may result from underivatized aminopropyl groups within the synthesis of this phase (Fig. 2.8).

2.4 CONCLUSIONS

Recent years have seen a maturation of the field, such that insights into the broad properties of each class of HILIC columns are now available. This allows grouping of columns of similar character and recognition of which columns will offer a different selectivity (and thus enable resolution of challenging samples). Nonetheless, it is anticipated that the field of HILIC will continue to see rapid growth in the number and types of columns commercially avail-able. To keep abreast of these developments, Ron Majors' annual preview of the new chromatographic columns introduced at Pittcon is highly recom-mended [124].

ACKNOWLEDGMENTS

This work was sponsored by the Natural Sciences and Engineering Research Council of Canada (NSERC) and the University of Alberta. Knut Irgum's generous gift of a preprint of his work is gratefully acknowledged.

REFERENCES

1. Alpert AJ. Hydrophilic-interaction chromatography for the separation of peptides, nucleic acids and other polar compounds. *J. Chromatogr.* 1990; **499**: 177–196.

2. Sielc Technologies. Sielc. http://www.sielc.com (accessed in October, 2012).

3. Jane I. The separation of a wide range of drugs of abuse by high-pressure liquid chromatography. *J. Chromatogr.* 1975; **111**: 227.

4. Paek IB, Moon Y, Ji HY, Kim HH. Hydrophilic interaction liquid chromatography–tandem mass spectrometry for the determination of levosulpiride in human plasma. *J. Chromatogr. B* 2004; **809**: 345–350.

5. Naidong W, Shou WZ, Eerkes A. Liquid/liquid extraction using 96-well plate format in conjunction with hydrophilic interaction liquid chromatography–tandem mass spectrometry method for the analysis of fluconazole in human plasma. *J. Pharm. Biomed. Anal.* 2003; **31**: 917–928.

6. Koh HL, Lau AJ, Chan ECY. Hydrophilic interaction liquid chromatography with tandem mass spectrometry for the determination of underivatized dencichine

(beta-*N*-oxalyl-L-α,β-diaminopropionic acid) in Panax medicinal plant species. *Rapid Commun. Mass Specrom.* 2005; **19**: 1237–1244.

7. Qi S, Junga H, Yong T, Li AC. Automated 96-well solid phase extraction and hydrophilic interaction liquid chromatography–tandem mass spectrometric method for the analysis of cetirizine (ZYRTEC®) in human plasma—with emphasis on method ruggedness. *J. Chromatogr. B* 2005; **814**: 105–114.

8. Su J, Hirji R, Zhang L, He C, et al. Evaluation of the stress-inducible production of choline oxidase in transgenic rice as a strategy for producing the stress-protectant glycine betaine. *J. Exp. Bot.* 2006; **57**: 1129–1135.

9. Qi S, Naidong W. Analysis of omeprazole and 5-OH omeprazole in human plasma using hydrophilic interaction chromatography with tandem mass spectrometry (HILIC–MS/MS)—Eliminating evaporation and reconstitution steps in 96-well liquid/liquid extraction. *J. Chromatogr. B* 2006; **830**: 135–142.

10. Kovarikova P, Stariat J, Klimes J, Hruskova K, Varova K. Hydrophilic interaction liquid chromatography in the separation of a moderately lipophilic drug from its highly polar metabolites—The cardioprotectant dexrazoxane as a model case. *J. Chromatogr. A* 2011; **1218**: 416–426.

11. Mcroczek H. Highly efficient, selective and sensitive molecular screening of acetylcholinesterase inhibitors of natural origin by solid-phase extraction-liquid chromatography/electrospray ionisation-octopole-orthogonal acceleration time-of-flight-mass spectrometry and novel thin-layer chromatography-based bioautography. *J. Chromatogr. A* 2009; **1216**: 2519–2528.

12. Hayama T, Yoshida H, Todoriki K, Nohta H, Yamaguchi M. Determination of polar organophosphorus pesticides in water samples by hydrophilic interaction liquid chromatography with tandem mass spectrometry. *Rapid Commun. Mass Spectrom.* 2008; **22**: 2203–2210.

13. Ji HY, Park EJ, Lee KC, Lee HS. Quantification of doxazosin in human plasma using hydrophillic interaction liquid chromatography with tandem mass spectrometry. *J. Sep. Sci.* 2008; **31**: 1628–1633.

14. Naidong W. Bioanalytical liquid chromatography tandem mass spectrometry methods on underivatized silica columns with aqueous/organic mobile phases. *J. Chromatogr. B* 2003; **796**: 209–224.

15. Hemstrom P, Irgum K. Hydrophilic interaction chromatography. *J. Sep. Sci.* 2006; **29**: 1784–1821.

16. Snyder LR, Kirkland JJ, Dolan JW. *Introduction to Modern Liquid Chromatography*, 3rd ed. Hoboken, NJ: John Wiley & Sons; 2010.

17. Kohler J, Chase DB, Farlee RD, Vega AJ, Kirkland JJ. Comprehensive characterization of some silica-based stationary phase for high-performance liquid chromatography. *J. Chromatogr. A* 1986; **352**: 275–305.

18. Nawrocki J. Silica surface surface controversies, strong adsorption sites, their blockage and removal. Part I. *Chromatographia* 1991; **31**: 177–192.

19. Nawrocki J. Silica surface surface controversies, strong adsorption sites, their blockage and removal. Part II. *Chromatographia* 1991; **31**: 193–205.

20. Engelhardt H, Low H, Gotzinger W. Chromatographic characterization of silica-based reversed phases. *J. Chromatogr.* 1991; **544**: 371–379.

21. Sindorf DW, Maciel GE. SI-29 NMR–Study of dehydrated rehydrated silica-gel using cross polarization and magic-angle spinning. *J. Am. Chem. Soc.* 1983; **105**: 1487–1493.

22. Neue UD. *HPLC Columns Theory, Technology and Practice.* New York: Wiley-VCH; 1997.

23. Sandoval JE, Pesek JJ. Synthesis and characterization of a hydride-modified porous silica material as an intermediate in the preparation of chemically bonded chromatographic stationary phases. *Anal. Chem.* 1989; **61**: 2067–2075.

24. Microsolv. Cogent Silica-C. http://www.microsolvtech.com/hplc/app_aromatic.asp (accessed in October 2012).

25. Kirkland JJ, Truszkowski FA, Dilks CH, Engel GS. Superficially porous silica microspheres for fast high-performance liquid chromatography of macromolecules. *J. Chromatogr. A* 2000; **890**: 3–13.

26. Kirkland JJ, Langlois TJ, DeStefano JJ. Fused core particles for HPLC columns. *Am. Lab.* 2007; **39**: 18.

27. Advanced Materials Technology. HALO HPLC columns. http://www.advanced-materials-tech.com/products.html (accessed in October 2012).

28. Gritti F, Pereira AD, Sandra P, Guiochon G. Efficiency of the same neat silica column in hydrophilic interaction chromatography and per aqueous liquid chromatography. *J. Chromatogr. A* 2010; **1217**: 683–688.

29. Tanaka N, Kobayashi H, Nakanishi K, Minakuchi H, Ishizuka N. Monolithic LC columns. *Anal. Chem.* 2001; **73**: 420A–429A.

30. Ikegami T, Tanaka N. Monolithic columns for high-efficiency HPLC separations. *Curr. Opin. Chem. Biol.* 2004; **8**: 527–533.

31. Miyabe K, Guiochon G. Characterization of monolithic columns for HPLC. *J. Sep. Sci.* 2004; **27**: 853–873.

32. Dawkins JV, Lloyd LL. Chromatographic characteristics of polymer-based high performance liquid chromatography packings. *J. Chromatogr.* 1986; **352**: 157–167.

33. Cabrera K. Applications of silica-based monolithic HPLC columns. *J. Sep. Sci.* 2004; **27**: 843–852.

34. Glenn KM, Lucy CA, Haddad PR. Ion chromatography on a latex-coated silica monolith column. *J. Chromatogr. A* 2007; **1155**: 8–14.

35. Pelletier S, Lucy CA. Achieving rapid low-pressure ion chromatography separations on short silica-based monolithic columns. *J. Chromatogr. A* 2006; **1118**: 12–18.

36. Ibrahim MEA, Zhou T, Lucy CA. Agglomerated silica monolithic column for hydrophilic interaction LC. *J. Sep. Sci.* 2010; **33**: 773–778.

37. Ikegami T, Tomomatsu K, Takubo H, Horie K, Tanaka N. Separation efficiencies in hydrophilic interaction chromatography. *J. Chromatogr. A* 2008; **1184**: 474–503.

38. Chambers SD, Glenn KM, Lucy CA. Developments in ion chromatography using monolithic columns. *J. Sep. Sci.* 2007; **30**: 1628–1645.

39. Pack BW, Risley DS. Evaluation of a monolithic silica column operated in the hydrophilic interaction chromatography mode with evaporative light scattering detection for the separation and detection of counter-ions. *J. Chromatogr. A* 2005; **1073**: 269–275.

40. Nawrocki JJ. The silanol group and its role in liquid chromatography. *J. Chromatogr. A* 1997; **779**: 29–71.

41. McCalley DV. Comparative evaluation of bonded-silica reversed-phase columns for high-performance liquid chromatography using strongly basic compounds and alternative organic modifiers buffered at acid pH. *J. Chromatogr.* 1997; **769**: 169–178.

42. McCalley DV. Reversed-phase HPLC of basic samples-An update. *LC-GC North Am.* 1999; **17**: 440–455.

43. Kirkland JJ, Henderson JW, DeStefano JJ, Straten MA, van Claessens HA. Stability of silica-based, endcapped columns with pH 7 and 11 mobile phases for reversed-phase high-performance liquid chromatography. *J. Chromatogr. A* 1997; **762**: 97–112.

44. Wyndham KD, O'Gara JE, Walter TH, Glose KH, Lawrence NL, Alden BA, Izzo GS, Hudalla CJ, Iraneta PC. Characterization and evaluation of C18 HPLC stationary phases based on ethyl-bridged hybrid organic/inorganic particles. *Anal. Chem.* 2003; **75**: 6781–6788.

45. Grumbach ES, Diehl DM, Neue UD. The application of novel 1.7 μm ethylene bridged hybrid particles for hydrophilic interaction chromatography. *J. Sep. Sci.* 2008; **31**: 1511–1518.

46. Orth P, Engelhardt H. Separation of sugars on chemically modified silica gel. *Chromatographia* 1982; **15**: 91–96.

47. Tomiya N, Awaya J, Kurono M, Endo S, Arata Y, Takahashi N. Analyses of N-linked oligosaccharides using a two-dimensional mapping technique. *Anal. Biochem.* 1988; **171**: 73–90.

48. Higashi H, Ito M, Fukaya N, Yamagata S, Yamagata T. Two-dimensional mapping by the high-performance liquid chromatography of oligosaccharides released from glycosphingolipids by endoglycoceramidase. *Anal. Biochem.* 1990; **186**: 355–362.

49. Takahashi N, Nakagawa H, Fujikawa K, Kawamura Y, Tomita N. Three-dimensional elution mapping of pyridylaminated N-linked neutral and sialyl oligosaccharides. *Anal. Biochem.* 1995; **226**: 139–146.

50. Yoshida T. Prediction of peptide retention time in normal-phase liquid chromatography. *J. Chromatogr. A* 1998; **811**: 61–67.

51. Yoshida T. Peptide separation in normal phase liquid chromatography. *Anal. Chem.* 1997; **69**: 3038–3043.

52. Ahmed WHA, Gonmori K, Suzuki M, Watanabe K, Suzuki O. Simultaneous analysis of alpha-amanitin, beta-amanitin, and phalloidin in toxic mushrooms by liquid chromatography coupled to time-of-flight mass spectrometry. *Forensic Toxicol.* 2010; **28**: 69–76.

53. Tomasek C. TSK-gel amide-80 HILIC columns for the analysis of melamine and cyanuric acid in milk by LC-MS-MS. *LC-GC North Am.* 2009; **27**: 44–45.

54. Risley DS, Yang WQ, Peterson JA. Analysis of mannitol in pharmaceutical formulations using hydrophilic interaction liquid chromatography with evaporative light-scattering detection. *J. Sep. Sci.* 2006; **29**: 256–264.

55. West C, Lesellier E. Characterisation of stationary phases in subcritical fluid chromatography with the solvation parameter model: III. Polar stationary phases. *J. Chromatogr. A* 2006; **1110**: 200–213.

56. Regnier FE, Noel R. Glycerolpropylsilane bonded phases in steric exclusion chromatography of biological macromolecules. *J. Chromatogr. Sci.* 1976; **14**: 316–320.

57. Brons C, Olieman C. Study of the high performance liquid chromatographic separation of reducing sugars, applied to the determination of lactose in milk. *J. Chromatogr.* 1983; **259**: 79–86.

58. Herbreteau B, Lafosse M, Morinallory L, Dreux M. High performance liquid chromatography of raw sugars and polyols using bonded silica gels. *Chromatographia* 1992; **33**: 325–330.

59. Tanaka H, Zhou XJ, Ohira M. Characterization of a novel diol column for high-performance liquid chromatography. *J. Chromatogr. A* 2003; **987**: 119–125.

60. Pazourek J. Monitoring of mutarotation of monosaccharides by hydrophilic interaction chromatography. *J. Sep. Sci.* 2010; **33**: 974–981.

61. Liu X, Pohl C. New hydrophilic interaction/reversed-phase mixed-mode stationary phase and its application for analysis of nonionic ethoxylated surfactants. *J. Chromatogr. A* 2006; **1191**: 83–89.

62. Jandera P, Hájek T, Skeríková V, Soukup J. Dual hydrophilic interaction-RP retention mechanism on polar columns: Structural correlations and implementation for 2-D separations on a single column. *J. Sep. Sci.* 2010; **33**: 841–852.

63. McClintic C, Remick DM, Peterson JA, Risley DS. Novel method for the determination of piperazine in pharmaceutical drug substances using hydrophilic interaction chromatography and evaporative light scattering detection. *J. Liq. Chromatogr. Relat. Technol.* 2003; **26**: 3093–3104.

64. Al-Tannak NF, Bawazeer S, Siddiqui T, Watson DG. The hydrophilic interaction like properties of some reversed phase high performance liquid chromatography columns in the analysis of basic compounds. *J. Chromatogr. A* 2011; **1218**: 1486–1491.

65. Dinh NP, Jonsson T, Irgum K. Probing the interaction mode in hydrophilic interaction chromatography. *J. Chromatogr. A* 2011; **1218**: 5880–5891.

66. Armstrong D, Jin H. Evaluation of the liquid chromatographic separation of monosaccharides, disaccharides, trisaccharides, tetrasaccharides, deoxysaccharides and sugar alcohols with stable cyclodextrin bonded phase columns. *J. Chromatogr.* 1989; **462**: 219–232.

67. Lai XH, Tang WH, Ng SC. Novel β-cyclodextrin chiral stationary phases with different length spacers for normal-phase high performance liquid chromatography enantioseparation. *J. Chromatogr. A* 2011; **1218**: 3496–3501.

68. Feng JT, Guo ZM, Shi H, Gu JP, Jin Y, Liang XM. Orthogonal separation on one beta-cyclodextrin column by switching reversed-phase liquid chromatography and hydrophilic interaction chromatography. *Talanta* 2010; **81**: 1870–1876.

69. Risley DS, Strege MA. Chiral separations of polar compounds by hydrophilic interaction chromatography with evaporative light scattering detection. *Anal. Chem.* 2000; **72**: 1736–1739.

70. Berthod A, Chang SSC, Kullman JPS, Armstrong DW. Practice and mechanism of HPLC oligosaccharide separation with a cyclodextrin bonded phase. *Talanta* 1998; **47**: 1001–1012.

71. Jiang W, Irgum K. Synthesis and evaluation of polymer based zwitterionic stationary phases for separation of tonic species. *Anal. Chem.* 2001; **73**: 1993–2003.

72. Jiang W, Irgum K. Covalently bonded polymeric zwitterionic stationary phase for simultaneous separation of inorganic cations and anions. *Anal. Chem.* 1999; **71**: 333–344.

73. Guo Y, Gaiki S. Retention behavior of small polar compounds on polar stationary phases in hydrophilic interaction chromatography. *J. Chromatogr. A* 2005; **1074**: 71–80.

74. Kane RS, Deschatelets P, Whitesides GM. Kosmotropes form the basis of protein-resistant surfaces. *Langmuir* 2003; **19**: 2388–2391.

75. Viklund C, Irgum K. Synthesis of porous zwitterionic sulfobetaine monoliths and characterization of their interaction with proteins. *Macromolecules* 2000; **33**: 2539–2544.

76. Jiang W, Awasum JN, Irgum K. Control of electroosmotic flow and wall interactions in capillary electrophosesis capillaries by photografted zwitterionic polymer surface layers. *Anal. Chem.* 2003; **75**: 2768–2774.

77. Rodriguez-Gonzalo E, Garcia-Gomez D, Carabias-Martinez R. Study of retention behaviour and mass spectrometry compatibility in zwitterionic hydrophilic interaction chromatography for the separation of modified nucleosides and nucleobases. *J. Chromatogr. A* 2011; **1218**: 3994–4001.

78. Kato M, Kato H, Eyama S, Takatsu A. Application of amino acid analysis using hydrophilic interaction liquid chromatography coupled with isotope dilution mass spectrometry for peptide and protein quantification. *J. Chromatogr. B* 2009; **877**: 3059–3064.

79. Di-Palma S, Boersema PJ, Heck AJ, Mohammed S. Evaluation of the deuterium isotope effect in zwitterionic hydrophilic interaction liquid chromatography separations for implementation in a quantitative proteornic approach. *Anal. Chem.* 2011; **83**: 3440–3447.

80. Van-Dorpe S, Vergote V, Pezeshki A, Burvenich C, Peremans K, DeSpiegeleer B. Hydrophilic interaction LC of peptides: Columns comparison and clustering. *J. Sep. Sci.* 2010; **33**: 728–739.

81. Pasakova I, Gladziszova M, Charvatova J, Stariat J, Klimes J, Kovarikova P. Use of different stationary phases for separation of isoniazid, its metabolites and vitamin B6 forms. *J. Sep. Sci.* 2011; **34**: 1357–1365.

82. Bengtsson J, Jansson B, Hammarlund-Udenaes M. On-line desalting and determination of morphine, morphine-3-glucuronide and morphine-6-glucuronide in microdialysis and plasma samples using column switching and liquid chromatography/tandem mass spectrometry. *Rapid Commun. Mass Spectrom.* 2005; **19**: 2116–2122.

83. Risley DS, Pack BW. Simultaneous determination of positive and negative counterions using a hydrophilic interaction chromatography method. *LC-GC North Am* 2006; **24**: 776–785.

84. Dorr FA, Rodriguez V, Molica R, Henriksen P, Krock B, Pinto E. Methods for detection of anatoxin-a(s) by liquid chromatography coupled to electrospray ionization-tandem mass spectrometry. *Toxicon* 2010; **55**: 92–99.

85. Lindegardh N, Hanpithakpong W, Phakdeeraj A, Singhasivanon R, Farrar J, Hien TT, White NJ, Day NP. Development and validation of a high-throughput zwitterionic hydrophilic interaction liquid chromatography solid-phase extraction–liquid chromatography–tandem mass spectrometry method for determination of the anti-influenza drug peramivir in plasma. *J. Chromatogr. A* 2008; **1215**: 145–151.

86. Hu WZ, Takeuchi T, Haraguchi H. Electrostatic ion chromatography. *Anal. Chem.* 1993; **65**: 2204–2208.

87. Lammerhofer M, Richter M, Wu JY, Nogueira R, Bicker W, Lindner W. Mixed-mode ion-exchangers and their comparative chromatographic characterization in reversed-phase and hydrophilic interaction chromatography elution modes. *J. Sep. Sci.* 2008; **31**: 2572–2588.

88. Beilmann B, Langguth P, Hausler H, Grass P. High-performance liquid chromatography of lactose with evaporative light scattering detection, applied to determine fine particle dose of carrier in dry powder inhalation products. *J. Chromatogr. A* 2006; **1107**: 204–207.

89. Linden JC, Lawhead CL. Liquid-chromatography of saccharides. *J. Chromatogr.* 1975; **105**: 125–133.

90. Que AH, Novotny MV. Separation of neutral saccharide mixtures with capillary electrochromatography using hydrophilic monolithic columns. *Anal. Chem.* 2002; **74**: 5184–5194.

91. Aturki Z, D'Orazio G, Rocco A, Si-Ahmed K, Fanali S. Investigation of polar stationary phases for the separation of sympathomimetic drugs with nano-liquid chromatography in hydrophilic interaction liquid chromatography mode. *Anal. Chim. Acta* 2011; **685**: 103–110.

92. Valette JC, Demesmay C, Rocca JL, Verdon E. Separation of tetracycline antibiotics by hydrophilic interaction chromatography using an amino-propyl stationary phase. *Chromatographia* 2004; **59**: 55–60.

93. Olsen BA. Hydrophilic interaction chromatography using amino and silica columns for the determination of polar pharmaceuticals and impurities. *J. Chromatogr. A* 2001; **913**: 113–122.

94. Lafosse M, Herbreteau B, Dreux M, Morinallorym L. Control of some high-performance liquid-chromatographic systems by using an evaporative light-scattering detector. *J. Chromatogr.* 1989; **472**: 209–218.

95. Jandera P. Stationary and mobile phases in hydrophilic interaction chromatography: a review. *Anal. Chim. Acta* 2011; **692**: 1–25.

96. Alpert AJ. Cation-exchange high-performance liquid chromatography of proteins on Poly(Aspartic acid)-silica. *J. Chromatogr.* 1983; **266**: 23–37.

97. Kiseleva MG, Kebets PA, Nesterenko PN. Simultaneous ion chromatographic separation of anions and cations on poly(aspartic acid) functionalized silica. *Analyst* 2001; **126**: 2119–2123.

98. Zhu BY, Colin CT, Hodges RS. Hydrophilic-interaction chromatography of peptides on hydrophilic and strong cation-exchange columns. *J. Chromatogr.* 1991; **548**: 13–24.

99. Tolstikov VV, Fiehn O. Analysis of highly polar compounds of plant origin: Combination of hydrophilic interaction chromatography and electrospray ion trap mass spectrometry. *Anal. Biochem.* 2002; **301**: 298–307.

100. Oyler AR, Armstrong BL, Cha JY, Zhou MX, Yang Q, Robinson RI, Dunphy R, Burinsky DJ. Hydrophilic interaction chromatography on amino-silica phases complements reversed-phase high-performance liquid chromatography and capillary electrophoresis for peptide analysis. *J. Chromatogr. A* 1996; **724**: 378–383.

101. Curren MS, King JW. New sample preparation technique for the determination of avoparcin in pressurized hot water extracts from kidney samples. *J. Chromatogr. A* 2002; **954**: 41.

102. Zywicki B, Catchpole G, Draper J, Fiehn O. Comparison of rapid liquid chromatography-electrospray ionization-tandem mass spectrometry methods for

determination of glycoalkaloids in transgenic field-grown potatoes. *Anal. Biochem.* 2005; **336**: 178–186.

103. Mihailova A, Lundanes E, Greibrokk T. Determination and removal of impurities in 2-D LC-MS of peptides. *J. Sep. Sci.* 2006; **29**: 576–581.

104. Orachrom Inc. STYROS™ Amino HILIC. http://www.orachrom.com/Product_Folder/Amino_HILIC.html (accessed in October 2012).

105. Klingenberg A, Seubert A. Comparison of silica-based and polymer-based cation exchangers for the ion chromatographic separation of transition metals. *J. Chromatogr.* 1993; **640**: 167–178.

106. Hargitai T, Reinholdsson P, Toernell B, Isaksson R. Functionalized polymer particles for chiral separation. *J. Chromatogr.* 1992; **630**: 79–83.

107. Samuelson O, Sjostrom E. Utilization of ion exchangers in analytical chemistry. XXIV. Isolation of monosaccharides. *Sven. Kem. Tidskr* 1952; **64**: 305–314.

108. Havlicek J, Samuelson O. Separation of oligosaccharides by partition chromatography on ion-exchange resins. *Anal. Chem.* 1975; **47**: 1854–1857.

109. Chambers TK, Fritz JS. Effect of polystyrene–divinylbenzene resin sulfonation on solute retention in high-performance liquid chromatography. *J. Chromatogr. A* 1998; **797**: 139–147.

110. Kawachi Y, Ikegami T, Takubo H, Ikegami Y, Miyamoto M, Tanaka N. Chromatographic characterization of hydrophilic interaction liquid chromatography stationary phases: Hydrophilicity, charge effects, structural selectivity, and separation efficiency. *J. Chromatogr. A* 2011; **1218**: 5903–5919.

111. TOSOH. Tosoh Bioscience LLC. http://www.separations.us.tosohbioscience.com (accessed in October 2012).

112. Guo Y, Gaiki S. Retention and selectivity of stationary phases for hydrophilic interaction chromatography. *J. Chromatogr. A* 2011; **1218**: 5920–5938.

113. Fountain KJ, Xu J, Diehl DM, Morrison D. Influence of stationary phase chemistry and mobile-phase composition on retention, selectivity, and MS response in hydrophilic interaction chromatography. *J. Sep. Sci.* 2010; **33**: 740–751.

114. Kimata K, Iwaguchi K, Onishi S, Jinno K, Eksteen R, Hosoya K, Araki M, Tanaka N. Chromatographic characterization of silica-C-18 packing materials—Correlation between a preparation method and retention behavior of stationary phase. *J. Chromatogr. Sci.* 1989; **27**: 721–728.

115. Neue UD. Stationary phase characterization and method development. *J. Sep. Sci.* 2007; **30**: 1611–1627.

116. Karlsson G, Winge S, Sandberg H. Separation of monosaccharides by hydrophilic interaction chromatography with evaporative light scattering detection. *J. Chromatogr. A* 2005; **1092**: 246–249.

117. Yanagida A, Murao H, Ohnishi-Kameyama M, Yamakawa Y, Shoji A, Tagashira M, Kanda T, Shindo H, Shibusawa Y. Retention behavior of oligomeric proanthocyanidins in hydrophilic interaction chromatography. *J. Chromatogr. A* 2007; **1143**: 153–161.

118. Daunoravicius Z, Juknaite I, Naujalis E, Padarauskas A. Simple and rapid determination of denaturants in alcohol formulations by hydrophilic interaction chromatography. *Chromatographia* 2006; **63**: 373–377.

119. di2chrom. Willkommen bei dichrom—vormals SeQuant. http://www.dichrom.com (accessed in October 2012).

120. Guo Y, Huang AH. A HILIC method for the analysis of tromethamine as the counter ion in an investigational pharmaceutical salt. *J. Pharm. Biomed. Anal.* 2003; **31**: 1191–1201.

121. Puy G, Demesmay C, Rocca JL, Iapichella J, Galarneau A, Brunel D. Electrochromatographic behavior of silica monolithic capillaries of different skeleton sizes synthesized with a simplified and shortened sol-gel procedure. *Electrophoresis* 2006; **27**: 3971–3980.

122. McCalley DV. Study of the selectivity, retention mechanisms and performance of alternative silica-based stationary phases for separation of ionised solutes in hydrophilic interaction chromatography. *J. Chromatogr. A* 2010; **1217**: 3408–3417.

123. Guo Y. Chapter 17. Retention and selectivity of polar stationary phases for hydrophilic interaction chromatography. In: Wang PG, He W, eds. *Hydrophilic Interaction Chromatography (HILIC) and Advanced Applications.* New York: CRC Press; 2011, pp. 401–426.

124. Majors RE. New chromatography columns and accessories at Pittcon 2010: Part I. *LC-GC North Am.* 2011; **29**: 218–235.

CHAPTER

3

HILIC METHOD DEVELOPMENT

YONG GUO

School of Pharmacy, Fairleigh Dickinson University, Madison, NJ

XIANDE WANG

Johnson & Johnson, Raritan, NJ

3.1 INTRODUCTION

Hydrophilic interaction chromatography (HILIC) has been widely recognized as a valid separation mode in modern liquid chromatography and increasingly selected to tackle analytical challenges in biomedical, environmental, and pharmaceutical fields [1–3]. Over the past decade, research in the HILIC area has generated substantial knowledge to enable us to better understand the fundamental aspects of HILIC, such as retention mechanisms, stationary phase properties, and effects of various chromatographic parameters (Chapter 1 and Chapter 2). In practice, chromatographic techniques are typically used to establish analytical procedures or methods for specific applications (e.g., assay or purity analysis). Therefore, the value of a chromatographic technique (e.g., HILIC) is ultimately realized through the methods developed for various applications. Method development is a process of establishing the analytical procedure for qualitative and quantitative determination of the analyte(s) of interest using the selected chromatographic technique. Many method development

Hydrophilic Interaction Chromatography: A Guide for Practitioners, First Edition.
Edited by Bernard A. Olsen and Brian W. Pack.
© 2013 John Wiley & Sons, Inc. Published 2013 by John Wiley & Sons, Inc.

strategies have been developed for well-established chromatographic techniques such as reversed-phase liquid chromatography (RPLC) [4,5]. However, systematic method development for HILIC has not been extensively discussed and documented in the literature [6–9]. Dejaegher et al. reviewed many biological and pharmaceutical assays based on HILIC and found that most method development employed the trial-and-error approach [10,11]. Chirita et al. proposed a decision tree to assist method development and optimization based on the effect of various chromatographic factors on the HILIC separation of neurotransmitters as well as a selectivity comparison of HILIC phases based on principal component analysis (PCA) [12].

In chromatographic method development, initial emphasis is often placed on achieving the desired separation. As a result, many development activities are centered around selecting appropriate columns and mobile phases. It should be noted that achieving the desired separation alone does not constitute a complete method, particularly for bioanalytical and pharmaceutical methods typically operated in a regulated environment. Method development should also include sample preparation, detector selection, system suitability, and quantitative calculation. This chapter focuses on method development employing HILIC as the chromatographic technique. Various aspects of method development are discussed including method objectives, sample considerations, column and mobile phase selection, and other operating parameters (e.g., column temperature, sample solvent, and detectors). The goal is to provide general guidance on HILIC method development based on a solid understanding of HILIC basics and the authors' experience with bioanalytical and pharmaceutical methods. Chapter 4, Chapter 5, Chapter 6, Chapter 7, Chapter 8, and Chapter 9 provide additional details for methods related to specific applications.

3.2   GENERAL METHOD DEVELOPMENT CONSIDERATIONS

Method development starts with a clear definition of method objective. With the objective in mind, it is important to build up a good understanding of the properties of the target analytes, such as the structure, molecular weight, pK_a, log P, and solubility. It is also important to understand the sample characteristics and be prepared for the sample preparation challenges. Selection of column and mobile phase is the core of method development and can be effectively conducted when a screening process is implemented using sound experimental design. Robustness evaluation is also an integral part of method development. Design of experiments (DOE) methodology can be applied throughout the method development, optimization, and validation to enhance the efficiency of the entire process, as illustrated in Figure 3.1.

3.2.1   Method Objectives

Chromatographic procedures and methods are developed to solve specific analytical problems or complete defined tasks. A clear definition of the problem

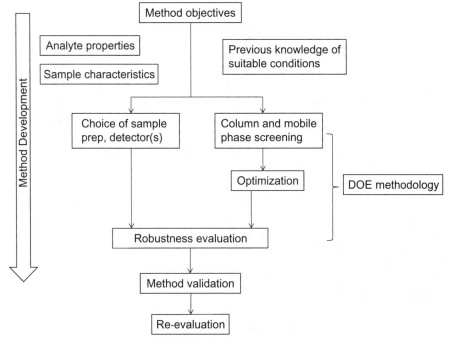

Figure 3.1. General scheme for HILIC method development.

or task is the first step toward successful method development. Similar to other chromatographic methods (e.g., RPLC methods), HILIC methods can be applied to various types of analytical problems in many fields including the following bioanalytical and pharmaceutical applications:

1. Single component assay, such as assay of active drug or metabolite in biological matrices, active pharmaceutical ingredient (API) in drug substance and dosage formulations, counterion in drug substance, and assay of special impurities (e.g., genotoxic impurity).

2. Analysis of multiple compounds of interest, such as synthesis impurities in drug substances, degradation products in drug products, and drug metabolites in biological matrices.

3. Stability-indicating methods for drug substance and drug product.

Although HILIC can be used in any of the above applications, it has some unique advantages in some specific areas over other chromatographic techniques. For example, the HILIC method can simplify sample preparation and also improve sensitivity when coupled with mass spectrometric (MS) detection [13]. The HILIC/MS method is also suitable for the analysis of impurities at low concentrations (e.g., genotoxic impurity) owing to its improved MS sensitivity. HILIC methods have also been successfully employed for the analysis of organic counterions in API [14,15]. These and other pharmaceutical applications are described further in Chapter 4, Chapter 5, and Chapter 7.

To ensure that the method development objectives are achieved with a high probability of success, a set of method attributes should be set early on to guide method development activities. Since the methods for bioanalytical and pharmaceutical applications are required to be validated, the International Conference on Harmonization (ICH) and Food and Drug Administration (FDA) guidances on method validation provide a general framework for the method attributes (e.g., accuracy, precision, linearity, robustness, limit of detection [LOD], and limit of quantitation [LOQ]) [16,17]. Methods for use in other fields or for different purposes may have different requirements from regulated pharmaceutical methods. In addition, other method attributes should also be considered based on the method objectives, such as desired resolution between critical pairs, required sensitivity for particular impurities, and run time. Development analysts should also consider other factors that have practical implications, for example, adaptability for automation (for high sample volume applications), suitability for routine application in quality control (QC) labs, and availability of instruments (e.g., detectors) when the methods need to be transferred to different QC labs.

3.2.2 Target Compounds Consideration

Understanding the targeted compounds (analytes) goes hand in hand with defining method development objectives. Although the structures of all targeted compounds may not be available at the start of method development, the structure of the active compound in the case of pharmaceutical analysis is typically known. It is important to gather as much information on the target compounds as possible (structure, molecular weight, pK_a, log P, solubility, etc.). This information will help the development analysts to assess the possibility of success in developing the HILIC method based on HILIC fundamentals. In general, HILIC is a chromatographic technique suitable for polar compounds due to stronger retention provided by HILIC than RPLC. Various types of compounds have been separated by HILIC methods, including amino acids, nucleic acids, organic acids, sugars, water-soluble vitamins, drugs and metabolites, and even peptides and proteins [1,3]. Typically the compounds suitable for HILIC separation are those with small positive or negative log P (P is the octanol–water partition coefficient) and relatively low molecular weight (excluding peptides and proteins). The possibility of a polar compound being retained in HILIC can be assessed using a web-based prediction model developed by Merck SeQuant (http://www.sequant.com). The prediction model is built upon the retention data of 40 polar compounds on a ZIC-HILIC column through quantitative-structure–retention relation (QSRR) and multivariate modeling. The structure of a particular compound in simplified molecular-input line-entry specification (SMILES) format is entered into the web application, and the prediction model generates a retention factor in the mobile phase that contains 70% acetonitrile and 30% 100 mM ammonium acetate (pH 5.6 or 6.7). Although the database on which the prediction model is built is relatively small and the mobile phase is relatively strong (30% aqueous buffer), it can still provide useful information regarding compound retention in HILIC at the beginning of method development.

Solubility is another compound property that needs special attention in developing HILIC methods. In HILIC, the mobile phase typically contains high organic solvent content (>60% by volume), which may lead to reduced solubility of some compounds. Therefore, it is critical for the development analyst to check the solubility of the target compounds not only in aqueous solutions, but also in the final mobile phase. Sample solvents also need to be carefully evaluated in relation to the final mobile phase. Although a sample solvent containing a high level of organic solvent (e.g., acetonitrile) may be desirable for better peak shape and efficiency, it may not provide sufficient solubility for the targeted compounds at the concentration required to achieve desired sensitivity. The samples may be prepared in a solvent stronger than the mobile phase; however, the difference in the solvent strength between the sample solvent and mobile phase may lead to distorted peak shapes and reduced efficiency (see Section 3.3.2).

3.2.3   Systematic Method Development

One-factor-at-a-time (OFAT) is a trial-and-error approach traditionally used in method development, where one chromatographic factor is tested while the other factors are set at rather arbitrary values. The OFAT approach may lead to a usable method, but the method may not have optimal performance. More importantly, the knowledge generated in the development process is very limited, and the method may not be robust and reproducible. A preferred approach is systematic method development, which is based on the understanding of HILIC basics and is critical to the success of method development and knowledge generation of the final method [6–9]. The systematic approach to method development is consistent with the quality-by-design (QbD) concept, which is defined as "a systematic approach to development that begins with predefined objectives and emphasizes product and process understanding and process control, based on sound science and quality risk management" [18]. Systematic method development takes a global view of various chromatographic factors on the desired separation and should begin with defined method objectives and rely on rational experimental design for initial screening and optimization (refer to Fig. 3.1). After the method is finalized, an assessment of method robustness needs to be performed. Method robustness is often evaluated at the end of method validation. However, if major issues are uncovered in robustness studies, this may result in redevelopment/validation and timeline delay. Method robustness should be evaluated before the start of method validation using appropriate experimental design (e.g., DOE) and method knowledge gained during development. The outcome of the robustness evaluation can help the development analyst define control strategies to ensure acceptable method performance throughout the duration of its use. The extent of robustness studies and validation will depend on the intended use of the method. Methods for quality control or other regulatory purposes typically require thorough robustness evaluation and validation.

Method development typically starts with screening potential stationary phases/columns. In systematic method development, column screening should be conducted in combination with various mobile phase conditions including organic solvent content and pH in a well thought-out design space. In HILIC, stationary phase chemistry and organic solvent content in the mobile phase are shown to be the most important factors for retention and selectivity [2,3]. The mobile phase pH is another important factor, particularly for ionizable compounds. These factors should be first screened in various combinations. Previous knowledge of suitable conditions can help in the choice of conditions to be screened for a new, but related application. The screening experiments can be based on a set protocol, or ideally a statistical approach (e.g., DOE). For example, acetonitrile content, salt concentration, and column temperature (three factors) were screened for aspirin and related compounds on selected stationary phases including amino, amide, silica, and sulfobetaine phases at five levels. A 3×5 design (with a center) yields 20 experiments including two dummies [19]. Figure 3.2 shows the response surface plot of salicyluric acid on

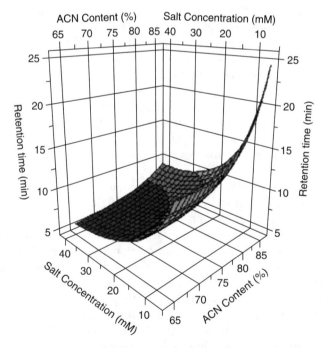

Figure 3.2. Response surface of salicyluric acid (salt concentration and acetonitrile content) on the amino phase (YMC-Pack NH₂ column). Reprinted from Reference 19 with permission from Springer.

the amino phase for acetonitrile content and salt concentration. The response surface plots based on the DOE results can not only help to identify optimal chromatographic conditions, but also provide information on the interactions of chromatographic factors.

It should be noted that systematic method development may require significant time and efforts upfront. In some cases, project timelines may not allow the time to fully implement the systematic approach. Nonetheless, the principles of systematic development should be followed as much as possible within the timeline and resource constraints. The following sections provide more detailed information on some important aspects of HILIC method development (e.g., columns, mobile phase, and detectors).

3.3 HILIC METHOD DEVELOPMENT

3.3.1 Systematic Approach to Column Screening

Column selection, to a large extent, depends on the method objectives. For a single component assay (e.g., biological sample assay, content uniformity, counterion analysis), the primary concern is peak shape, run time, and column reproducibility. Simple column chemistry, such as a bare silica phase, is advantageous since the performance of the bare silica column does not gradually deteriorate due to "bleeding" of the bonded stationary phase. In addition, the peak shape of basic compounds on silica columns often improves in HILIC in comparison to RPLC [20]. However, silica stationary phases are subject to dissolution at pH > 8 and some compounds can adsorb irreversibly to silica. For impurity analysis using stability-indicating methods, selectivity is more important to ensure that all impurities and/or degradation products are well separated from each other. When the method is intended to determine the impurities and degradation products at very low levels, sufficient resolution, stable baseline, and low column bleeding are often critical to achieve optimal sensitivity. As shown in Figure 3.3, a stable baseline and sufficient resolution allowed 5-fluorouracil to be detected at levels below 0.1% in 5-fluorocytosine on an amino column [21].

A large number of columns, based on a wide variety of stationary phase chemistry, are now commercially available for HILIC applications and are shown to have different selectivity toward polar compounds (see Chapter 2 for details). In HILIC, water shields the silica surface but does not completely prevent the interaction of polar analytes with surface silanol groups. In order to achieve reproducible separation, ultra-high purity silica should be selected for HILIC. In addition to the bare silica phase, bonded polar phases (e.g., amide, diol, and sulfobetaine phases) can effectively reduce the interactions between charged analytes and residual silanol groups as well as provide different selectivity. A survey of published reports, as shown in Figure 3.4, reveals that the bare silica phase is most popular in HILIC, followed by sulfobetaine

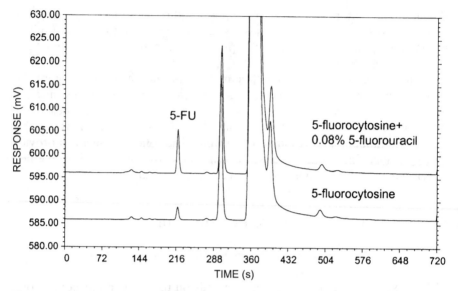

Figure 3.3. Determination of 5-fluorouracil in 5-fluorocytosine. Zorbax NH₂ column; mobile phase acetonitrile–25 mM potassium phosphate, pH 6.5 (80:20); 275 nm detection; fluorocytosine (1.0 mg/mL). Reprinted from Reference 21 with permission from Elsevier.

and amide phases [22]. Diol and mixed-mode phases are less frequently used, probably due to lower hydrophilicity, but are still favored in certain applications, particularly for the mixed-mode columns, such as ion-exchange phases. In comparison, the amino phase is least favored but can provide significant retention for anionic compounds. The columns listed in Figure 3.4 account for more than 90% of the applications among all the stationary phases that are commercially available [22].

A typical systematic approach to method development employs a number of stationary phases with different selectivity. The selected stationary phases are screened in order to identify the optimal stationary phase for desired selectivity, peak shape, and overall performance. Based on retention and selectivity characteristics, five types of polar stationary phases, as shown in Table 3.1, are recommended for initial screening. The list of the stationary phases can be expanded or shortened, depending on the available knowledge of the target analytes and method objectives. For example, Liu et al. screened seven different columns in an effort to develop a HILIC method for the analysis of 4-(aminomethyl) pyridine (4-AMP) and its related compounds [23]. Figure 3.5 shows the chromatograms on seven different columns. Significant differences in retention and selectivity were observed on the stationary phases screened. The silica phase was selected for the final method based on the screening results, but some differences were observed among the silica phases from different manufacturers, as shown in Figure 3.6.

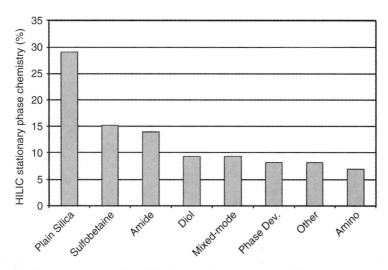

Figure 3.4. Relative number of published scientific papers using different stationary phases chemistries for HILIC separations. Data from literature survey performed in 2009. Adapted from Reference 22 with permission from Chromatography Today.

Table 3.1. Recommended Stationary Phases for Initial Screening

Stationary Phase	Functional Group	Example Columns[a]
Silica	Silanol	Atlantis HILIC
Zwitterionic	Sulfobetaine	ZIC HILIC
Amide	Carbamoyl	TSKgel Amide 80
Diol	Hydroxyl	YMC-pack Diol
Amino	Amino	Zorbax NH_2

[a]Acceptable columns for similar phases may also be obtained from other suppliers.

In addition to the stationary phase, the type of organic solvent and the content of organic solvent in the mobile phase are also very critical to HILIC methods. Acetonitrile is the most commonly used organic solvent in the mobile phase, and retention is very sensitive to the change of acetontrile content in the mobile phase, especially above 85% (v/v). Mobile phase pH and buffer/salt concentration are also critical to HILIC methods. Ammonium acetate or formate is commonly used to control mobile phase pH and is also compatible with detectors that require a volatile buffer such as MS, or evaporation-based detectors such as the charged aerosol detector (CAD). In the systematic approach, column screening should be performed in combination with various mobile phase conditions. Considering the impact of organic solvent and mobile phase pH, we recommend a range of acetonitrile content (e.g., 60%, 80%,

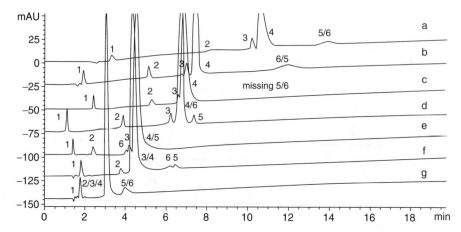

Figure 3.5. Separation of 4-AMP and its related compounds on different HILIC stationary phases. Column temperature is at 30°C. Mobile phases A and B are acetonitrile and 50 mM ammonium formate (pH 3.0), respectively. Gradient is from 10% B to 50% B in 20 min. Flow rate is 1.5 mL/min and detection is at 254 nm. Compounds 1–6 are 4-MP, 2-AMP, 3-AMP, 4-AMP, Degradant-1 and Degradant-2, respectively. Columns: (a) TSKgel Amide-80, (b) SeQuant ZIC-HILIC, (c) Developsil 100Diol-5, (d) PolyHydroxylEthyl A, (e) Phenomenex Luna-NH₂, (f) Phenomenex Luna CN, and (g) Hypersil Gold PFP. Reprinted from Reference 23 with permission from Elsevier.

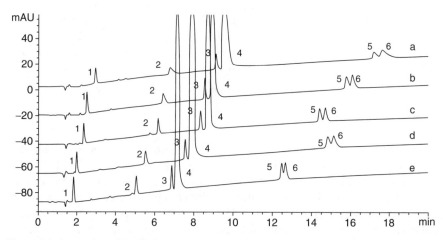

Figure 3.6. Separation of 4-AMP and its related compounds on bare silica HILIC columns from different manufactures. Column temperature is at 30°C. Mobile phases A and B are acetonitrile and 50 mM ammonium formate (pH 3.0), respectively. Gradient is from 10% B to 50% B in 20 min. Flow rate is 1.5 mL/min and detection is at 254 nm. Compounds 1–6 are 4-MP, 2-AMP, 3-AMP, 4-AMP, Degradant-1 and -2, respectively. Columns: (a) Phenomenex Luna Silica (2), (b) Phenomenex Kromasil Silica, (c) Waters Atlantis HILIC, (d) Alltech Alltima HP HILIC, and (e) Thermo Electron HyPurity Silica. Reprinted from Reference 23 with permission from Elsevier.

95%) and mobile phase pH (e.g., pH 3.0 and 6.5) in the screening experiment. In addition, the mobile phase should also contain 10 mM buffer, for example, 10 mM ammonium formate for pH 3.0 and 10 mM ammonium acetate for pH 6.5. These combinations produce six screening runs (isocratic) for each stationary phase listed in Table 3.1. Under the proposed screening conditions (pH 3.0 and 6.5), the recommended columns include both neutral and charged phases and may provide a variety of interactions necessary for HILIC separations. Initial screening can also be conducted in gradient mode using a gradient range from 5% to 40% water. It should be noted that gradient elution often requires extra re-equilibration time in HILIC, and the baseline in gradient elution is often unsteady, particularly for CAD.

Many instruments for column and mobile phase screening are available with various configurations and software. Most configurations allow the screening of a combination of multiple mobile phase compositions and columns automatically. The screening results can be processed automatically or manually to identify the column and mobile phase combination that provides the desired separation and targeted performance. Since many hydrophilic compounds are ultraviolet (UV) transparent, we recommend that a UV detector and a CAD, connected in series, be used in the screening experiment to ensure all species are detected. This is particularly important if not all analytes of interest are known, such as with impurity investigations or forced degradation studies. The UV/CAD combination can also provide valuable information on response factors and mass balance, which could be critical in the method optimization at a later stage [24]. For example, the UV and CAD detectors were used in combination during method development for a marketed drug product and its degradation products (unpublished data). The HILIC method is being developed as an orthogonal method to a validated RPLC method. Figure 3.7 shows the chromatograms of a typical sample generated by both UV and CAD detectors. The CAD detector provides a more stable baseline than the UV detector, and also detects other species in the sample matrix that are not detected by the UV detector (e.g., Na^+/K^+ peak at 34.8 min).

3.3.2 Optimization of Method Parameters

3.3.2.1 Final Column Selection

The initial screening experiments generate the separation of the target compounds on various stationary phases under different mobile phase conditions. The development analyst can typically select one stationary phase as the primary column based on desired selectivity, peak shape, retention time, and other method objectives. The screening results also provide the opportunity to select another stationary phase with different selectivity in case an orthoganol method is needed.

After the type of stationary phase is selected through the screening experiments, final column selection should also consider parameters such as particle

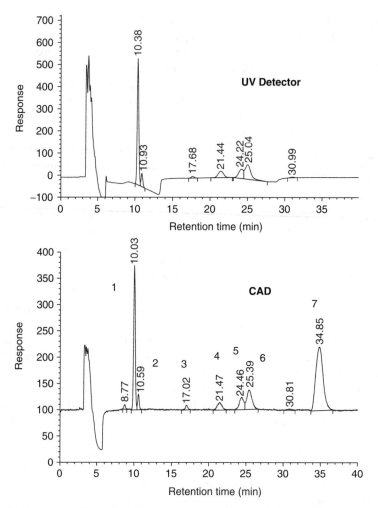

Figure 3.7. Separation of a marketed drug product and its degradation products using both UV and CAD detectors. Column: TSKgel Amide-80, 150 × 4.6 mm, 3 μm particle size. Mobile phase: ACN/water (85/15, v/v) containing 10 mM ammonium acetate pH 5.0. Column temperature: 30°C. Flow rate: 0.5 mL/min. UV detection at 240 nm. Peak identification: (1) active compound, (2) α-anomer, (3) unknown impurity, (4) degradation products 1 and 2, (5), degradation product 3, (6) degradation product 4, and (7) Na⁺/K⁺.

size and column dimension. Typical HILIC columns are packed with 3- or 5-μm particles commonly used in analytical separations. Recently, HILIC columns packed with sub-2 μm particles (e.g., 1.7 μm) became available for ultra-high performance liquid chromatography (UHPLC) applications. A HILIC method employing sub-2 μm particles provides the opportunity to speed up separations without compromising separation efficiency, and is particularly suitable for the applications where fast analysis is needed [25].

Together with the particle size, appropriate column length should be selected to achieve sufficient efficiency and desired resolution while balancing the run time (particularly for isocratic methods) and the backpressure. The backpressure is typically lower in HILIC than in RPLC due to lower viscosity of the mobile phase containing high organic solvents. The columns with 4.6-mm ID are commonly used in the analytical flow rate range (about 0.5 to about 3 mL/min) and 2.1-mm ID columns are more appropriate for HILIC-MS and UHPLC applications.

3.3.2.2 Organic Solvents

Water is the essential component of the mobile phase (minimum of 3–5% in the mobile phase) as it maintains the stagnant aqueous layer on the surface of the stationary phase necessary for HILIC separation. A weaker solvent is also needed to decrease the overall polarity of the mobile phase so that the solutes can partition into the aqueous layer and be retained. Acetonitrile is the most common organic solvent in HILIC. Due to its aprotic nature, the use of acetonitrile encourages stronger hydrogen bonding between the analytes and stationary phases, leading to increased retention. Alcohols are alternative solvents for HILIC, and the retention time typically increases with the carbon number in the order of methanol < ethanol < isopropyl alcohol (IPA) [3]. IPA is less hydrophilic and also has less hydrogen bonding capability, thus resulting in a stronger retention of polar analytes. Methanol can form hydrogen bonds with polar analytes and thereby decreases hydrophilic interactions leading to reduced retention. Other organic solvents such as acetone and tetrahydrofuran (THF) can also be used. The overall solvent strength increases in the order of acetone < acetonitrile < IPA < ethanol < methanol < water [3]. The organic solvents can also be used in various combinations to adjust retention and selectivity.

Method development typically starts with acetonitrile as the organic solvent, as employed in the screening step. Other solvents may be evaluated under certain circumstances. For example, if shortage of acetonitrile or solvent cost is a concern, the sample is not soluble in acetonitrile, or the desired selectivity cannot be achieved using acetonitrile, alcohols may be substituted if retention and selectivity are not negatively affected.

HILIC mobile phases should generally contain at least 60% organic solvent to ensure sufficient hydrophilic interaction. It should be noted that a small change of acetonitrile content can cause a significant change in retention when the acetontrile content in the mobile phase exceeds 85%. Therefore, the selection of final organic content in the mobile phase can have a significant impact on the reproducibility and robustness of the HILIC method.

3.3.2.3 Mobile Phase pH

Mobile phase pH has a significant effect on retention and selectivity by changing the ionization state of both analytes and stationary phases in HILIC [2,3].

However, precautions must be taken since most HILIC silica-based columns are not stable at extreme pHs (less than 2 or greater than 8). Ionized analytes (e.g., acidic and basic compounds) typically have stronger retention due to increased hydrophilic interactions. For the stationary phases, not only can the mobile phase pH affect ionizable functional groups of the stationary phase (e.g., amino and trizaole phase), but it can also have a significant impact on the bare silica and silica-based neutral phases as well by ionizing surface silanol groups. Normal silanol groups are weak acids and a small fraction, particularly when activated by metal impurities embedded in the silica substrate, demonstrates much higher acidity. The silanol groups become deprotonated and negatively charged at higher pHs. Positively charged basic compounds would have increased electrostatic attractions with negatively charged silanol groups in addition to hydrophilic interaction, resulting in stronger retention. On the other hand, acidic compounds would have less retention at higher pHs due to stronger electrostatic repulsion from the silanol groups. The mobile phase pH can also impact selectivity and usually has more influence on acidic compounds than basic compounds. One example is shown in Figure 3.8 to demonstrate the effect of mobile phase pH on the selectivity of selected compounds [26].

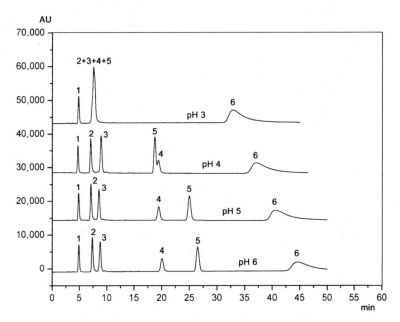

Figure 3.8. Effect of pH on separation of selected water-soluble vitamins. Conditions: column, Inertsil diol (150 mm × 4.6 mm, particle size 5 μm); column temperature, 25°C; flow rate, 0.6 mL/min; detection wavelength, 272 nm; mobile phase, ACN-H$_2$O 90:10 v/v containing ammonium acetate 10 mM, with the aqueous buffer adjusted at various pH values. Peak assignment: (1) nicotinamide, (2) pyridoxine, (3) riboflavin, (4) nicotinic acid, (5) L-ascorbic acid, and (6) thiamine. Reprinted from Reference 26 with permission from Wiley.

For simplicity purposes, aqueous solutions of organic acids such as formic and acetic acids can be directly used to obtain low pHs at low concentrations (e.g., below 0.2%) without pH adjustment. The mobile phases at higher pHs (less than pH 7) are commonly prepared using buffer salts with pH titration, such as ammonium formate or acetate. It should be noted that the apparent pH value of the mobile phase containing high organic solvents is different from that of the aqueous acid or buffer solutions [20,27]. At acidic pHs, the apparent pH of the mobile phase is likely higher than that of the aqueous acid or buffer solution used to prepare the mobile phase. The pK_a values of the analytes are also affected by the high organic content in the mobile phase, particularly for basic compounds [27]. Therefore, it is important to consider the impact of the organic solvent on the pH of the mobile phase and pK_a of the analytes so that the ionization of the analytes is correctly estimated.

3.3.2.4 *Buffer Types and Concentration*

Buffers are commonly employed to adjust the mobile phase pH in chromato-graphic methods, and buffer capacity can prevent pH fluctuation of the mobile phase, resulting in more reproducible and robust methods. In HILIC, buffer salts soluble in the mobile phase containing high organic content are required to provide the desired pH and ionic strength. Ammonium acetate or formate is typically selected for acidic pHs, and ammonium hydroxide and carbonate are good alternatives for high pHs. Phosphate buffers should be generally avoided due to their low solubility in the mobile phase typically used in HILIC and their incompatibility with MS or CAD detectors. However, phosphate buffers at low concentration can be useful when UV detection at low wave-length is desired.

Appropriate buffer concentration (5–100 mM) is typically recommended for HILIC to maintain an acceptable peak shape. The impact of buffer con-centration on the retention and selectivity is dependent on the nature of the analyte's interaction with the stationary phase in HILIC. For nonionizable compounds, partitioning between the stagnant aqueous layer on the stationary phase and hydrophobic mobile phase is the dominant force that determines the retention of analytes. In such cases, the retention time increases moder-ately when the buffer/salt concentration increases. It is proposed that higher salt concentration increases the volume of the stagnant aqueous layer, which in turn provides more partitioning of analytes, leading to longer retention times [28]. For ionizable species, electrostatic interactions play an important role and can modify the retention behavior ([2], also see Chapter 1). Depending on the charges on the solutes and stationary phases, the electrostatic inter-actions between the charged solutes and stationary phases can be either attractive or repulsive. The charged solutes, in the absence of a salt or buffer, have a much greater tendency to be associated with the oppositely charged groups on the stationary phase (e.g., protonated amino or deprotonated silanol groups) through electrostatic attraction, resulting in strong or permanent retention on the column. In this case, addition of salt/buffer becomes necessary for elution and good peak shape, and the retention time decreases when the buffer/salt concentration increases as the electrostatic attraction is weakened at higher ionic strength. Similarly, the retention behavior of negatively charged analytes (e.g., acidic compounds) on silica-based phases is influenced by the electrostatic repulsion between negatively charged analytes and surface silanol groups. As the buffer concentration increases, weakened electronic repulsion leads to stronger retention. It has been observed that higher salt concentrations can improve the peak shape of charged analytes as well as column loadability [20]. The change in the buffer/salt concentration may also result in significant changes in selectivity. For example, the elution pattern of a group of acidic compounds on a silica column showed a marked change by simply changing the ammonium acetate concentration from 10 to 20 mM, as shown in Figure 3.9 [29].

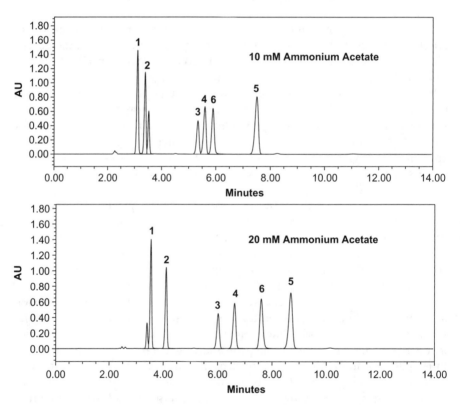

Figure 3.9. Separation of the acidic model compounds on the silica phase. Mobile phase, acetonitrile/water (85/15, v/v) containing 10 and 20 mM ammonium acetate. Column dimension, 250 mm × 4.6 mm ID, 5 μm particle size. Column temperature: 30°C. Flow rate: 1.0 mL/min. UV detection at 228 nm. Peak label: (1) salicylic acid, (2) gentisic acid, (3) acetylsalicylic acid, (4) salicyluric acid, (5) hippuric acid, and (6) α-hydroxyhippuric acid. Reprinted from Reference 28 with permission from Elsevier.

In case of electrostatic interactions, the counterions of the buffer salt (typically cations) may also have an impact, to a lesser extent, on the retention of charged analytes due to different elution strength. For example, Liu et al. found that the retention times of hydrazines decreased in the order of $Na^+ <$ $NH_4^+ <$ triethylamine (TEA^+) for phosphate buffers with different cations [30]. In addition, it is possible that the presence of an excess concentration of counterions might promote the formation of ion pairs with the charged solutes. Formation of ion pairs reduces the hydrophilicity of the analytes and hence results in shorter retention times [31].

Based on the above discussion, the buffer salt in the mobile phase needs to be carefully optimized when the column and mobile phase are selected based on the initial screening results. It is critical to identify the type of electrostatic interactions (attractive or repulsive) between the charged analytes and the stationary phases, so the buffer/salt type and concentration can be optimized to achieve the desired retention time and selectivity.

3.3.2.5   Column Temperature

Column temperature is sometimes used to adjust retention and selectivity of the target analytes in chromatographic methods. For most polar compounds in HILIC, an increase in column temperature usually leads to a decrease in retention when hydrophilic interaction is the primary retention mechanism. However, deviation from this behavior may occur when other retention mechanism becomes dominant. For example, the retention of acetylsalicylic acid increases as the column temperature increases on the amino phase [28]. It should be noted that the column temperature is found to have less impact on the retention of acidic compounds than the content of organic solvent or the buffer/salt concentration in HILIC [19]. Marrubini's study on nucleic acid bases and nucleosides also showed a relatively small change in retention when the column temperature increased from 20°C to 50°C. Significant changes were only observed when the column temperature reached 80°C [32]. Although the column temperature plays a relatively limited role in HILIC, its impact on retention and selectivity should be clearly understood during method development, particularly in the case of aberrant behavior. The column temperature can be optimized to fine-tune the method in the final step of the method development. Controlling column temperature can also lead to greater robustness of a separation and is especially necessary when temperature is known to impact retention.

3.3.2.6   Sample Solvents

Acetonitrile is often the first choice of sample solvent for small molecule compounds, but other organic solvents (e.g., methanol, ethanol, and IPA) can also be used for sample preparation in HILIC. When the sample solubility in pure acetonitrile is an issue, a mixture of acetonitrile and IPA (e.g., 50:50, v/v) may be used as sample solvent. In some cases, dimethyl sulfoxide (DMSO) may be considered due to its ability to dissolve a wide variety of organic compounds at high concentrations. Water or buffer solutions at certain pHs are commonly mixed with the organic solvent to improve sample solubility. However, the amount of water in sample diluents should be carefully adjusted since too much water in the sample diluents can lead to a deteriorated peak shape (broadened or split) [33]. To achieve the desired peak shape and targeted sensitivity, the amount of water in the sample diluted should also be balanced with appropriate injection volumes. As demonstrated in Figures 3.10 and 3.11, large injection volumes of the sample solution containing a large amount of water cause peak deterioration, but reducing injection volume can restore the peak shape without reducing water content in the sample solvent [34].

3.4 DETECTION FOR HILIC METHODS

For analytes with one or more chromophores, UV detection is probably the most widely used detection method in liquid chromatography due to its broad linear range, relatively low cost, ease of use, and the fact that it is compatible with most solvents used in the mobile phase in isocratic or gradient elution mode. As alternatives to UV/Vis detection, particularly for compounds lacking strong UV absorbance, MS, and charged aerosol detection (CAD) are advantageous. Other detectors based on eluent evaporation (evaporative light-scattering [ELSD] and nanoquantity analyte detection [NQAD] are also options. Less commonly used detectors such as refractive index (RID), chemiluminescent nitrogen (CLND), electrochemical, and fluorescence may also be compatible with HILIC conditions and may be appropriate for specific applications. However, ELSD and RID detectors have certain limitations in precision, sensitivity, and dynamic range, and RID is also not compatible with gradient elution. Due to the presence of high organic content in the mobile phase, MS and CAD detectors have special advantages in HILIC.

3.4.1 MS Detector

Electrospray ionization (ESI) or atmospheric pressure chemical ionization (APCI) is commonly used with MS detectors when coupled with liquid chromatography. ESI or APCI-MS requires volatile buffer salts in the mobile phase. The mobile phase used for HILIC is a perfect match for ESI or APCI-MS. Moreover, the high organic content in the mobile phases facilitates eluent vaporization and increases ionization efficiency in the electrospray interface, which leads to improved sensitivity of ESI-MS detection. These advantages have been exploited for HILIC-MS determination of polar drugs in bioanalytical samples as described in Chapter 7. Volatile buffers should be used at reasonable concentrations (5–20 mM) since high buffer concentrations can suppress ESI signals. Signal enhancement can also be achieved by postcolumn

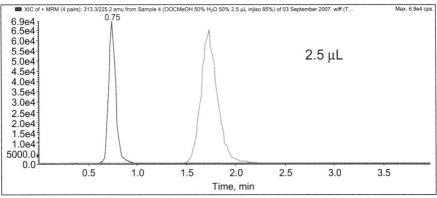

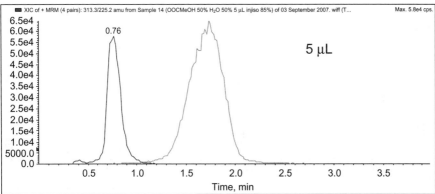

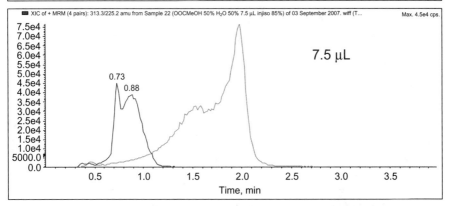

Figure 3.10. Effect of water in sample solvents and injection volume. Column: 50 mm × 2.1 mm, particle size 5 μm. Mobile phase: acetonitrile/ammonium acetate 10 mM (90/10). Flow rate: 0.5 mL/min. Sample solvent: methanol/ammonium acetate 50 mM (50/50). Figure adapted from W. Jiang, Reference 34, personal communication.

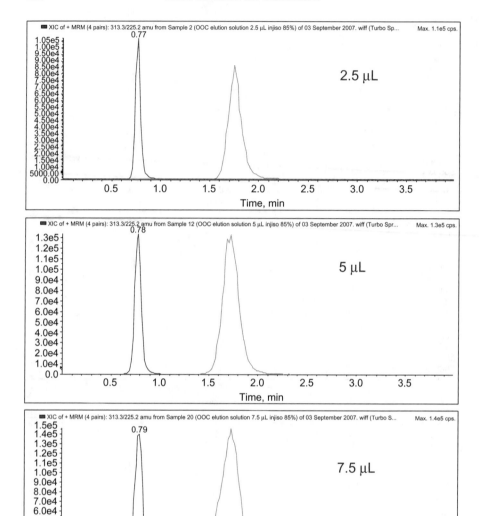

Figure 3.11. Effect of water in sample solvents and injection volume. Sample solvent: methanol/ammonium acetate 50 mM (90/10). Figure adapted from W. Jiang, Reference 34, personal communication.

addition of aprotic solvents or long-chain alcohols, following HILIC separation, although sensitivity may be reduced due to dilution. It should be noted that ionization suppression caused by mobile phase additives (e.g., TFA) or the sample matrix may also occur with ESI-MS detection for HILIC. The APCI source is less vulnerable to matrix ionization suppression effects than the ESI source, and the recovery time for the APCI response to return to its presample injection level is also shorter. In order to achieve high-quality analytical results by HILIC-MS methods, matrix effects and other chromatographic characteristics such as peak shape, carry-over, cross-talk, and mass spectrometric interference are important to evaluate [35].

3.4.2 CAD

In a CAD detector the eluent from a chromatographic system is nebulized using a flow of nitrogen, and the aerosol is transported through a drift tube where the volatile components and solvents are evaporated. Then the dried particle stream is charged with a secondary stream of nitrogen after passing a high-voltage platinum wire; the resulting charged particle flux is measured by an electrometer. Unlike UV detection, CAD is a mass-dependent detector and its response does not depend on the spectral or physicochemical properties of the analytes as long as the analytes are nonvolatile. In theory, this means that CAD can generate similar responses for identical amounts of different analytes. CAD has higher sensitivity than ELSD, as the signal generated by CAD is much less influenced by the aerosol droplet size or its size distribution. However, linearity of the CAD response is typically observed over a narrow concentration range. Since the droplet diameter is related to several other factors including density and viscosity of the mobile phase, the response in CAD is also dependent on the mobile phase composition. Higher organic content in the mobile phase leads to greater transport efficiency of the nebulizer, which results in a larger number of particles reaching the detector chamber and a higher signal. Therefore, the response factor will vary significantly with the mobile phase composition in gradient elution, although the responses can be normalized with a postcolumn inverse gradient [36]. For this reason, isocratic elution is preferred in HILIC-CAD, particularly for impurity or degradation product analysis when authentic materials are not available to establish response factors by other means.

3.5 CONCLUSIONS

Developing HILIC methods is not fundamentally different from other chromatographic methods, such as RPLC methods, and should be based on a solid understanding of HILIC basics. Wherever possible, a systematic approach should be applied to identify the column and mobile phase conditions that provide desired method performance. Five commercially available stationary phases with different selectivity are recommended for initial screening in combination with various mobile phase conditions. Following initial screening, other aspects of the method should be carefully evaluated including sample considerations, detector selection, system suitability, and quantitation schemes. In addition to mobile phase pH, a buffer in the mobile phase is particularly important to HILIC methods. The types and concentrations of buffer salts need to be optimized based on the interactions between the analytes and stationary phases. To achieve greater universality for detecting unknown analytes, a CAD or MS detector should be used in tandem with UV detection during HILIC method development since the mobile phase in HILIC is uniquely compatible with CAD or MS detection and many polar compounds lack strong UV chromophores. In addition, more HILIC columns packed with sub-2 μm particles will be developed and used with UHPLC instruments. Developing methods for ultra-high pressure instruments brings benefits but may present additional challenges to HILIC method development.

REFERENCES

1. Wang PG, He W, eds. *Hydrophilic Interaction Liquid Chromatography (HILIC) and Advanced Applications*. Boca Raton, FL: CRF Press; 2011.
2. Guo Y, Gaiki S. Retention and selectivity of stationary phases for hydrophilic interaction chromatography. *J. Chromatogr. A* 2011; **1218**: 5920–5938.
3. Hemstrom P, Irgum K. Hydrophilic interaction chromatography. *J. Sep. Sci.* 2006; **29**: 1784–1821.
4. Dong MW. *Modern HPLC for Practicing Scientists*. Hoboken, NJ: John Wiley and Sons; 2006.
5. Snyder LR, Kirkland JJ, Glajch JL. *Practical HPLC Method Development*, 2nd ed. New York: John Wiley and Sons; 1997.
6. Olsen BA, Castle BC, Myers DP. Advances in HPLC technology for the determination of drug impurities. *Trends Analyt. Chem.* 2006; **25**: 796–805.
7. Hewitt EF, Lukulay P, Galushko S. Implementation of a rapid and automated high performance liquid chromatography method development strategy for pharmaceutical drug candidates. *J. Chromatogr. A* 2006; **1107**: 79–87.
8. Li Y, Terfloth GJ, Kord AS. A systematic approach to RP-HPLC method development in a pharmaceutical QbD environment. *Am. Pharm. Rev.* 2009; **12**: 87–95.
9. Vogt FG, Kord AS. Development of quality-by-design analytical methods. *J. Pharm. Sci.* 2011; **100**: 797–812.
10. Dejaegher B, Vander Heyden Y. HILIC methods in pharmaceutical analysis. *J. Sep. Sci.* 2010; **33**: 698–715.
11. Dejaegher B, Mangelings D, Vander Heyden Y. Method development for HILIC assays. *J. Sep. Sci.* 2008; **31**: 1438–1448.
12. Chirita R, West C, Finaru A, Elkafir C. Approach to hydrophilic interaction chromatography column selection: Application to neurotransmitters analysis. *J. Chromatogr. A* 2010; **1217**: 3091–3104.

13. Hsieh YS. Potential of HILIC-MS in quantitative bioanalysis of drugs and drug metabolites. *J. Sep. Sci.* 2008; **31**: 1481–1491.

14. Guo Y, Huang A. A HILIC method for the analysis of tromethamine as the counter ion in an investigational pharmaceutical salt. *J. Pharm. Biomed. Anal.* 2003; **31**: 1191–1201.

15. Risley DS, Pack BW. Simultaneous determination of positive and negative counterions using a hydrophilic interaction chromatography method. *LC-GC N. Am.* 2006; **24**: 776–785.

16. International Conference on Harmonization Q2B Guideline, Validation of Analytical Procedures: Methodology, 1996.

17. Food and Drug Administration, CDER Reviewer Guidance, Validation of Chromatographic Methods, 1994.

18. International Conference on Harmonization, ICH Annex to Q8, Step 2 version, November 2007.

19. Guo Y, Srinivasan S, Gaiki S. Investigating the effect of chromatographic conditions on retention of organic acids in hydrophilic interaction chromatography using a design of experiment. *Chromatographia* 2007; **66**: 223–229.

20. McCalley DV. Is hydrophilic interaction chromatography with silica column a viable alternative to reversed-phase liquid chromatography for the analysis of ionisable compounds? *J. Chromatogr. A* 2007; **1171**: 46–55.

21. Olsen BA. Hydrophilic interaction chromatography using amino and silica columns for the determination of polar pharmaceuticals and impurities. *J. Chromatogr. A* 2001; **913**: 113–122.

22. Hemstrom P, Jonsson T, Appelblad P, Jiang W. HILIC after the hype: A separation technology here to stay. *Chromatography Today*. 2011; May/June.

23. Liu M, Chen EX, Ji R, Semin D. Stability-indicating hydrophilic interaction liquid chromatography method for highly polar and basic compounds. *J. Chromatogr. A* 2008; **1188**: 255–263.

24. Vehoveca T, Obrezab A. Review of operating principle and applications of the charged aerosol detector. *J. Chromatogr. A* 2010; **1217**: 1549–1556.

25. Orentiene A, Olsauskaite V, Vickackaite V, Padarauskas A. UPLC a powerful tool for the separation of imidazolium ionic liquid cations. *Chromatographia* 2011; **73**: 17–24.

26. Karatapanis AE, Fiamegos YC, Stalikas CD. HILIC separation and quantitation of water-soluble vitamins using diol column. *J. Sep. Sci.* 2009; **32**: 909–917.

27. Subirats X, Roses M, Bosch E. On the effect of organic solvent composition on the pH of buffered HPLC mobile phases and the pK_a of analytes—A review. *Sep. Purif. Rev.* 2007; **36**: 231–255.

28. Guo Y, Gaiki S. Retention behavior of small polar compounds on polar stationary phases in hydrophilic interaction chromatography. *J. Chromatogr. A* 2005; **1074**: 71–80.

29. Guo F. Retention and selectivity of polar stationary phases for hydrophilic interaction chromatography. In Wang PG, He W, eds., *Hydrophilic Interaction Liquid Chromatography (HILIC) and Advanced Applications*. Boca Raton, FL: CRF Press; 2011, pp. 401–425.

30. Liu M, Ostovic J, Chen EX, Cauchon N. Hydrophilic interaction liquid chromatography with alcohol as a weak eluent. *J. Chromatogr. A* 2009; **1216**: 2362–2370.

31. Wang X, Li W, Rasmussen H. Orthogonal method development using hydrophilic interaction chromatography and reversed phase high performance liquid chromatography for the determination of pharmaceuticals and impurities. *J. Chromatogr. A* 2005; **1083**: 58–62.

32. Marrubini G, Mendoza BEC, Massolini G. Separation of purine and pyrimidine bases and nucleosides by hydrophilic interaction chromatography. *J. Sep. Sci.* 2010; **33**: 803–816.

33. Ruta J, Rudaz S, McCalley DV, Veuthey JL, Guillarme D. A systematic investigation of the effect of sample diluent on peak shape in hydrophilic interaction liquid chromatography. *J. Chromatogr. A* 2010; **1217**: 8230–8240.

34. Jiang W. HILIC Method Development and Troubleshooting, HILIC Day USA, New Jersey; 2011.

35. Nguyen HP, Schug KA. The advantages of ESI-MS detection in conjunction with HILIC mode separations: Fundamentals and applications. *J. Sep. Sci.* 2008; **31**: 1465–1480.

36. Gorecki T, Lynen F, Szucs R, Sandra P. Universal response in liquid chromatography using charged aerosol detection. *Anal. Chem.* 2006; **78**: 3186–3192.

CHAPTER

4

PHARMACEUTICAL APPLICATIONS OF HYDROPHILIC INTERACTION CHROMATOGRAPHY

BERNARD A. OLSEN

Olsen Pharmaceutical Consulting, LLC, West Lafayette, IN

DONALD S. RISLEY and V. SCOTT SHARP

Pharmaceutical Sciences Research and Development, Lilly Research Laboratories, A Division of Eli Lilly and Company, Indianapolis, IN

BRIAN W. PACK and MICHELLE L. LYTLE

Analytical Sciences Research and Development, Lilly Research Laboratories, A Division of Eli Lilly and Company, Indianapolis, IN

Hydrophilic Interaction Chromatography: A Guide for Practitioners, First Edition.
Edited by Bernard A. Olsen and Brian W. Pack.
© 2013 John Wiley & Sons, Inc. Published 2013 by John Wiley & Sons, Inc.

4.1 INTRODUCTION

Reversed-phase high performance liquid chromatography (RP-HPLC) has been arguably the primary technique used for analysis of pharmaceuticals for over 25 years. It has applicability for a wide range of chemical structures and has high selectivity for separating compounds similar in structure. Improvements in RP column reproducibility, method development tools, and protocols coupled with robust instrumentation have made RP-HPLC the technique of choice for most investigators concerned with methods for pharmaceutical development and quality control for drug substances as well as final drug products.

There are situations, however, where RP-HPLC is not well suited for pharmaceutical analysis. These situations most often arise when a drug, or impurities of interest in the drug, is relatively polar and is therefore not retained well nor separated from other sample components using typical RP-HPLC conditions. Other situations include quantitative determination of a polar counterion in a pharmaceutical salt or a polar excipient in a drug product.

Hydrophilic interaction chromatography (HILIC) has emerged as a useful tool for the determination of polar compounds of pharmaceutical interest. As described in the following sections, several other approaches have been taken to overcome the problem of polar compound retention even before the rise in popularity of HILIC. These approaches should also continue to be considered for pharmaceutical analysis, but will not be discussed at length.

The development of the ion-pairing mode of RP-HPLC was an early response to the problem of poor retention/separation of charged polar analytes [1]. In this mode, the pH of the mobile phase is buffered to maintain a positive or negative charge on the analyte and a salt containing a counterion capable of forming an ion pair with the analyte is added. This process neutralizes the charge, and greater hydrophobicity of the ion pair leads to greater retention of the analyte. The ion-pairing reagent type and concentration are parameters that may be optimized to yield the desired results. Although ion-pairing separations have been widely used, this separation mode is not without problems. Impurities present in the ion-pairing reagent can produce artifacts in the analysis, particularly if gradient elution is employed [2]. The need to restrict a method to a certain supplier of reagent is usually not desired from a robustness standpoint. Many ion-pairing reagents are not volatile and thereby preclude the use of mass spectrometric (MS), evaporative light scattering (ELSD), or charged aerosol detection (CAD). Some fluorinated ion-pairing reagents that are volatile have been used to overcome this limitation [3–5].

Reducing the elution strength of the mobile phase in RP-HPLC can increase the retention of polar compounds. Totally aqueous (no organic modifier) mobile phases have been used for this purpose but have led to problems with robustness and reproducibility. The initial cause of retention loss or slow equilibration observed with totally aqueous systems was thought to be caused by "phase collapse" of the alkyl chains onto the silica particles. More recent work has shown that retention loss is more likely due to exclusion of the polar aqueous mobile phase from stationary phase pores containing the alkyl chains [6,7]. This mechanism leads to poor partitioning of analytes into the stationary phase and irreproducible chromatography. This phenomenon can also be observed with gradient elution where the initial conditions are totally aqueous and require extended equilibration times.

In order to alleviate problems with aqueous mobile phases, some column suppliers have offered columns purportedly designed to function well with highly aqueous mobile phases. Many of these columns are based on the introduction of a polar group within the alkyl chain that is bonded to the silica support. The polar group helps prevent the retention time loss phenomenon observed with regular alkyl phases and can provide greater retention for ionizable and polar compounds. Information from column manufacturers can be consulted to obtain the characteristics of these columns and suggestions for their use.

Although the use of polar stationary phases with water-containing mobile phases has been known for many years, very few applications of this separation mode had been described for pharmaceutical analysis. Even after Alpert first coined the term hydrophilic interaction chromatography (HILIC) [8] to describe this mode, it took several years to become more widely used. The wide applicability of RP-HPLC and familiarity of chemists with RP method development techniques undoubtedly contributed to the slow uptake of HILIC. Resistance to adopting a new technique also perhaps led to "force-fitting" RP-HPLC to applications for which it is poorly suited. Over the last 10 years, however, research to better understand HILIC retention mechanisms and information on HILIC applications has grown dramatically. This growth has been accompanied and facilitated by a rapid increase in the availability of HILIC stationary phases from column suppliers.

In this chapter, we will describe several applications for HILIC analysis related to pharmaceuticals. Most examples will be related to synthetic organic molecule drugs, sometimes referred to as "small molecules," but some concepts could apply to biomolecules as well. The focus of the chapter will be on "chemistry, manufacturing, and control" issues in pharmaceutical development. Understanding that many of these issues and methodologies may be unique to the pharmaceutical industry, each section in this chapter first outlines the importance of the analytical measurement that is performed. As discussed later and in Chapter 3, definition of the problem is the key first step in deciding upon a separation technique. Without prior knowledge of the pharmaceutical challenge, the reader may be left wondering the importance of the assay. For example, cleaning validation and dissolution assays are specific to the pharmaceutical industry. As a result, we have outlined the importance of the assay, the execution of the assay, and subsequently how HILIC may serve as the appropriate analytical tool. The uses of HILIC in drug discovery, biochemical, and bioanalytical (drugs in biological samples) applications are described in Chapter 5, Chapter 6, and Chapter 7, respectively. Basic HILIC method development considerations are given in Chapter 3. This chapter will build on those topics as specific applications of interest to pharmaceutical development chemists are considered. These applications include not only drug substance and drug product analysis, but also analysis of raw materials, intermediates, and reagents that are involved in the synthesis of the drug substance. These precursors may be smaller, more polar molecules than the drug substance and, as a result, are often amenable to HILIC analysis. Excipients used in drug formulations may be amenable to HILIC analysis. In addition, chiral separations using HILIC conditions will be discussed. Rather than providing a comprehensive review of pharmaceutical applications we will focus on considerations involved in method development using selected applications as examples.

4.1.1   Definition of the Problem

As with any analytical problem, the goals of the analysis should be considered in order to choose an appropriate analytical technique. If HILIC is chosen as a promising technique, an understanding of the separation mechanisms and experimental parameters involved is important for optimizing a separation. These concepts are discussed in detail in Chapter 1 on mechanisms, Chapter 2 on stationary phases, and Chapter 3 on general method development. Various pharmaceutical analysis problems and how HILIC as a separation mode can be applied to them are delineated and discussed below.

The goals of analysis at different stages of pharmaceutical development can vary significantly. Some of the types of analytical problems and their associated analytes are listed in Table 4.1.

As suggested in Table 4.1, method requirements can often be classified into two main categories: determination of the drug substance and determination of impurities. In many cases, it is desirable to be selective for both. The method requirements can greatly impact the method conditions and type of development necessary to achieve the goal. For example, when only the drug substance concentration in a sample is needed without concern for low-level impurities, as in solubility studies or dissolution testing, the main focus will be on separation of the drug substance from the sample matrix and detection at the necessary levels. In other cases, separation and detection of specific impurities will be the goal without concern about determination of the drug substance. The purpose of the method, especially with regard to whether it will be used for a short-term investigation or long-term quality control, will impact the extent of development, optimization, and validation performed. Factors such as column lifetime, robustness, prior experience with similar conditions, and individual laboratory preferences will be important in the choice of a final method, especially if it is to be used over an extended period of time or transferred to other laboratories.

Table 4.1. Common Pharmaceutical Development Studies and Analytes

Analytical Need	Analytes
Solubility studies	Drug substance
Permeability studies	Drug substance
Toxicology formulations	Drug substance
Salt selection	Drug substance, counterion
Counterion determination	Counter ion
Stress testing/forced degradation studies	Drug substance, degradation products
Synthetic process development	Starting materials, intermediates, drug substance, synthesis by-products, degradation products
Excipient compatibility studies	Drug substance, degradation and excipient interaction products
Formulation development	Drug substance, degradation and excipient interaction products
Excipient analysis	Main component and/or impurities
Regulatory stability studies	Drug substance, degradation, and excipient interaction products
Dissolution tests	Drug substance
Equipment cleaning tests	Drug substance, cleaning agents

4.1.2 Selection of Conditions

Some basic considerations for analysis of pharmaceuticals using HILIC are:

Is the sample appropriate for HILIC analysis—what other options are available?

What components need to be separated in the sample?

What level of detection is required?

What type of detector will be appropriate for the compound(s) of interest?

What is the solubility of the sample in various solvents?

Are there sample throughput and analysis time goals?

How robust does the method need to be?

As described previously, there are options other than HILIC for the analysis of polar compounds. These options should be considered together with the questions outlined above to determine the choice of technique. The sample complexity in terms of other components present will determine the level of selectivity needed for analysis. The sample matrix will also determine the need for sample pretreatment before analysis. The nature of the sample components may also influence the choice of HILIC stationary phase. For example, use of amino or diol columns would not be recommended for samples containing compounds such as acylating agents that could undergo reactions with the bonded phase functional groups. Chiral separations will obviously require a chiral stationary phase or chiral mobile phase additive to achieve separation of enantiomers. Detectability requirements and nature of the analyte will determine the type of detector that should be used. The detector choice can influence the choice of mobile phase, especially if volatile buffers are needed when employing detection by MS, CAD, or ELSD. The level of detection needed can impact the amount of sample that must be applied to the column, which, in turn, may be impacted by sample solubility. The nature of the sample solvent, especially regarding the amount of water present, can influence the analysis through effect on peak shape.

The choice of stationary and mobile phases for HILIC analysis of pharmaceuticals is a basic decision when it has been determined that HILIC is an appropriate mode for the analysis. Useful information concerning method development and case studies can be obtained from HILIC column manufacturers, although in most cases the information is focused toward that manufacturer's brand of column as described in Chapter 2. Columns can be divided between bare silica and bonded-phase silica columns. Polymeric or monolithic columns may be considered based on the analysis and the developer's experience, but most published pharmaceutical applications have employed silica-based columns. The bonded-phase columns can be further grouped into those with polar neutral moieties bonded to silica and those with ionic character. It is important to keep in mind that all silica-based columns will have some degree of ionic character due to silanol groups that can ionize depending on their acidity and the mobile phase pH. The degree of ionic character of different silicas will vary with the silica supplier and manufacturing process used. As shown in examples below, ionization of analytes and stationary phase can be manipulated in many cases to achieve the desired selectivity.

Aqueous acetonitrile mobile phases are most often used for HILIC with the aqueous component ranging from about 3% to 40%. In some cases, methanol or other alcohols have been used as the weak solvent instead of acetonitrile [9]. Unbuffered mobile phases may suffice for some applications but several column vendors recommend the use of at least 20 mM ammonium formate at about pH 3.2 or ammonium acetate at about pH 5.2. Buffering of the mobile phase provides pH control and rapid equilibration of both the mobile and stationary phases. Volatile buffers and high organic composition are also advantageous for MS detection (i.e., sensitivity is typically improved). Mobile phase gradients ranging from 5% to 40% aqueous content can help determine the retention properties of the sample analytes. Retention can then be adjusted further using different gradient conditions or perhaps isocratic elution.

Parameters such as solvent strength, buffer concentration, and pH can all affect retention and selectivity in HILIC. The impact of some parameters may be difficult to predict given the multiple retention mechanisms that can be operative under HILIC conditions. For example, a change in pH can simultaneously affect silanol ionization as well as the ionization state of weak acids and bases in the sample. Empirical optimization of such parameters can be facilitated by using statistical design of experiments (DOE). Several parameters can be studied efficiently and their interactive effects on responses determined. A few such studies have been described for HILIC optimization, but these techniques are well known for other modes (see Chapter 3 and the example below for further discussion of DOE optimization) [10–12].

The robustness requirements of the method should also be considered during method development. Methods intended for long-term use in quality control laboratories and/or in multiple laboratories should be thoroughly investigated to determine the impact of small changes in operating parameters on the quality of the separation. The DOE techniques referred to above can aid in efficient evaluation of robustness for several parameters.

4.1.3   Validation of the Method

After suitable method conditions have been established, method validation appropriate for the intended use of the method is necessary. For example, methods used for quality control purposes require more rigorous validation than those used to obtain information strictly for development. The phase of drug development for a particular drug is also usually a factor in the extent of validation that is performed.

4.1.4   General References

Reviews [13,14] and several articles have described specific examples of HILIC applications for pharmaceutical analysis. Many of these have focused on determination of drugs in biological fluids, which is the subject of Chapter 7 in this book. Some papers have described general method development [15,16] or impurity determination [17]. Others have used pharmaceutical compounds to investigate HILIC retention mechanisms (see Chapter 1) but have not looked at specific drug development problems.

4.2   DETERMINATION OF COUNTERIONS

4.2.1   Salt Selection and Options for Counterion Determination

Quantitative analytical tests for counterions of drug salts are frequently performed in support of drug substance and drug product development in the pharmaceutical industry. Counterions are often used to form salts of drug substances that have acidic or basic functional groups. Sodium and chloride are examples of a positive and negative inorganic counterion, respectively, while organic acids and bases can also be used for basic and acidic drugs, respectively. It should be noted that one of the key reasons for making pharmaceutical salts rather than working with free-base or free-acid forms of pharmaceutical compounds is that the salt forms often have physicochemical properties that provide greater bioavailability, stability, and manufacturability of drug products. Therefore, salt selection for acids or bases is a critical part of the drug development process and selecting a developable salt form early in the development process can avoid timeline delays due to chemistry, manufacturing, and control (CMC) issues (such as repeating a stability study) or clinical trials (bridging salt form studies).

The most common pharmaceutical salt forms are sodium salts of acids and hydrochloride salts of amines [18]. However, even the most common salt forms do not always possess the best physicochemical properties needed for a successful drug candidate. The initiation of the salt selection process generally takes place for all ionizable compounds that have successfully passed some form of toxicology screening. Automated salt screening systems are increasingly being used to perform comprehensive exploration of numerous counterions with a multitude of different crystallization solvent systems to find viable salt forms. Crystalline "hits" are discovered and then manually scaled up to produce enough material for characterization. The most promising salt forms are evaluated to determine if they possess physicochemical properties suitable for future development often with respect to hygroscopicity, solubility, thermal properties, stability, bioavailability, impurity rejection, and formulatability. Analytical methodologies are needed to provide qualitative and quantitative measurements of counterions.

The key reasons counterion analyses are performed in the pharmaceutical industry include:

1. An identification assay for a counterion is used to confirm that the correct drug substance salt form was synthesized.
2. A quantitative assay for a counterion is used to confirm that the correct salt stoichiometry was achieved.
3. Determination of whether the integrity of the salt form of the molecule is maintained as the drug substance or drug product is subjected to normal or accelerated storage conditions.
4. Determination of the free acid or base potency is needed for a pharmaceutical salt or characterization of a reference standard.
5. Determination of an unwanted impurity or residue is needed during drug processing or synthesis.

Many techniques exist for the analysis of counterions including wet chemistry titration, electrochemical, spectroscopic, and separation-based methods (see Table 4.2). Titration and spectroscopic techniques are relatively inexpensive, easy-to-use techniques that can provide good precision and accuracy comparable to chromatography. However, chromatography offers the advantages of selectivity, automated sample injections, and online sample analysis that allows for a higher throughput of samples and faster generation of results. Capillary electrophoresis (CE) has been used frequently in ion analyses. Ultraviolet (UV) detection (direct or indirect) is the most common type of detection used for CE analysis. CE with indirect UV detection has been used

Table 4.2. Techniques for Counterion Analysis

Technique	Mode of Counterion Analysis
Wet Chemistry	A. Titration (1) acid–base (2) precipitation (i.e., chloride-silver nitrate, calcium–EDTA) B. Gravimetric (volatilization)—sample weight loss
Electrochemical	A. Ion-selective electrodes (ISE) B. pH electrode (H$^+$) C. Inorganic (F$^-$, Cl$^-$, Na$^+$, etc.)
Spectroscopic	A. Atomic absorption spectroscopy (AAS) B. Inductively coupled plasma (ICP) with atomic emission or mass spectrometry C. NMR—organic molecules

Chromatography	A. Reversed-phase (RP)
	B. Ion pairing (IPC)
	    Alky sulfonates (cation)
	    Tetra-alkyl quaternary amines (anion)
	C. Ion chromatography (IC)
	    Conductivity detection
	D. Capillary electrophoresis (CE)
	    Direct
	    Indirect
	E. HILIC
	    ELSD, NQAD, CAD (aerosol detectors)
	    UV

in the analysis of inorganic ions. This type of detection is based on the use of an absorbing co-ion as the main component in the electrophoretic buffer. Because the highest sensitivity for indirect UV detection is achieved when the analyte has an effective mobility similar to that of the absorbing co-ion, the mobilities of both the analyte and the absorbing co-ion must be known in order to select the most appropriate absorbing co-ion to use for the separation. Also, the linear dynamic range of indirect UV detection is not as wide as that for direct UV detection and sample injection reproducibility for CE in general is often not as good as other techniques. Ion exchange chromatography (IEC) is the most popular technique used in the analysis of inorganic ions such as sodium or nitrate. Conductivity is the primary type of detection used for ion chromatography (IC). Conductivity detection can be performed in either suppressed mode or nonsuppressed mode. Suppressed conductivity detection involves a suppressor unit that modifies the eluent and the analyte in order to increase detection of the solute. Traditional IC methods are plagued with issues arising from the lack of solubility in the sample solvent or mobile phase that is typically 100% aqueous. In using IC columns, separation of the drug substance is nearly guaranteed from the counterion since positive or negative ions are selectively retained.

HILIC has become a viable option for the analysis of positive and negative counterions of pharmaceutical salts in part because these analytes typically have no retention or very short retention times in the RP mode, which is often the technique of choice for the drug substance. The HILIC mode offers a cost-effective advantage over other techniques because it uses existing HPLC instrumentation and detectors that are common in most pharmaceutical laboratories. Aerosol detectors, such as ELSD, CAD, and nano-quantity analyte (NQAD) detectors have been used successfully for the detection of counterions. Aerosol detectors are not affected by the UV spectral properties of the mobile phase components but only by the volatility of these components; therefore, analytes that are less volatile than the mobile phase can be detected and gradient elution can also be used. Peterson and Risley first used ELSD for the determination of sodium [19] and then later of chloride [20] for pharmaceutical drug substance salt forms, although strong anion exchange (SAX) and strong cation exchange (SCX) columns were used with high organic content in the mobile phase.

In general, aerosol detectors rely on three main processes: nebulization, evaporation, and detection. The nebulization process transforms the mobile phase into an aerosol of fine droplets that can be more easily evaporated. As a gas carries the droplets through a heated drift tub, the evaporation step removes the volatile mobile phase components, leaving particles of the nonvolatile analytes. These nonvolatile analytes then proceed to a detection cell. In the case of ELSD and NQAD, the detector is based on light scattering, whereas CAD is based on charge detection. Chloride as HCl is volatile and ordinarily would not be detected by aerosol detectors; however, by adding a volatile buffer such as ammonium acetate or ammonium formate, chloride can be detected because it will form ammonium chloride upon evaporation of the mobile phase. Likewise, sodium detection will also be enhanced because it will form sodium acetate during the detection process with aerosol detectors.

4.2.2 Specific Counterion Analysis

References for the determination of specific counterions are given in Table 4.3 and Table 4.4 with the following details given for some examples. Pack and Risley used an isocratic HILIC-ELSD method for the separation and detection of lithium, sodium, and potassium from drug substances and chloride [21]. A Chromolith silica column with a mobile phase composed of 80:20 acetonitrile:50 mM ammonium acetate buffer (pH 3.75) was used to retain and separate these positive counterions. Due to the porous nature of this monolith stationary phase and subsequently low backpressure, a flow rate of 5.0 mL/min was used, resulting in a run time under 3 min (Fig. 4.1). McClintic et al. incorporated a cyanopropyl (CN) stationary phase in the HILIC mode for the analysis of piperazine, which was used as a counterion during salt selection for drug substances. An ELSD was used since piperazine lacks a UV chromophore, with a mobile phase of 95:5 acetonitrile:0.15% nitric acid in water [22].

Table 4.3. HILIC Methodology References for Common Organic Acids and Inorganic Counterions of Basic Pharmaceutical Compounds

Acids	Structure	Reference
Fumaric acid		[29]
Glucuronic acid		[29]
Glutaric acid		[29]
Glycolic acid		[29]
Hippuric acid		[24]
Maleic acid		[29,32]
Malic acid		[29]
Mandelic acid		[29]
Naphthalene-1,5-disulfonic acid		[29]
Succinic acid		[29]

(Continued)

Table 4.3. (Continued)

Acids	Structure	Reference
Tartaric acid		[29]
Toluenesulfonic acid		[29]
Citric acid		[29]
Hydrobromic acid	Br^-	[27,29,30]
Hydroiodic acid	I^-	[30]
Hydrochloric acid	Cl^-	[26,27,29,30]
Nitric acid	NO_3^-	[26,27,29,30]
Methane sulfonic acid		[29]
Ethanesulfonic acid		[29]
Isethionic acid		[29]
Ethane-1,2-sulfonic acid		[29]
Phosphoric acid		[27,29,30]
Sulfuric acid		[27,29,30]

Table 4.4. HILIC Methodology References for Common Organic Bases and Inorganic Counterions of Acidic Pharmaceutical Compounds

Bases	Structure	Reference
Arginine		[29]
Piperazine		[22,28,29]
Benzylamine		[29]
Choline chloride		[25,29]
Diethanolamine		[29]
Lysine		[29]
Potassium hydroxide	K^+	[21,26,27,29,30]
Sodium hydroxide	Na^+	[21,26,27,29,30]
Lithium hydroxide	Li^+	[21,30]
Tromethamine (TRIS)		[23,29]
Zinc hydroxide	Zn^{2+}	[29]
Calcium hydroxide	Ca^{2+}	[27,29]
Magnesium hydroxide	Mg^{2+}	[27,29,62]

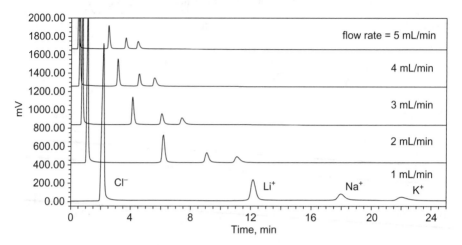

Figure 4.1. HPLC-ELSD chromatograms showing the effect of flow rate from 1 to 5 mL/min on the retention time of lithium, sodium, and potassium. A Chromolith Performance SI silica column (100 × 4.6 mm monolithic) with a mobile phase of 90% acetonitrile/10% 50 mM ammonium acetate (pH 4.45 with acetic acid) was used.

Guo and Huang developed a HILIC method utilizing refractive index (RI) detection for the tromethamine salt form of an investigational drug substance [23]. Tromethamine is very polar, lacks a UV chromophore, and is not retained by the conventional RP mode; however, an aminopropyl column operated in the HILIC mode with 80:20 acetonitrile:water adequately retained and separated tromethamine from the drug substance. The high organic content of the mobile phase offered another advantage in this case because this salt form has extremely poor aqueous solubility and therefore a more suitable sample solvent could be used. However, one of the disadvantages of RI detection is the incompatibility with gradient elution. Guo also compared the separation of six organic acids, including hippuric acid, which can be used as a counterion, on five different HILIC stationary phases with UV detection at 228 nm [24]. Statistical experimental design was used to evaluate the effect on retention of various parameters such as acetonitrile content, buffer concentration, and column temperature. A mobile phase consisting of 85:15 acetonitrile:10 mM ammonium acetate column was chosen as optimal. Furthermore, Guo studied five quaternary amines, including choline, which can also be used as a counterion, using similar chromatographic parameter variations to above with TSKgel Amide-80 and YMC-Pack amino columns [25]. When using MS, 0.4 fmol of choline could be detected with this HILIC methodology. Liu and Pohl investigated the HILIC effect on a Trinity Acclaim P1 column, which is a mixed-mode phase containing reversed-phase, weak anion exchange, and strong cation exchange characteristics [26]. Sodium and chloride retention was tracked across various organic contents, buffer concentrations, and pH values of the mobile phase. Additionally, the separation of penicillin G from its potassium counterion was demonstrated with ELSD detection using a mobile phase of 90:10 acetonitrile:20 mM ammonium acetate pH 5.2. The potassium peak eluted before the penicillin G peak with 60% or 70% organic modifier but eluted after the penicillin G peak when the organic modifier was 80% or greater. Thus, the retention of potassium was affected more by the HILIC mobile phases than was that of penicillin G. Huang et al. developed two different isocratic HILIC methods, one for monovalent counterions and another for multivalent counterions, using a ZIC-HILIC stationary phase with CAD [27]. The method for monovalent ions used a 15-cm column with 75:25 acetonitrile:100 mM ammonium acetate, pH 7.0, as the eluent, and the multivalent ion method used a 5-cm column with 70:30 acetonitrile:100 mM ammonium formate, pH 3.5, as the eluent. Six drug substance salt forms were analyzed for the counterion by this HILIC-CAD technique (relative standard deviation [RSD]% < 2) and the results were also compared with IC. One drug substance was not amenable to analysis by IC due to the poor aqueous solubility of the sample in the IC sample solvent, although it could be analyzed by HILIC-CAD because of the high organic content in the HILIC sample solvent. Additionally, the HILIC-CAD method was shown to have two other advantages over the IC method, including simultaneously being able to quantify anions and cations and requiring minimal training of the operator because existing HPLC equipment is used (rather than IC systems). Cohen et al. extended CAD detection to volatile bases by addition of a suitable acid in the mobile phase which formed a nonvolatile salt with the base [28]. The conditions could be suitable for counterion analysis or determination of volatile bases as impurities.

4.2.3 Counterion Screening with Gradient Elution

Gradient elution has also been used in the HILIC mode for the analysis of counterions by starting the gradient with a high organic concentration and gradually increasing the aqueous component of the mobile phase. This type of methodology is very useful for screening purposes or when analyzing samples with unknown counterions and other charged or polar analytes. Risley et al. simultaneously retained, separated, and detected 33 counterions (12 cations and 21 anions) using gradient elution on a ZIC-HILIC column with ELSD and a starting mobile phase of 85:15 acetonitrile:75 mM ammonium acetate (pH 4.8 with acetic acid) with a 2-min hold and then a linear gradient to 10:90 acetonitrile:75 mM ammonium acetate (pH 4.8) in 22 min [29]. The effects of organic content, buffer strength, and pH were evaluated. Increasing the organic concentration in the mobile phase increased the retention time for both cations and anions (Fig. 4.2). In general, increasing pH (3.0–6.5) resulted in retention time increases for cations but retention time decreases for anions. Increasing the buffer concentration (10–200 mM) resulted in retention time decreases for cations but retention time increases for anions. Both of these observations are consistent with the repulsion effect outlined in the discussion of electrostatic

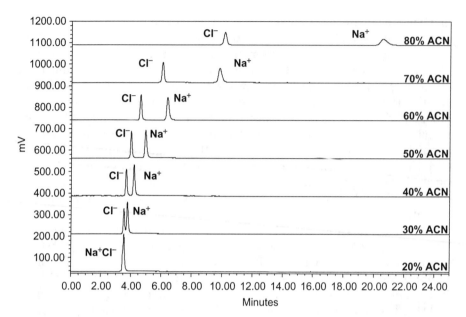

Figure 4.2. Effect of mobile phase aqueous content on sodium and chloride retention. These chromatograms were generated on a ZIC-HILIC column (250 × 4.6 mm) using 75 mM ammonium acetate, pH 4.8 acetic acid as the aqueous mobile phase component with ELSD. The organic content ranged from 20% to 80% acetonitrile.

repulsion hydrophilic liquid interaction chromatography (ERLIC) from Chapter 1. In the case of pH, the effect on cation retention is presumably due to the H^+ interacting with the negatively charged part of the zwitterions (SO_3^-), which ultimately shields the cation from having a strong interaction at a lower pH. With regard to buffer concentration, the NH_4^+ interacts strongly with the SO_3^- fixed negative charges and as the buffer increases, access to these fixed charges is diminished. As a result, cations do not interact with SO_3^- and are not significantly retained. Anion retention is affected in the opposite manner. As the buffer cation concentration increases, anions do not experience the typical repulsion forces of the SO_3^- functionality and can then access the tertiary amine for ion exchange. With better access (less repulsion) to the fixed positive charge of this stationary phase, anions can be retained more strongly. Both anions and cations had improved peak shape with increasing buffer concentration. This gradient HILIC-ELSD method was also used to quantify the counterions from 10 different pharmaceutical drug substances with RSD% ($n = 3$) ranging from 0.26% to 2.03%. It should be noted that one of the drug substance salts, a phosphate salt form, had poor reproducibility when using high organic content of the sample solvent due to the poor solubility of phosphate in organic solvents. However, reducing the organic content in the sample solvent (to 50:50) resulted in typical reproducibility (Figs. 4.3, 4.4, 4.5). Crafts et al.

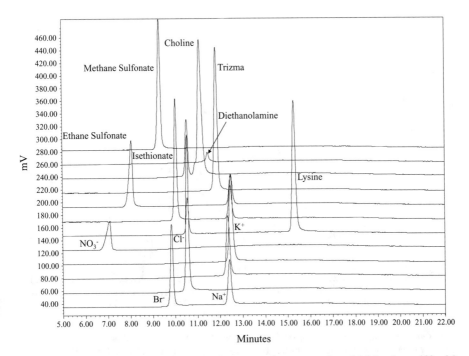

Figure 4.3. Retention of positive and negative monovalent counterions. Mobile phase: 100 mM ammonium acetate buffer with gradient from 85% ACN to 10% ACN in 20 min. A ZIC-HILIC column (250 × 4.6 mm) with ELSD was utilized.

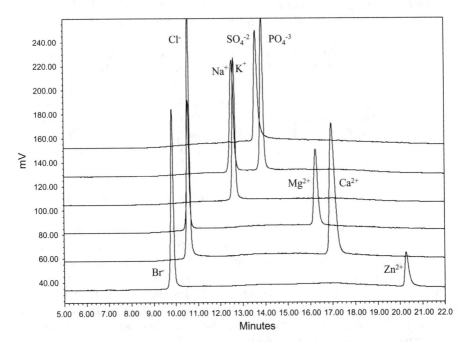

Figure 4.4. Retention of multivalent counterions. Mobile phase: 100 mM ammonium acetate buffer with gradient from 85% ACN to 10% ACN in 20 min. A ZIC-HILIC column (250 × 4.6 mm) with ELSD was utilized.

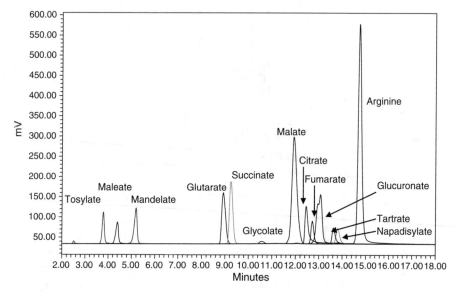

Figure 4.5. Retention of common organic ions used for pharmaceutical salt selection. Mobile phase: 75 mM ammonium acetate buffer (pH 3.8) with gradient from 85%ACN to 10% ACN in 20 min. A ZIC-HILIC column (250 × 4.6 mm) with ELSD was utilized.

also employed HILIC-CAD for counterions using the ZIC-HILIC column with gradient elution [30]. The mobile phases A (15:65:20:5) and B (50:25:20:5) were mixtures of 100 mM ammonium acetate (pH 4.7), acetonitrile, isopropanol, and methanol with a gradient starting at 45% B, at 15 min 65% B, holding to 20 min, at 25 min 40% B, holding to 26 min and then 45% B at 30 min. This HILIC-CAD method demonstrated high sensitivity (ppb level) and excellent reproducibility and was thus a competitive alternative to IC with conductivity detection. The use of HILIC for counterion analysis is becoming more popular because of its use of conventional HPLC systems, ease of use, solubility advantages for poorly aqueous soluble samples, good accuracy and precision, and excellent reproducibility.

4.2.4 Suitable Reference Standards for Counterion Analysis

Commercial reagents that have 98% or greater purity are recommended for reference standard materials used for quantitative analyses of samples containing counterions. The standard may be purchased as a solid chemical or stock solution of known concentration. For solid chemicals, it is always important to find the certificate of analysis (COA) for the specific lot number to identify the specified purity. For example, p-toluenesulfonic acid monohydrate can be listed as 98.5% on the bottle label; however, this value is the minimum requirement, whereas the purity listed on the COA may be 99.9% for the specific lot. Also for solid chemicals, be aware of the hygroscopicity of the solid and the potential need to dry the standard at 105°C prior to use, or apply a correction for water content. Some counterions are available in different forms. For example, tartaric acid can be used for the analysis of tartrate salts. However, if solubility of this form is an issue, then sodium tartrate dibasic dihydrate could also be used as an alternate standard form. Several references have provided examples of accurate quantitative methods for various counterions from pharmaceutical salts [19–23,27,29,30].

4.3 MAIN COMPONENT METHODS

As mentioned in the introduction, when only the drug substance concentration in a sample is needed without concern about low-level impurities, a method may be utilized where the main focus will be on separation of the drug substance from the sample matrix. Such methods are often used for drug substance or drug product potency measurements, for drug product content uniformity determinations, to support equipment cleaning analysis, for dissolution testing, and for solubility studies. In these cases, rapid isocratic methods are often desired to support the screening of a large number of samples. While separation of impurities is necessary for potency methods to ensure accuracy, detection and quantitation of impurities and degradation products are often not needed. HILIC is especially advantageous for these types of methods where highly aqueous RP-HPLC may not provide adequate selectivity for polar compounds and typical drug product excipients, which are often also not well retained.

4.3.1 Potency/Assay Methods

Potency/assay methods are the cornerstone of the pharmaceutical industry. It is of utmost importance to ensure that the patient is receiving the expected dose. For the active pharmaceutical ingredient, a typical assay specification might be 97.0–102.0%, where 102.0% on the high side would allow for assay variability and 97.0% would allow for impurities, assay variability, and degradation on storage. For the drug product (e.g., a tablet, capsule, or parenteral product), a typical specification early in development would be 90–110% of label claim. As more experience is gained with the formulation and the analytical methodology, more stringent acceptance limits may be developed. In order to ensure tablet-to-tablet or capsule-to-capsule consistency, content uniformity per United States Pharmacoepia (USP) general chapter <905> is determined [31]. In this determination, 10–30 unit doses are analyzed to ensure that each unit is within a predefined range of potency, with a range of 85–115% being acceptable per USP. The average potency of the lot is considered as well as the variability in order to assess if a product lot should be released to the market or used in a clinical trial if the compound is in development.

In contrast to the number of HILIC methods used to determine drug substances for bioanalytical purposes (see Chapter 7), there have been relatively few publications on the use of HILIC for a drug substance/product potency assay. In general, if a method can be developed for the intent of measuring impurities in the presence of an active compound, that method can be validated for potency because the requirements of an impurity assay are more stringent than those of a potency assay. The main differences arise with the selectivity and sensitivity requirements. In an impurity assay, all impurities should be separated from the main component and the impurities must be detectable at 0.05% of the main component or less; in a potency method, the impurities only have to be separated from the main component and sensitivity for impurity levels is not required.

One example of a potency method is the determination of dimethindene maleate in a topical gel [32]. Dimethindene maleate is a polar basic drug that can be formulated in a gel matrix. The samples were prepared by sonication with 100% acetonitrile as the solvent (HILIC compatible as the weak solvent). A ZIC-HILIC (50 mm × 2.1 mm; 5 μm) analytical column was used with a flow rate of 0.3 mL/min. The run time was less than 3 min and specificity was demonstrated for the maleate counterion, diltiazem hydrochloride (internal standard), and dimethindene. There was no particular interest in quantifying the maleate counterion; however, it responded at 258 nm and therefore selectivity had to be demonstrated. The method was validated per International Conference on Harmonization (ICH) Guideline Q2 (R2) [33] for accuracy linearity, and precision and was found to be suitable for routine analysis.

In another application, a HILIC method was developed for the simultaneous determination of pseudoephedrine hydrochloride (PSH), diphenhydramine hydrochloride (DPH), and dextromethorphan hydrobromide (DXH) in a cough-cold syrup [34]. For this application, a Supelcosil LC-Si, 25 cm × 4.6 mm, 5 μm was used as stationary phase. A chromatogram is shown in Figure 4.6. This application is interesting because PSH, DPH, and DXH are considerably hydrophobic. The authors demonstrate that the compounds undergo the HILIC effect (increased retention with increased organic content); however, the ion exchange interaction is dominant with these compounds. A thorough investigation of the impact of pH and buffer modifier and organic eluent was performed. In this case, all three compounds are weak bases with pK_a values greater than 8.3. Therefore, the compounds are nearly completely ionized in the pH range investigated (pH 4.0–5.5) and pH was demonstrated to have only a small impact on retention. The charge state of some stationary phase silanols changes in this pH range and, as a result, retention times of all three compounds are shifted slightly longer due to cationic interactions.

Jia et al. evaluated two aerosol-based detectors (CAD and ELSD) for the analysis of gabapentin in pharmaceutical formulations. Gabapentin is a highly polar compound without a chromophore, which precluded the use of UV detection [35]. In this work, they evaluated four HILIC columns: ZIC-HILIC (zwitterionic), ZIC-pHILIC (zwitterionic polymer), Luna HILIC (cross-linked diol), and Atlantis HILIC (silica). Mobile phase composition and pH were evaluated. The authors reported increased baseline noise attributed to column

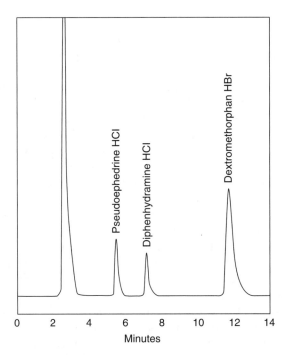

Figure 4.6. Chromatogram of a cough-cold syrup sample solution showing separated peaks of PSH, DPH, and DXH (initial wavelength 254 nm; changed at 9 min to 280 for DXH detection). Note that nominal salt forms of analytes are indicated. Reprinted from Reference 34 with permission from Elsevier.

bleeding with the Atlantis HILIC and the ZIC-HILIC under highly aqueous conditions with CAD detection; therefore, the authors recommended ELSD as it provided a much smoother baseline. Others have utilized the ZIC-HILIC column with ELSD and CAD and have not noted column bleed or increased noise with their method conditions [29,30]. The impact of pH was evaluated for each column and the results were as expected. Gabapentin has two pK_a values at 3.7 and 10.7, and could therefore exist as a cation, anion, or zwitterion in solution. For both ZIC-HILIC columns, the retention time could be altered by a simple change in pH. Thus, pH could be used as a method optimization parameter.

4.3.2 Equipment Cleaning Verification Assays

The Food and Drug Administration (FDA) regulations on cleaning and maintenance state that "Equipment and utensils shall be cleaned, maintained, and sanitized at appropriate intervals to prevent malfunctions or contamination that would alter the safety, identity, strength, quality, or purity of the drug product beyond the official or other established requirements." [36] ICH has also issued recommendations for compliance and safety that include very similar requirements with more elaboration on specific details [37]. This cross-contamination concern applies not only to active pharmaceutical ingredients but includes residual cleaning solvents and detergents.

Equipment cleaning verification is a demonstration for each manufacturing run (typically by submission of cleaning swabs to the analytical laboratories) that the active pharmaceutical ingredient used in the previous manufacturing run has been removed from the equipment to a level below the pre-established acceptance limit. Although the cleaning verification procedure is not submitted to regulatory authorities, it is critical in order to support a clinical trial or commercial manufacturing area. Cleaning verification is usually accomplished by one of two methods. In the first method, the active ingredient is measured in an equipment rinse solution that may be organic or aqueous in nature. In the case where an organic rinse is utilized, HILIC may be advantageous for determination of the analyte, in particular when the analyte is polar. The other method utilized to ensure equipment cleanliness is through the analysis of swab samples that have been applied to the equipment to check for residues. Cleaning verification methods for swabs are typically validated by demonstrating recovery from a "test coupon." This coupon is made of a material that is representative of the manufacturing equipment with regard to material type (e.g., 316L stainless steel) and surface finish. A known amount of an analyte is spiked onto the coupon and the coupon is swabbed in a methodical manner. Typically, a swab is submersed in methanol or water in order to wet it prior to swabbing the surface. A swabbing solvent of methanol has the inherent advantage in that it is volatile. When the equipment is swabbed, the residual methanol will evaporate, leaving behind a clean surface. Water takes longer to dry and may leave spots or streaks, necessitating a subsequent wipe down of the equipment. During swab method validation, methanol is commonly used to dissolve the analyte (both polar and nonpolar) for spiking onto the test coupon. Again, methanol promotes evaporation from the test coupons in order to mimic dried residues on equipment surfaces. A solvent is needed to wet the swab and extract residue from the equipment to which the swab is applied. Depending on the compound, methanol may provide better solubility than water even for polar compounds and therefore better recovery of analyte from the equipment surface may be obtained. After swabbing, the swab is extracted in solvent and analyzed. Better analyte recoveries are also possible when using methanol to extract the analyte from the swab as part of the sample preparation process. If HILIC is chosen as the chromatographic technique, water is undesirable as a sample solvent as it is the strong solvent. This solvent mismatch may produce chromatography disruption and result in peak tailing, splitting, and shifting. Methanol is an excellent choice for sample extraction solvent as it is compatible in solvent strength with a typical HILIC mobile phase. Thus, HILIC could be a good separation choice because it is compatible with the organic solvent used throughout the swabbing process.

A HILIC method was developed for the analysis of cleaning swabs for a pharmaceutical tablet formulation [38]. The active ingredient in the tablet was a polar, weakly basic, low molecular weight molecule that is not retained on typical reversed-phase stationary phases. The HILIC method employed a Chromolith ZIC column (a monolithic zwitterionic stationary phase) with a mobile phase of 95% acetonitrile and 5% 10 mM ammonium acetate. The weakly basic compound was retained using the zwitterionic stationary phase. During swab validation, one of the most important parameters to evaluate is spiked recovery. The average recovery of the polar analyte was 87%, with a relative standard deviation of 2.9%. Another desirable attribute of a cleaning verification method is a short run time. A short run time will allow for rapid sample throughput so that the cleanliness of the equipment can be verified and the equipment put back into service. Monoliths have the inherent advantage of very low backpressure, which allows for high mobile phase flow rates. At a flow rate as high as 5 mL/min, peak efficiency was maintained and the run time was only 1.5 min for this analysis.

4.3.3 Dissolution Methods

Dissolution testing is a common characterization method employed in the pharmaceutical industry to assess product quality and performance and to design formulations. Most regulatory authorities require a dissolution method as part of the control strategy for solid oral dosage forms. USP guidance specifies the type of method that should be employed for various types of dosage forms, including immediate release, extended release, and delayed release formulations [39]. Dissolution testing determines the rate of drug solubilization of a dosage form and is used to monitor how storage conditions and manufacturing processes might impact drug release.

HILIC would usually not be the first choice for a dissolution method as aqueous media are typically used and disruption of chromatography in a polar organic mode is likely when injecting highly aqueous solutions. However, it can be advantageous when a polar molecule cannot be retained by any other means and direct UV spectroscopy is not an option for sample analysis. The sample injection volume can be reduced to minimize the impact of sample solvent mismatch, but sensitivity is also reduced.

In the dissolution method example provided below, it was desired to monitor a drug substance and two potential degradation products that can be formed through hydrolysis after dissolution of the dosage form [38]. The possible degradation products are extremely polar and not well-retained on standard RP-HPLC stationary phases. The dissolution method was used to monitor the drug substance and understand the level of degradation products that could form under varying pH conditions. A HILIC method was developed to separate rapidly the three components as shown in Figure 4.7A. To address the sample solvent–mobile phase mismatch, dissolution samples

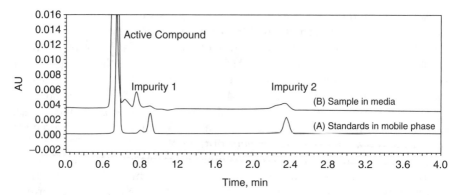

Figure 4.7. Dissolution analysis of active pharmaceutical ingredient and two potential degradation products. Chromolith ZIC column; mobile phase acetonitrile–10 mM ammonium acetate, pH 4.0 (95:5 acetonitrile:ammonium acetate), flow rate = 3.0 mL/min, injection volume = 5 μL; UV detection at 265 nm. (A) Aqueous samples diluted in mobile phase. (B) Sample from 40% ethanol/0.1 N HCl dissolution media injected without dilution.

and standards were prepared by performing a twofold dilution into mobile phase. The less polar drug molecule and more polar impurities (denoted Impurity 1 and 2 in Fig. 4.7) are resolved and sufficiently retained for quantitation of dissolution samples.

While HILIC might be problematic for most dissolution applications due to differences in solvent strength between dissolution media and HILIC mobile phases, an application where HILIC might be advantageous is in the evaluation of dissolution in ethanolic media. For delayed release and extended release dosage forms, recent warnings by regulatory bodies have led to an interest in understanding the potential impact of alcohol consumption on the release rate of non-immediate release dosage forms [40,41]. Analysis of release in media containing up to 40% ethanol can be problematic for methods designed for aqueous sample solutions. For the example provided above, sample solutions consisting of 40% ethanol in 0.1 N HCl were evaluated and demonstrated to provide acceptable chromatography, as shown in Figure 4.7B. While band-broadening is observed compared with Figure 4.7A, the method conditions are suitable for evaluation of dissolution samples using high ethanol levels and would not require further sample dilution to address the sample solvent incompatibility. Alternatively, a method was available where standard RP-HPLC was employed using a highly aqueous mobile phase to retain the polar impurities. When samples containing 40% ethanol were injected on this system, retention and resolution of the polar impurities were lost. Further dilution was attempted into a more aqueous sample solvent, but quantitation of the impurities was not possible due to loss of sensitivity. The HILIC method provided adequate retention and sensitivity for all three components.

4.4   DETERMINATION OF IMPURITIES

4.4.1   Impurity Screening and Orthogonal Separations

Impurity screening is usually conducted during pharmaceutical development to search for impurities that are not already known or readily predictable. Gradient elution RP-HPLC is usually used as a general screening technique to investigate impurities in pharmaceuticals [42]. In some cases, however, the parent drug compound and/or potential impurities may be very polar and HILIC may be appropriate. HILIC offers an attractive option for impurity screening due to the dramatic differences in selectivity compared with RP-HPLC. HILIC can also be advantageous for drug product analysis when excipients or color/flavor ingredients interfere with impurities under RP conditions. To accomplish more thorough orthogonal screening for impurities, both RP and HILIC modes can be used. HILIC can be an especially useful complement to RP-HPLC if there is concern about potential polar impurities that might be eluted in the solvent front with RP-HPLC even when gradient elution starting with a highly aqueous mobile phase is used. As described below, orthogonal screening can be done with separate RP and HILIC methods, serial RP and HILIC columns, column switching, or full two-dimensional systems (see Chapter 9).

An example of orthogonal selectivity for polar impurities using HILIC is shown in Figure 4.8 for moxonidine drug substance (peak 5) [43]. The RP-HPLC method required ion-pairing conditions to provide retention for moxonidine and related compounds. The more polar hydroxylated impurities were still not well retained and the dihydroxy impurity (peak 1) eluted in the same region as system peaks due to the injection. Under HILIC conditions, however, the three polar hydroxylated impurities were eluted in reversed order as compared with the RP-HPLC separation and well after the drug substance. The polar impurities could be determined easily without interference from system peaks using HILIC conditions. In this example, separate RP-HPLC and HILIC methods could also provide advantages for unknown-impurity screening.

In another example, Wang et al. developed a HILIC method for a polar pharmaceutical to complement an RP-HPLC method and further demonstrate adequate selectivity [44]. The HILIC method was validated and could be used for drug substance assay and purity determinations.

Gavin et al. utilized separate HILIC and RP-HPLC methods to screen for impurities in a polar drug substance starting material [45]. Samples from different manufacturers prepared by different synthetic routes were screened with multiple methods to reveal differences in impurity profiles. In this example, the unoptimized HILIC method provided alternate selectivity but suffered from poor peak shape compared with an RP ion-pairing method. Both HILIC and RP methods were used for impurity screening, but the RP method was chosen for further development.

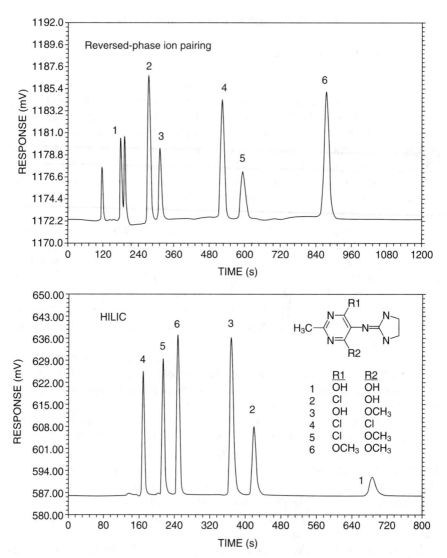

Figure 4.8. Comparison of selectivity provided by reversed-phase ion paring and HILIC separation modes. Reversed-phase conditions: LiChrospher RP-Select B, 250 mm × 4.6 mm ID, 5 μm column; mobile phase acetonitrile–20 mM sodium pentanesulfonate (14:86), pH 3.5 with sulfuric acid; 230 nm detection; 1.2 mL/min; column temperature 40°C. HILIC conditions: Zorbax NH2 column; mobile phase acetonitrile–5 mM potassium phosphate, pH 6.5 (65:35); 230 nm detection. Compound numbering is the same for both chromatograms. Reprinted from Reference 43 with permission from Elsevier.

For complex samples containing many components, two-dimensional chromatography is a powerful screening technique. HILIC can offer orthogonal selectivity advantages as the separation mode in one of the dimensions. This topic is covered in depth in Chapter 9. In a variation of two-dimensional separation, Wang et al. applied column switching to achieve an orthogonal analysis [46]. They coupled a HILIC column followed by an RP column with a switching valve and RP transfer column between the two analytical columns. Nonpolar components that were not retained on the HILIC column were mixed with an aqueous solution to allow focusing on the RP transfer column. The nonpolar components were then backwashed onto the RP analytical column using a separate pumping system while the original analysis of polar components continued on the HILIC column. Separate chromatograms resulted from the HILIC and RP separations since separate detectors were used in their experimental setup. Orthogonal selectivity was demonstrated for test solutes and natural product extracts.

Louw et al. coupled a 2-mm ID RP column directly to a 4.6-mm inner diameter (ID). HILIC column with a T-junction where acetonitrile could be added to the eluent of the RP column to achieve HILIC conditions for the second column (Fig. 4.9) [47]. The differences in optimal flow rate for the different diameter columns provided the basis for coupling the columns in series with the addition of acetonitrile for the larger diameter HILIC column. While providing suitable solvent strength for HILIC, this dilution decreased sensitivity. Unlike the system described by Wang, this system produced a single RP-HILIC chromatogram unless the flow was split and directed to both CAD and MS detectors. Unretained components in the RP dimension were retained and

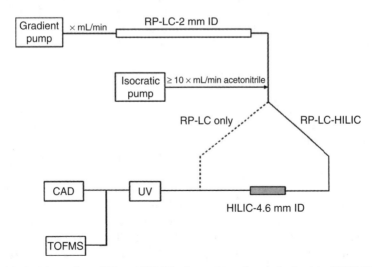

Figure 4.9. Serial coupling of RP and HILIC columns for orthogonal selectivity. TOFMS = time-of-flight mass spectrometer. Reprinted from Reference 47 with permission from Elsevier.

separated on the HILIC column, while the separation of components obtained by RP was maintained through the second column since these components simply passed through in the void of the HILIC column. The separation was demonstrated for polar sugars and nonpolar sulfonamides. The system could be employed as a generic screen for both polar and nonpolar impurities in a single chromatographic run.

4.4.2   Impurity Identification

LC-MS has become a very powerful tool for impurity identification in pharmaceutical development. Assignment of a molecular ion and fragmentation ions coupled with chemistry knowledge can often provide a high degree of confidence for structural identification. As discussed in depth in Chapter 7, HILIC conditions can promote analyte ionization and aid qualitative as well as quantitative analysis. Liu et al. employed HILIC with online hydrogen-deuterium (H-D) exchange and tandem mass spectrometry to elucidate the structures of 4-aminomethylpyridine degradation products [48]. The H-D exchange provided information beyond the typical molecular ion and fragmentation that allowed more complete assignment of structures.

4.4.3   Specific Impurity Determination

In contrast to screening for unknown impurities, some analyses involve determination of a specific known polar impurity in the presence of the drug substance, which itself may or may not be polar. Excipients present in drug products may also impact the selectivity needed for determination of a polar impurity. Conditions for separation of impurities may also be used for drug potency or content analysis as described previously. As with RP-HPLC, there may be several combinations of HILIC conditions that meet the goals of the analysis. Specific examples related to drug impurities are given below.

4.4.3.1   Pyrimidines, Purines, and Nucleosides

HILIC has been used frequently for the analysis of pyrimidine and purine bases and nucleosides. In addition to their importance in biological systems, some of these compounds are used as drugs or intermediates in drug synthesis. Strege et al. investigated the separation of cytosine from uracil and a process impurity on amino and TSKgel Amide-80 (silica with bonded carbamoyl group) columns [49]. Gradient elution from 5% to 25% aqueous sodium or ammonium formate, pH 3.5, was utilized. The method did not suffer from problematic baseline artifacts that were present when a RP ion-pairing method was used and MS detection could be used for identification of impurities when ammonium formate buffer was employed. As shown in Figure 4.10, excellent separation was obtained on all three columns investigated. The method was validated using the TSKgel Amide-80 column for impurities over a range of

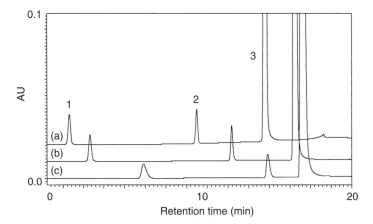

Figure 4.10. HILIC separation of 0.01 mg/mL uracil, 0.01 mg/mL 7-ADOP (a major impurity in commercial cytosine), and 0.5 mg/mL cytosine obtained with (a) Spherisorb amino, (b) Zorbax amino, and (c) TSKgel Amide-80 columns. Peaks: 1 = uracil, 2 = 7-ADOP, 3 = cytosine. Chromatogram courtesy of M.A. Strege.

0.02–4.2% and was shown to be robust with respect to small changes in column temperature, mobile phase pH, buffer concentration, injection volume, and flow rate. Also, good reproducibility was obtained for cytosine retention time (RSD = 0.12%) and peak area (RSD = 0.73%) from over 100 injections on multiple days.

Guo and Gaiki examined the retention behavior of nucleobases and nucleosides including cytosine and uracil on four different columns: amino, TSKgel Amide-80, ZIC-HILIC, and silica [50]. Similar selectivity was obtained using the amino and TSKgel Amide-80 columns under the conditions examined. Of the columns used, silica provided the most different selectivity. All columns gave excellent separation of cytosine and its degradation product, uracil. The authors also studied the retention of acidic compounds on these columns and found that amino and silica columns gave the same retention order.

Marrubini et al. also studied the separation of purines, pyrimidines, and nucleosides on TSKgel Amide-80 and ZIC-HILIC columns [51]. Partition and adsorption mechanisms were both operative using aqueous buffer/acetonitrile mobile phases. Retention of the compounds studied was balanced by ionization of silanols and analytes on the TSKgel Amide-80 and less affected on ZIC-HILIC in the pH 3–5 range. The effects of acetonitrile percentage, salt concentration, pH, and temperature were investigated and gradient conditions were identified for separation of the 12 compounds studied on the ZIC-HILIC column.

In early work with HILIC, Olsen demonstrated the determination of 5-fluorouracil in 5-fluorocytosine on both silica and amino columns [43]. In this case, the main component, 5-fluorocytosine, was more polar than the 5-fluorouracil impurity. The HILIC method was mentioned as a possible improvement over the USP monograph method that used thin layer chromatography (TLC). As shown in Figure 3.2 (see Chapter 3), unknown impurities were also observed in the samples. Substituting a volatile buffer would allow MS investigation of the unknown structures. The determination of guanine in acyclovir was also demonstrated using silica or amino columns as potential alternatives to RP-HPLC with highly aqueous mobile phases [43].

4.4.3.2 Hydrazines with Ethanol as Weak Solvent

Liu et al. determined hydrazine and 1,1-dimethylhydrazine at trace levels in a drug substance intermediate without derivatization using a zwitterionic ZIC-HILIC stationary phase and a chemiluminescent nitrogen detector (CLND) [9]. Because this detector responds to nitrogen-containing compounds, acetonitrile could not be used in the mobile phase and ethanol was used as the weak solvent. Eluting strength of alcohols in the mobile phase decreased in the order of methanol > ethanol > isopropyl alcohol. Trifluoroacetic acid (TFA), acetic acid, and formic acid modifiers were evaluated. TFA provided the best peak shape and shortest retention time. In addition to the ZIC-HILIC column, TSKgel Amide-80, diol, and amino columns were evaluated. The ZIC-HILIC column gave the longest retention, which is consistent with electrostatic interactions between the positively charged hydrazines and the negatively charged sulfonate groups on the stationary phase. With the amino column, the hydrazines were eluted before the void volume, presumably due to electrostatic repulsion from the positively charged amine groups. Retention decreased with increasing ionic strength, indicating involvement of ion exchange. Temperature did not have a significant effect on retention. Figure 4.11 shows

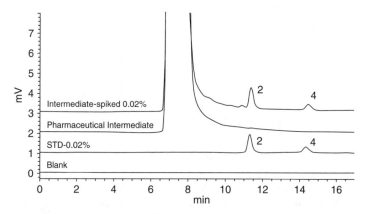

Figure 4.11. Determination of 1,1-dimethylhydrazine and hydrazine in a pharmaceutical intermediate on a ZIC-HILIC column. Mobile phase: TFA/water/ethyl alcohol (0.1/30/70, v/v/v). Flow rate: 0.4 mL/min with a splitter and CLND. Injection volume: 20 μL. Sample concentration: 10 mg/mL of the intermediate in DMSO/ethyl alcohol (30/70, v/v). The spiked intermediate sample consisted of 10 mg/mL intermediate, 2 μg/mL 1,1-dimethylhydrazine, and 2 μg/mL hydrazine in DMSO/ethyl alcohol (30/70, v/v). The standard (STD) solution had 2 μg/mL each of 1,1-dimethylhydrazine and hydrazine in water/ethyl alcohol (30/70, v/v). 2: 1,1-dimethylhydrazine, and 4: hydrazine. Reprinted from Reference 9 with permission from Elsevier.

data for a standard, sample, and spiked sample, which demonstrates good recovery of the analytes in the presence of the sample matrix. Note that DMSO used in the solvent to dissolve the sample did not appear to impact the separation.

4.4.3.3 Neutral and Charged Polar Impurities in a Drug Substance

Olsen exploited both hydrophilic partitioning and ion-exchange mechanisms using a propyl amine column to determine low levels of oxamide (neutral), oxamic acid (monoacid), and oxalic acid (diacid) in a nonpolar drug substance [43]. These impurities were all very poorly retained and not separated from each other under RP-HPLC conditions. Yang et al. had found previously that these analytes could be determined in the drug substance using ion exclusion chromatography, a mode often used for organic acids [52]. However, the drug substance was not eluted from the column in the ion exclusion method, which led to a concern about buildup of the drug substance on the column after repeated injections.

As indicated in Figure 4.12, hydrophilic partitioning into a water layer on the amino stationary phase took place because increasing the aqueous content of the mobile phase while holding the buffer strength constant decreased the

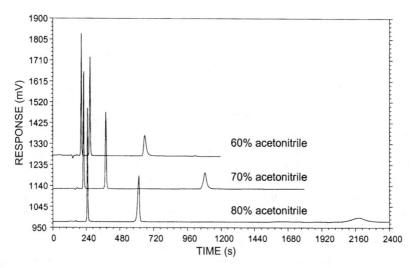

Figure 4.12. Effect of aqueous–acetonitrile ratio at constant buffer strength on the separation of oxamide, oxamic acid, and oxalic acid. Conditions: Zorbax NH2 column; mobile phase percentage acetonitrile as indicated with aqueous phosphate buffer, pH 7.0, to give an overall mobile phase buffer concentration of 10 mM; 205 nm detection; elution order: oxamide (0.03 mg/mL), oxamic acid (0.06 mg/mL), oxalic acid (0.15 mg/mL). Reprinted from Reference 43 with permission from Elsevier.

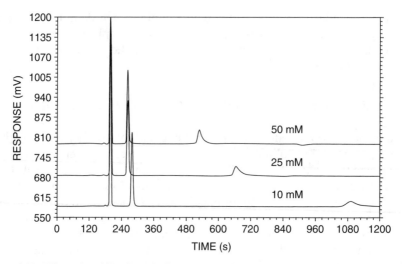

Figure 4.13. Effect of mobile phase buffer concentration on separation of of oxamide, oxamic acid, and oxalic acid. Conditions: Zorbax NH2 column; mobile phase acetonitrile–potassium phosphate buffer (concentration before mixing with acetonitrile, as given in figure), pH 7.0 (60:40); 205 nm detection; elution order: oxamide (0.08 mg/mL), oxamic acid (0.08 mg/mL), oxalic acid (0.06 mg/mL). Reprinted from Reference 43 with permission from Elsevier.

retention of all three analytes, including the neutral oxamide. Ion exchange was also taking place as shown in Figure 4.13 where the buffer strength was varied while holding the aqueous concentration of the mobile phase constant. In this case, the retention of only the charged analytes, oxamic acid and oxalic acid, was affected, with the diacid retention being impacted to a greater extent. Both the aqueous concentration and buffer strength were optimized to allow determination of all three analytes in the drug substance (Fig. 4.14). The aqueous pH was buffered at 7.0 to maintain the acidic analytes with a negative charge while still keeping a positive charge on the stationary phase amino groups, thereby providing the basis for ion exchange. Phosphate buffer was used because of transparency at 205 nm, although solubility limited the buffer concentration. Note that the solubility of buffers in the mobile phase should always be checked during development, especially if the aqueous and organic components are being mixed by the HPLC pump rather than premixed. A drawback of phosphate buffers at around neutral pH is the potential for shortened column lifetimes. The drug substance in the above application was suitably soluble in aqueous acetonitrile to give good sensitivity (~0.1%) for the analytes. Also, the nonpolar drug substance was eluted in the void volume, which alleviated concern about buildup on the column or the need to remove it after completing an analysis.

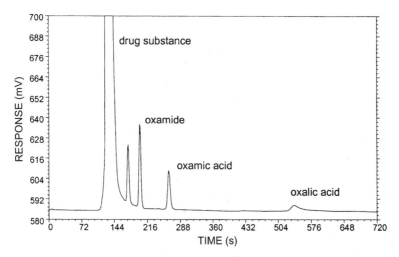

Figure 4.14. Determination of oxamide, oxamic acid, and oxalic acid in a drug substance. Conditions: Zorbax NH2 column; mobile phase acetonitrile–50 mM potassium phosphate, pH 7.0 (60:40); 205 nm detection; sample concentration: 4.9 mg/mL; standard concentrations: oxamide (0.01 mg/mL), oxamic acid (0.01 mg/mL), oxalic acid (0.03 mg/mL). Reprinted from Reference 43 with permission from Elsevier.

4.4.3.4 Polar Basic Compounds and Impurities

In another mixed retention mode example, Ali et al. developed a HILIC method for metformin drug substance and the impurities melamine and cyano-guanidine. A silica column was used, where mobile phase pH impacted ionization of silanols and two of the basic, polar analytes [53]. The authors referred to the separation mechanism as mixed-mode HILIC because the separation was not wholly explained by HILIC, but also relied on an ion-exchange mechanism. A mobile phase of pH 3.0 aqueous buffer:acetonitrile in a 16:84 ratio was used. The impact of organic content and buffer ionic strength was investigated. As mentioned previously, when developing a HILIC method for the pharmaceutical industry, it is important to understand the ruggedness of the method. When the organic composition was varied, it was observed that metformin and melamine retention (and thus resolution) was significantly impacted, whereas cyanoguanidine retention did not change. These interactions were explained by the fact that cyanoguanidine is the most hydrophobic of the three compounds. Mobile phase pH was also considered an important variable in this work across the range of 3.7–5.7. Due to the pK_a values of melamine (5.1) and metformin (2.8 and 11.5), and the increase in negative charge density of the active silanols, it was demonstrated that pH impacted the retention of melamine and metformin, but not the nonionizable cyanoguanidine. This type of multivariate interaction may give researchers the impression that HILIC separations lack robustness; however, with the proper design of experiments and fundamental understanding of the important separation factors and mechanisms, including the ionization states of the analytes, ionization state of the column, and analyte interactions with the column (hydrophilic or ionic in many cases), a very robust method can be developed and validated.

In an interesting comparison, Al-Rimawi described a method for cyanoguanidine in metformin HCl tablets. In this paper there was an indication of misunderstanding of HILIC conditions since concern was mentioned about IIILIC methods requiring special columns and analyst expertise [54]. Although not described as a HILIC method, a silica column was used with a mobile phase of pH 5.0 ammonium dihydrogen phosphate buffer:methanol, 21:79. The separation of metformin and cyanoguanidine was nearly the same as that obtained by Ali et al. [53], suggesting that the HILIC mode is very robust for separation of these compounds. The method was validated for linearity, range, accuracy, precision, specificity, limit of detection, and limit of quantitation. The method was judged to be stability indicating based on its ability to separate metformin from cyanoguanidine.

Not only is melamine a potential impurity in metformin hydrochloride production, it has also received much interest in the pharmaceutical industry recently as an adulterant that potentially can be added to excipients in order to deceive typical compendial testing. Melamine, which is low in molecular weight and high in nitrogen content, was fraudulently added to protein-containing products (such as dairy products and animal feeds) to increase the apparent protein content [55]. As a result, the FDA issued guidance for all excipients containing a nonspecific nitrogen test to implement a specific test for melamine [56]. The acceptable limit for melamine defined in this guidance is 2.5 ppm. In order to develop a method specific for melamine, HILIC was evaluated. The method outlined by Ali et al. [53] would have probably been suitable from a selectivity perspective. However, multiple HILIC columns were also investigated to determine melamine. A method developed by X.-L. Zheng et al. utilized an amino column (150×4.6 mm, 5 µm) as opposed to using ion-pairing reagents for RP separation [57]. The mobile phase was acetonitrile:phosphate buffer (88:12 v:v) at a flow rate of 1.0 mL/min. The FDA validated a method for melamine and cyanuric acid in infant formula [58]. This method employed the ZIC-HILIC column and used mass spectrometric detection. The method is useful for both the food and pharmaceutical industry, and the conditions are referenced in the melamine guidance for industry [56].

4.4.4 Statistical DOE for Optimization

The multiple mechanisms that may be involved when using HILIC conditions can lead to interactions of mobile phase parameters; that is, the value of one parameter can impact the optimum value of another parameter. In such cases, one-factor-at-a-time (OFAT) optimization may lead to a suboptimal overall separation. In a 2008 review, Dejaegher et al. mentioned that all literature examples of HILIC methods at that time had been developed using the OFAT approach [15]. As mentioned in Chapter 3, statistical DOE techniques are helpful for identifying interactions and selection of parameter combinations that produce the best overall result [59,60]. A detailed discussion of DOE techniques is beyond the scope of this book, but an example can illustrate their application and utility. Hatambeygi et al. described a DOE study for the separation of aspirin, salicylic acid (hydrolytic degradation product), and ascorbic acid in a tablet formulation [11]. They employed a Box-Behnken response surface design to investigate the following parameters and ranges: acetonitrile concentration (70–80%), ammonium acetate concentration (10–50 mM), pH (3–6), and temperature (25–35°C). A silica column was used and the parameters were examined at three levels each. Twenty-seven experiments, including three center points at nominal conditions, were conducted, with retention and resolution determined as the responses. Models were developed for the responses and significant effects on all responses were observed for buffer strength and the interaction of buffer strength and pH. Results for the retention time of aspirin as a function of pH and buffer concentration are shown in Figure 4.15. Hydrophilic and electrostatic retention mechanisms were both involved as pH changes affected analyte and stationary phase ionization. At high pH, increasing buffer concentration led to greater retention due to buffer

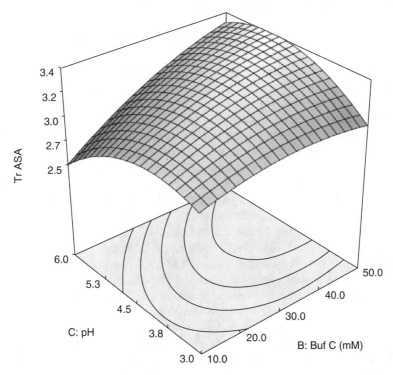

Figure 4.15. Response surface plots representing the retention time of aspirin (ASA) as a function of pH and buffer concentration with acetonitrile content and temperature constant at 75% and 30°C, respectively. Reprinted from Reference 11 with permission from Elsevier.

ion pairing with ionized silanols, thereby reducing electrostatic repulsion of the aspirin and increasing its retention. Derringer's desirability function was used to obtain a global desirability based on analysis goals such as resolution and retention time. Parameters could then be optimized based on overall goals of the analysis. The models could also be used to test the robustness of the method by checking predicted responses when parameters were changed slightly from the optimum conditions.

4.5 EXCIPIENTS

Excipients and other additives (inactive ingredients) are typically used in drug products to enhance stability, enhance solubility, control release, increase absorption, improve flavor, and aid in the manufacturing process. Analytical methodologies are needed during formulation development, excipient compatibility testing, in-process monitoring, homogeneity testing, and drug product stability testing to quantify the drug product components to ensure mixture or drug product integrity. These methods are often used to determine excipients in drug product-related samples. However, the development of analytical methods can be especially challenging for very polar excipients, which also often lack sufficient chromophores for UV detection.

4.5.1 Parenteral and Solution Formulations

For parenteral formulations, mannitol can be used as a bulking agent and stabilizer, as well as for maintaining lyophilized plug elegance and for ensuring an amorphous form of the drug for rapid dissolution upon reconstitution. Risley et al. developed and validated a HILIC-ELSD method for the quantitation of mannitol in a gemcitabine hydrochloride parenteral formulation [12]. This method employed a TSKgel Amide-80 column using 75% acetonitrile/25% water with 0.1% TFA, which was capable of separating sodium, gemcitabine (active ingredient), and mannitol. The authors also incorporated a gradient starting at 90% acetonitrile to show the feasibility of separating other potential excipients used in parenteral formulations (Fig. 4.16).

Another excipient used in parenteral or solution formulations is cyclodextrin (CD). Captisol® (SBE7-β-CD), an anionically charged sulfobutyl ether β-CD, can be used to increase the solubility, stability, and bioavailability of poorly aqueous soluble drugs in solution or intravenous (IV) formulations. Two products have been successfully commercialized using Captisol: Vfend (voriconazole) I.V. and Zeldox/Geodon (ziprasidone HCl) for injection. Captisol is also frequently used in early development to enhance solubility in toxicology formulations. Captisol can be assayed using HILIC-CAD and a ZIC-HILIC column with gradient elution starting at 90% acetonitrile (Fig. 4.17). In this profile, Captisol elutes as a distribution of sulfobutyl-ether-substituted species ($n = 1$ to $n = 10$) between 15.0 and 20.5 min. The sodium

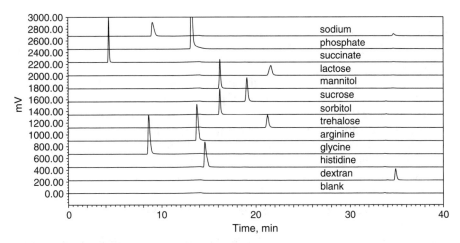

Figure 4.16. HPLC-ELSD overlaid chromatograms of polar excipients using a starting mobile phase of 90:10 (ACN:0.1% TFA in water) was used for 5 min with a linear gradient to 70:30 from 5 to 10 min followed by a 15-min hold and then another linear gradient to 10:90 from 25 to 40 min. Flow rate was 1.0 mL/min with 10-μL injections. Column was a TSKgel Amide-80 (250 × 4.6 mm, 5 u) maintained at 35°C. ELSD operated at 40°C, 3.5 bar nitrogen, and gain setting of 1.

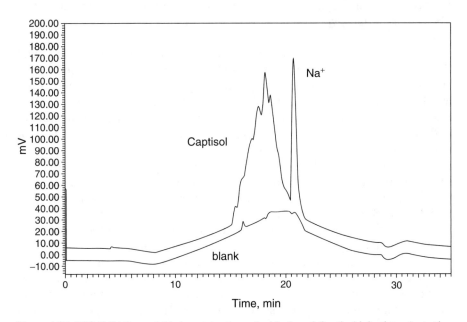

Figure 4.17. HPLC-CAD overlaid chromatograms of a blank and Captisol injections. A starting mobile phase of 90:10 (acetonitrile:10 mM ammonium acetate buffer) with a linear gradient to 40:60 (acetonitrile:10 mM ammonium acetate buffer) in 22 min, hold for 3 min, and then return to starting mobile phase. Flow rate was 1.0 mL/min with 5 μL injection volume (sample concentration 2 mg/mL in starting mobile phase). The column was a SeQuant ZIC-HILIC 250 × 4.6 mm, 5 um and the CAD setting range was 500 pA. Chromatogram courtesy of M.A. Strege, unpublished results.

counterion is the large peak at 21.0 min. LC-MS analyses performed using this method confirmed that hydrophilicity within the Captisol distribution (and therefore HILIC retention) increased with an increase in the degree of substitution.

PEG 400 (a low molecular weight grade of polyethylene glycol) is another very polar excipient used in a variety of pharmaceutical formulations. Webster et al. used a HILIC-like strategy for the analysis of PEG 400 in perfusate samples. The authors used a Cogent Diamond Hydride (silicon hydride surface) column with gradient elution starting at 100% acetonitrile and ending with 50% acetonitrile in 20 min. The ELSD was also used for quantitation due to lack of a suitable chromophore for PEG 400 [61].

4.5.2 Tablets, Capsules, and Inhalation Products

Lactose is commonly used in tablets and capsules as a filler/filler-binder due to its low cost, stability, low hygroscopicity, and compatibility with other drug product components, but it is also found in other types of drug products. Beilmann et al. used HILIC-ELSD for the analysis of lactose from a dry inhalation product where lactose and ipratropium bromide were components of the drug product. Retention of lactose was accomplished by using an APS-3 Hypersil aminopropyl stationary phase with a mobile phase of 80:20 acetonitrile:water [62].

Magnesium stearate is often added as a lubricant in capsule or tablet drug product manufacturing to prevent blended formulation ingredients from sticking to the manufacturing equipment. It can also be used in formulations as a binder. In this example, a HILIC-NQAD method was used for the retention, separation, and quantitation of magnesium from magnesium stearate in a drug product capsule formulation. Magnesium stearate has very poor aqueous solubility, but a sample solvent of 49:49:2 acetone:methanol:trifluoroacetic acid provided high solubility. A ZIC-HILIC column was used for the separation and retention of magnesium using a mobile phase composed of acetone:methanol:150 mM ammonium formate pH 4.5 with formic acid, 32.5:32.5:35. Stearic acid can also be quantified using this method, although the peak elutes quickly under these conditions [63]. This method could serve as an alternative to atomic spectroscopy for magnesium.

Croscarmellose (carboxymethylcellulose or CMC) sodium is often used as a disintegrant additive for tablets to increase dissolution although it has other applications in pharmaceutical formulations. In addition to monitoring the drug substance during stability studies, sometimes it can be important to also monitor the excipients to ensure product integrity. HILIC-ELSD can be used to assay for CMC using conditions shown in Figure 4.18, which also shows the separation and detection of polysorbate-80, sodium (from CMC sodium), mannitol, and CMC.

Povidone is a water-soluble polymer that can be added as a binder, or as a dispersing or suspending agent to pharmaceutical formulations. In another

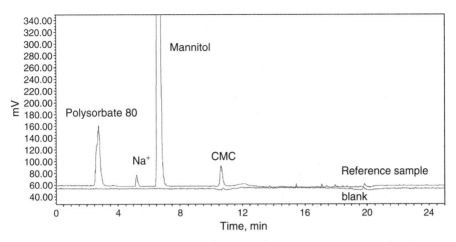

Figure 4.18. HPLC-ELSD chromatograms of blank and parenteral diluent components. The mobile phase was 60:40 (acetonitrile:0.1% trifluoroacetic acid in water) for 6 min followed by a gradient to 100% 0.1% trifluoroacetic acid in water from 6.1 to 15 min. The flow rate was 0.8 mL/min with a 10 μL injection volume. The column was a TSKgel Amide-80, 250 × 4.6 mm, 5 μm. ELSD detection was used. Chromatogram courtesy of M.B. Arif, unpublished results.

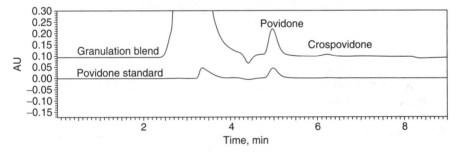

Figure 4.19. HPLC-UV (205 nm) chromatograms for the determination of povidine in a pharmaceutical granulation blend. Column: TSKgel Amide-80 column. Mobile phase: 70% acetonitrile:30% 0.1% trifluoroacetic acid in water. Flow rate: 0.8 mL/min. Sample concentration: 1.0 mg/mL. Detection: UV at 205 nm. Chromatogram courtesy of M.L. Lytle, unpublished results.

example, a method was developed for povidone using HILIC-UV to assess the uniformity of povidone in granulation blends, as part of DOE studies to understand the design space for the manufacturing process of the tablet formulation. The ingredients of the formulation included the drug substance, lactose, polysorbate-80, microcrystalline cellulose, and magnesium stearate. This HILIC method retained and separated the polar components of the granulation blend (povidone and crospovidone) while the less polar excipients and drug substance eluted in the void. Thus, the method provided good specificity for the quantitation of povidone in the blends using a TSKgel Amide-80 column with 70% acetonitrile for elution (Fig. 4.19).

4.5.3 Sugars

Glucose is used in chewable tablets as a flavor enhancer or sweetener additive and is also common in glucose IV administration bags that can have a drug added during therapy. Liu et al. used an Atlantis silica column with an aceto-nitrile:water mobile phase and ELSD detection for the analysis of three com-pounds: glucose, glucosamine, and 1-deoxynojirimycin [64]. Although this method was used as an impurity assay, the chromatography conditions are viable for more complex formulations containing glucose. Sucrose, a disaccha-ride, has also been added to protein solutions as a stabilizer. Karlsson et al. employed HILIC-ELSD for the analysis of sucrose from antithrombin preparations [65]. The method also separated and detected fructose, glucose, and Triton X-100. The separation and retention of sucrose was accompli-shed using a Poly LC poly-2-hydroxyethylaspartamide HILIC column with a mobile phase composed of 75% acetonitrile:25% water. This author also developed HILIC-ELSD methodology for the separation of several monosac-charides including L-fucose, D-galactose, D-mannose, N-acetyl-D-glucosamine, N-acetylneuraminic acid, and D-glucuronic acid using a TSKgel Amide-80 column with 82% acetonitrile for elution [66]. In a different sugar example, Figure 4.20 illustrates the analysis of dextrose, lactose, mannitol, sorbitol, sucrose, trehalose, and xylitol using eight different HILIC stationary phases. Dextrose (D-glucose) and lactose are known to form α- and β-anomers in solu-tion, and three of the columns were capable of resolving these forms while other columns exhibited poor peak shape for these analytes attributed to ano-merization. It is important to note that sugars do not need buffers in the mobile phase for HILIC elution; however, buffers or other mobile phase additives may be needed if other ionic components are present in the matrix that also need to be analyzed or eluted from the column. Although the eight columns tested did not completely resolve all of the sugars in the HILIC mode, the chromato-grams in Figure 4.20 can be used as a guide for the desired sugar analysis. As outlined in Chapter 3, this approach is typical of column screening performed in systematic method development for a particular set of analytes.

4.5.4 Stabilizers and Antioxidants

Amino acids, such as glutamine, can be added to pharmaceutical preparations as stabilizers to improve product stability. For example, 2.4% of glutamine and 0.9% of glycine can be added during the preparation of vancomycin lyophi-lized vials to increase shelf life through improved drug product stability [67]. The amino acid L-cysteine can be added as an antioxidant (oxidized to cystine) to improved product stability by inhibiting oxidation of the active ingredient. A HILIC-ELSD method was used to separate L-cysteine and cystine in a tablet matrix using a ZIC-HILIC column and starting mobile phase of 85% acetonitrile (Fig. 4.21). Additionally, there are several HILIC applications for the evaluation of amino acids [68–72]. Amino acids are also often used as

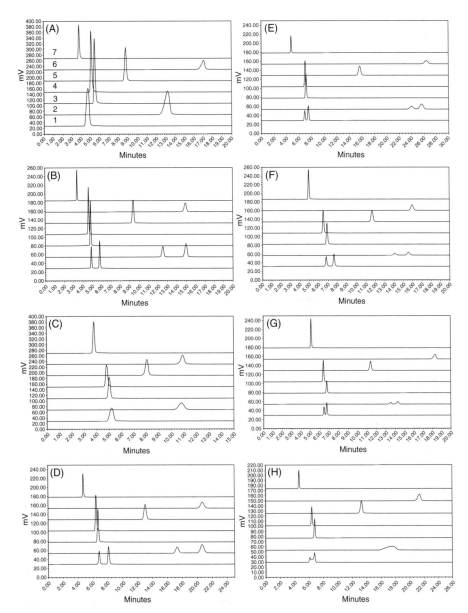

Figure 4.20. HPLC-ELSD overlaid chromatograms using (A) Dionex Trinity Acclaim P1, (B) Monolithic ZIC, (C) Phenomenex Luna Amino (150 × 4.6 mm), (D) SeQuant ZIC-HILIC (150 × 4.6 mm), (E) Tosoh Bioscience TSKgel Amide-80 (100 × 4.6 mm), (F) SeQuant ZIC-cHILIC (150 × 4.6 mm), (G) Phenomenex Luna HILIC (250 × 4.6 mm, diol) with mobile phase of 90% acetonitrile:10% water (G), 92% acetonitrile:8% water (A), 85% acetonitrile:15% water (B, C, D, E), or 80% acetonitrile:20% water (F) for the analysis of (1) dextrose, (2) lactose, (3) mannitol, (4) sorbitol, (5) sucrose, (6) trehalose, and (7) xylitol. Flow rates were 1.5 mL/min, except for (A), which was 0.4 mL/min with ELSD detection. Two peaks for dextrose (1) and lactose (2) in (B, D, E, F, G) are attributed to α- and β-anomers. Timescales: 15 min (C); 20 min (A, B, F, G); 25 min (D); 26 min (H); 30 min (E). Chromatograms courtesy of D.S. Risley, unpublished results.

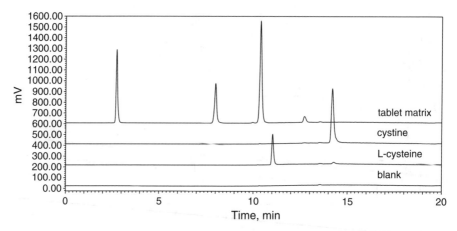

Figure 4.21. HPLC-ELSD chromatograms for the determination of L-cysteine (anti-oxidant) in a tablet formulation. A starting mobile phase of 85:15 (acetonitrile:50 mM ammonium acetate) for 2 min and then a linear gradient to 10% acetonitrile in 22 min. Flow rate was 1.0 mL/min with 20 μL injection volume. The column was a SeQuant ZIC-HILIC, 250 × 4.6 mm, and the Alltech 800 ELSD settings were temperature 55°C, nitrogen 3.7 bar, gain = 4. Chromatogram courtesy of D.S. Risley and B.W. Pack, unpublished results.

counterions to make salt forms of drug substances; therefore, additional chromatographic conditions are also discussed in the counterion section of this chapter.

4.6   CHIRAL APPLICATIONS

As described in Chapter 1, a unique factor inherent in HILIC is the formation of a specific microenvironment at the stationary phase–eluent interface. In certain situations, this stationary–mobile phase microenvironment is amenable to the separation of enantiomers when using chiral stationary phases (CSPs). Not every CSP, however, can be subjected to HILIC conditions. Many of the more established "coated" CSPs are not compatible with aqueous systems, no matter how minimal the water content [73]. Other CSP ligands are not of sufficient polarity to allow for the formation of the necessary water layer inherent with HILIC. Furthermore, the resulting interface must allow for the sufficient combination of partitioning and adsorption to produce adequate retention. Only when these conditions are met can stereoselective interaction become viable.

Although the minimal presence of water within the eluent system is a trademark of HILIC, this solvent is not necessarily beneficial concerning chiral recognition. The presence of a small proportion of water in the mobile phase can be particularly valuable for the solubilizing power of water for polar analytes, along with the marked effect of water on analyte retention. Reports of chiral separations in HILIC mode indicate an inverse relationship between eluent water content and the observed analyte retention and selectivity [74,75]. Achieving acceptable compound retention factors (k) within HILIC requires varying amounts of water, usually ranging from a few percent up to as high as 50% [74]. This value is very dependent on the polarity of both the CSP and the analyte. The increased polarity of either requires more water in the mobile phase for compound elution. Additional aqueous content in the eluent will continue to reduce retention to a point where reversed-phase processes begin to predominate, resulting in a shift to increasing retention. Plots of the well-known U-shaped retention curves demonstrate this phenomenon [76].

4.6.1   Chiral Selectors and HILIC

Despite the relatively limiting set of conditions necessary for HILIC, a few classes of chiral CSPs have proven successful in this mode. These selectors are characterized by the presence of highly polar moieties on the surface of the stationary phase that are presumably accessible within the stagnant water layer. CSPs based on CDs, macrocyclic antibiotics, chiral crown ethers, and cyclofructans have proven quite successful in separating enantiomers in the HILIC mode.

4.6.1.1   Cyclodextrins

The enantiomeric separation potential of CDs has been known for at least 30 years. These starch-based, naturally occurring oligosaccharides consist of six (alpha), seven (beta), or eight (gamma) glucopyranose units bound in a circular shape. Both rims of the CD cavity are lined with hydroxyl groups. This molecular arrangement results in a toroidal shape composed of a relatively hydrophobic cavity and a hydrophilic rim (see Fig. 2.5) [77].

Traditionally, CDs have been employed in run buffers for CE analyses or bonded to silica or added to the mobile phase for HPLC analyses. Additionally, multiple derivatized CDs, involving the modification of various rim hydroxyl groups, have greatly expanded the functionality within this class of chiral selectors. Although CDs exhibit chiral discrimination in multiple modes, their most notable success for HPLC has been with reversed-phase separations. The primary selectivity mechanism in this mode is inclusion complexation. While a "right-sized" portion of the analyte passes into the toroidal cavity, other analyte ligands interact selectively with the rim hydroxyls or derivatized hydroxyls. Normal phase separations, while achievable with some derivatized CDs, are not as numerous due to the preferential filling of the toroidal cavity with the nonpolar constituent of the mobile phase [77].

CD chiral separations in the HILIC mode are thought to occur via a different mechanism from that of either the reversed or normal phase modes. In HILIC, it is believed that the much more predominant constituent of the mobile phase (usually acetonitrile) preferentially partitions within the hydrophobic CD cavity, leaving the polar hydroxyl groups on the rim of the cavity to interact with the analytes [78,79]. At this interface, a sort of competition ensues between polar water molecules (part of the aqueous layer inherent within HILIC) and polar moieties of the analyte. According to this model, HILIC separations on CDs are not technically achieved by inclusion complexation. The analytes instead lie across the surface of the CD cavity. The degree of analyte to CD rim hydroxyl interaction is controlled in part by the eluent water content. Since water is actually deleterious to this interaction, it is not surprising that many HILIC chiral separations occur with very low aqueous content, approaching polar organic mode conditions. In fact, where analyte aqueous solubility is not an issue, and for analytes of moderate polarity, polar nonaqueous eluents can be of significant value in chiral separations [80,81].

Acidic and basic modifiers in the mobile phase can be used to optimize chiral separations with CDs. Since these additives are thought to mostly affect the ionization state of the analyte [73], with a lesser impact on the stationary phase, this topic is discussed in detail concerning the macrocylic antibiotic CSPs where the same phenomenon may occur. Reports are available demonstrating the success of separating enantiomers employing CDs in the HILIC mode at various pH values for cyclic carboxylic acids [74], nicotine analogs [82], and dansyl amino acids [83].

4.6.1.2   Macrocyclic Antibiotics

Compared with CDs, macrocyclic antibiotic (MA) chiral selectors are relatively new to the stereoisomer separation arena. First introduced in 1994 as enantioselective agents [84], macrocyclic glycopeptides (some used clinically as antibiotics) have become widely used for the separation of enantiomers of many compound classes within multiple chromatographic modes. Unlike the CD market, which offers dozens of unique native and derivatized selectors, only four basic MA structures are currently commercially available. These are based on vancomycin, teicoplanin, teicoplanin aglycone, and ristocetin A. The structure of vancomycin, a representative MA, is shown in Figure 4.22 and illustrates the chiral nature of MAs and the multiplicity of sites available for interactions.

MAs present a more complicated landscape concerning the microenvironment of the HPLC chiral separation. Their large molecular weights and complex structures offer multiple sites for stereoselective action that are not fully understood. The ionizable groups within MAs are of special interest, where simple pH changes (reversed-phase and HILIC modes) can markedly alter the spatial arrangement of the selector. Similar to CDs, MAs possess inclusion cavities, although these are thought to be shallow and offer weaker interaction [85]. MAs also lack the spatial rigidity inherent with CDs. In this spatial flexibility, the antibiotics function in ways similar to protein selectors, although MAs are significantly more stable [78].

The presence of water within HILIC provides the same advantages for MAs as it does for CDs. The main advantage is enhanced eluent solubility for highly

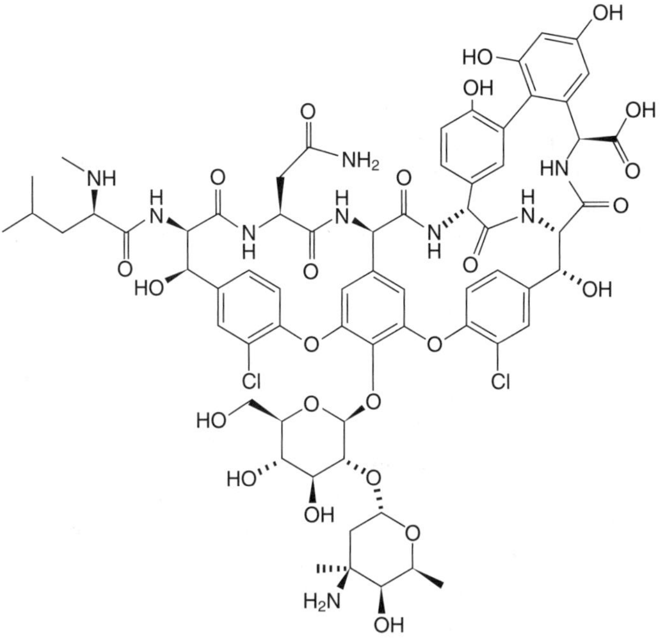

Vancomycin

Figure 4.22. Structure of the macrocyclic antibiotic vancomycin.

polar analytes. Another advantage lies with retention control. Within HILIC, an increase in eluent aqueous content results in decreased analyte retention, from less to more hydrophilic molecules. The phenomenon is well documented in MA separations [74,77,86–89].

With macrocyclic glycopeptide CSPs, depending on the analyte structure, the use of combinations of acidic and basic mobile phase additives can prove paramount in achieving an enantioseparation in HILIC and when using neat polar solvents as the mobile phase. These modifiers, frequently a tailored combination of acetic acid (HOAc) and triethylamine (TEA), appear to affect interaction between the ionizable groups of the MA and similar ionic moieties in the analyte. Additives provide no benefit when separating neutral compounds. Combinations of these modifiers with neat polar solvents such as methanol, ethanol, or acetonitrile, or co-solvents of these, have been dubbed the polar ionic mode (PIM) [85,90]. References supporting the use of the PIM for the chiral separation of polar analytes with MA CSPs abound [88–97].

Although not technically within the scope of HILIC, the interplay of mobile phase additive used in PIM with MAs warrants further discussion for a few reasons. First, the success of the PIM was accomplished using polar analytes that are likely also amenable to HILIC. Second, at the risk of oversimplification,

HILIC is simply a PIM eluent plus a small amount of water. If the aqueous content in HILIC mobile phases is primarily a retention tool only (such as reported with CDs), then the retention and resolution characteristics of PIM could certainly apply to a degree in HILIC. Finally, even if a chiral separation is obtained using an MA CSP in PIM, HILIC may still be a viable alternative if the analytes are less soluble in the solvents (usually alcohols or acetonitrile) employed within PIM systems.

Since HILIC and PIM systems appear closely related with respect to their interactions on MA stationary phases, further consideration of retention and resolution characteristics for PIM is useful. Whereas retention of analytes in binary systems (including HILIC) can be controlled by varying solvent ratios, this option does not exist in PIM systems where the eluent usually consists of a neat solvent with additives. Retention is controlled instead by adjusting the total additive amount (acid + base or volatile salt) within the mobile phase, up to about 2%. The larger the total amount employed, the less overall interaction that occurs between analyte and MA, resulting in reduced retention. Enantiomer resolution is modified by varying the ratio of additives, usually with higher acid content for acidic analytes, or with an excess of base for basic analytes [85]. This ratio controls the degree of ionization of the analytes, a significant factor in enantioselectivity [86].

The introduction of small amounts of water into PIM mobile phases by definition transforms them to HILIC. Such aqueous content, when added to the sample diluents, can enhance analyte solubility, improving diluent and eluent compatibility. Concerning column performance, increasing the water composition in the mobile phase usually decreases analyte retention in the same manner as is typically seen by increasing total acid/base modifier content. As observed in the separation of sotalol enantiomers (Fig. 4.23), increasing aqueous content in the eluent reduces retention and resolution, although baseline resolution was still obtained with 15% water [98].

In instances where isomer resolution is more than adequate (>2), augmenting a PIM eluent system with limited water content, rendering it HILIC, could prove to be a powerful tool to promote elution of polar analytes, without the need for excessive amounts of acid and base modifiers. It is conceivable that in such cases, a dynamic interplay between analyte charge (controlled by acid/base additive ratio) and partitioning and/or preferential adsorption (with the introduction of the HILIC water layer) provide powerful options for altering retention and enantioselectivity.

4.6.1.3　Chiral Crown Ethers

Chiral crown ethers (CCEs) are so named due to their inherent structural configuration. These macrocyclic polyethers are formed by linking multiple alternating oxygen and methyl moieties in a cyclic crown-like structure. Chirality is introduced to the crown ethers through the addition of chiral moieties to selected methylene groups on the cyclic molecules. In this respect, CCEs resemble CDs

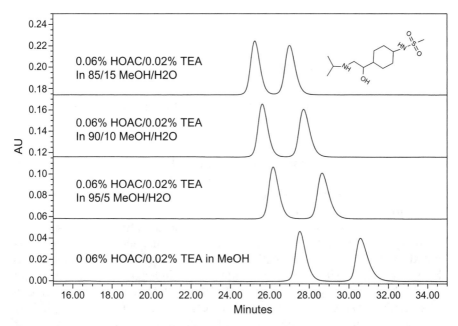

Figure 4.23. Effect of aqueous eluent content on the chiral separation of sotalol enantiomers: Transferring from polar ionic to HILIC chromatographic modes. Column: Chirobiotic T, 4.6 × 250 mm, 5-μm particles.

in that various selectors can be synthesized with differing enantioselective properties depending on the identity of the attached ligand [78].

The charged ring environment of the CCE is ideal for stereogenic interaction. Unfortunately, this charge and spatial combination is apparently so specialized that chiral separations employing CCEs are almost completely limited in scope to primary amines. Specifically, enantioselectivity is driven by interaction between the crown ether electron-donating oxygen atoms and the protonated amino group of the analyte [99]. The need for a charged amino group usually requires an acidic mobile phase environment.

Although not specifically identified as such, a significant proportion of HPLC chiral separations using CCEs reported in the literature can be considered HILIC in nature. In his extensive review of crown ethers, Huyn explains analyte-to-CSP interaction within the HILIC and non-HILIC environments as it relates to CSP polarity and mobile phase aqueous content. He does not, however, identify the appropriate separations as HILIC [99].

One basic requirement for the HILIC chromatographic mode, the presence of a polar stationary phase, is certainly met in a specific class of CCEs built on a (+)-(18-crown-6)-2,3,11,12-tetracarboxylic acid unit (Fig. 4.24). With this CCE type, when employing hydrorganic mobile phases, polar analyte retention increases with increasing organic content. As indicated previously with

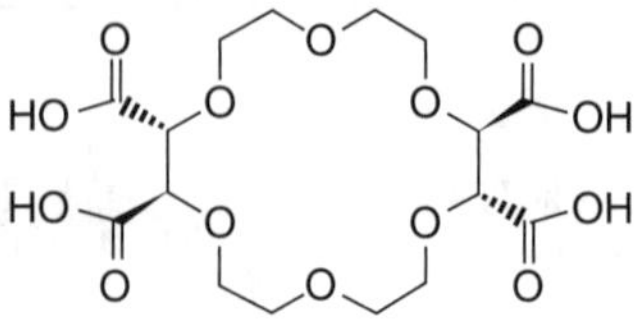

Figure 4.24. Structure of (+)-(18-crown-6)-2,3,11,12-tetracarboxylic acid.

CDs and MAs, this phenomenon is classic with HILIC, where polar analytes "compete" with water for interaction sites on the CSP. As the amount of water in the eluent decreases, greater CSP-to-analyte interaction becomes possible, leading to greater retention [99–104].

The effect of the degree of analyte polarity on retention in an 18-crown-6 HILIC system is well documented by Hyun and colleagues [105]. They investigated the retention and resolution behavior of a series of five homologous alpha amino acids with varying aliphatic side chain lengths. Within a given set of HILIC conditions, the capacity factor of the analyte increased in correlation to its decreasing side chain length, and corresponding increased hydrophilicity [105]. Hyun highlights the opposite retention behavior observed with nonpolar compounds in an otherwise HILIC environment, which further emphasizes the requirement for a hydrophilic analyte to partition between the aqueous and nonaqueous mobile phase layers [106]. The importance of stationary phase polarity in HILIC is confirmed when a structurally different and relatively nonpolar CCE based on a lipophilic 1,1'-binaphthyl is subjected to similar hydroorganic eluents and polar analytes. With CCEs of this nature, capacity factors decreased with increasing organic content in the eluent, typical of a reversed-phase system bereft of the stagnant aqueous eluent layer inherent with HILIC [107].

4.6.1.4 Cyclofructans

Cyclofructans, along with the previously discussed CDs, are classified as macrocyclic oligosaccharides. These molecules are similar in structure to crown ethers, possessing an 18-crown-6 ether central cavity. Whereas certain ring oxygens are derivatized in the case of CCEs, these atoms remain as unsubstituted ether linkages with cyclofructans. Instead, multiple β-(2→1) linked D-fructofuranose units, typically 6–8, are bound to ring carbons. The nomenclature of these cyclofructans is designated as CF6, CF7, or CF8, depending on the number of bound units in the crown (Fig. 4.25). The physical properties and orientation of the CF6–CF8 cavities have been extensively studied [108].

Despite the apparent similarities inferred from two-dimensional representations of CDs, crown ethers, and cyclofructans, the spatial orientation of the respective cyclic selectors has been shown to vary significantly. Specifically, the nature of underivatized CF6, with its internal hydrogen bonding, is not

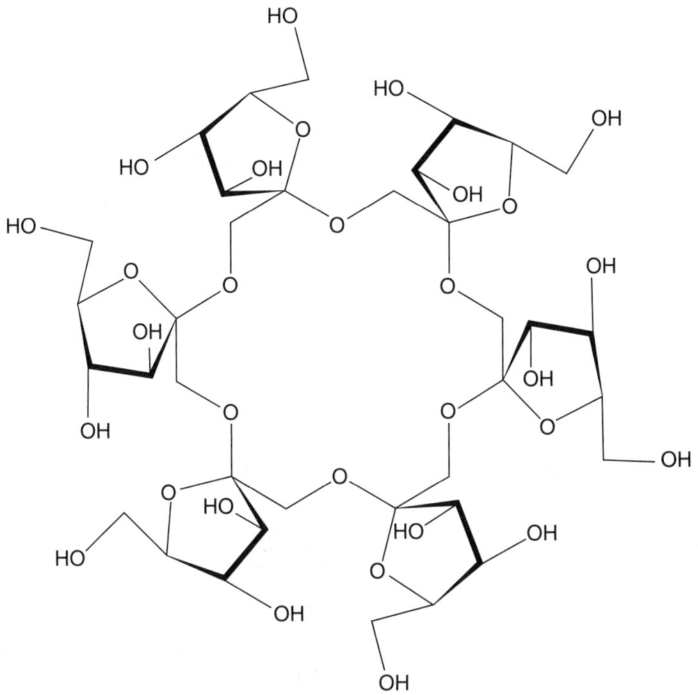

Figure 4.25. Structure of an underivatized cyclofructan macrocycle. Reprinted from Reference 110 with permission from Elsevier.

amenable to analyte access to the crown cavity. Armstrong and colleagues demonstrated that the macrocyclic structure of the CF6 could be "relaxed" via the derivatization of a limited number of hydroxyl groups within the fructo-furanose units (Fig. 4.25) [108]. When derivatized using selected moieties, CF6 becomes an impressive chiral selector for the resolution of multiple classes of chiral compounds. The enantioselective capabilities of the aliphatic derivatized CF6 selectors are of special interest, as these molecules preferentially separate analytes in PIM [108], only a small step from potential HILIC functionality.

Among the subset of aliphatic derivatized CF6 macrocycles, the isopropyl carbamate (IP-CF6) has been identified as a premier chiral selector. The CSP has indeed been commercialized as the LARIHC™ column, a unique alterna-tive to CCEs for the enantioseparation of primary amines [109]. In a recent report, 93% of a set of 119 chiral primary amines were separated using this CSP, most employing a PIM mobile phase consisting of various combinations of acetonitrile and methanol containing 0.3% and 0.2% of HOAc and TEA, respectively [110].

While the presence of water in a low pH environment has been shown to be necessary for CCE separations, the same apparently cannot be said when using the IP-CF6 CSP in PIM. Figure 4.26 demonstrates the effects on

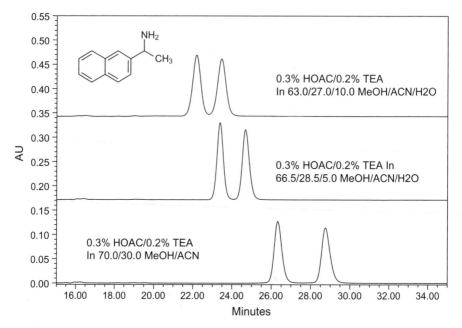

Figure 4.26. Effect of aqueous eluent content on the chiral separation of 1-(2-naphthyl)ethylamine enantiomers: Transferring from polar ionic to HILIC chromatographic modes. Column: CFC-P-06 (IP-CF6 prototype) 4.6 × 250 mm. Flow rate = 0.2 mL/min.

resolution and retention of 1-(2-naphthyl)ethylamine enantiomers with the addition of small amounts of water to the mobile phase, rendering it a HILIC system. Increasing the aqueous portion in the eluent from 0% to 10% (while maintaining the HOAc and TEA content and the acetonitrile-to-methanol ratio) resulted in decreased resolution and analyte retention [111]. One possibility for the negative effect of water on enantioselectivity is potential competition of water molecules for the limited hydrogen bonding sites on the CSP [108]. Nevertheless, in cases where analyte solubility is greatly enhanced in aqueous-based solvents, and where sample diluent and eluent compatibility is paramount, a limited loss of enantioselectivity imposed by the HILIC mode could be tolerated.

4.6.2   Conclusions for Chiral Separations

The dynamics of any HPLC chiral separation are complex. Analyte partitioning with the stationary phase must allow not only for retention, but also for stereoselective retention. CSP–eluent compatibility imposes yet additional challenges. As a result, the HILIC mobile phase environment is not readily applicable to many CSPs. Cyclodextrins, macrocyclic antibiotics, chiral crown ethers, and cyclofructans are notable exceptions, where the parameters of

HILIC mesh beautifully with the nature of the chiral selector to provide a desirable environment in which HPLC chiral separations are attainable.

4.7 CONCLUSIONS

HILIC is not likely to replace RP-HPLC as the workhorse technique for pharmaceutical analysis, but it has been shown to have broad applicability for polar compounds. Polar drugs, drug substance salt counterions, polar excipients, and drug or excipient impurities are amenable to HILIC analysis. The separation of enantiomers may also be conducted in the HILIC mode with an appropriate CSP.

Multiple combinations of HILIC stationary and mobile phases have been shown to be suitable for some applications. In other cases, specific consideration of the analytical goals and the sample and/or analyte properties were used to guide method development and choice of analytical conditions. Most applications employ silica or a polar moiety bonded to silica as the stationary phase. Buffered mobile phases are recommended particularly when ionic interactions between analytes and the stationary phase are involved. General guidelines for method development will continue to advance as further experience is gained with HILIC applications.

REFERENCES

1. Snyder LR, Glajch JL, Kirkland JJ. *Practical HPLC Method Development*, 2nd ed. New York: Wiley; 1997, pp. 317–339.

2. Meyer V. *Pitfalls and Errors of HPLC in Pictures*. Heidelberg: Hüthig; 1997, p. 56.

3. Deshpande GR, Roy AK, Rao NS, Rao BM, Reddy JR. Rapid screening of volatile ion-pair reagents using UHPLC and robust analytical method development using DoE for an acetyl cholinesterase inhibitor: Galantamine HBr. *Chromatographia* 2011; **73**: 639–648.

4. Aruda WO, Aruda KO. Review of volatile perfluorocarboxylic acids as ion pair reagents in LC: Part I. *LCGC N. Am.* 2009; **27**: 626–636.

5. Aruda WO, Aruda KO. Review of volatile perfluorocarboxylic acids as ion pair reagents in LC: Part II. *LCGC N. Am.* 2009; **27**: 916–929.

6. Walter TH, Iraneta P, Capparella M. Mechanism of retention loss when C8 and C18 HPLC columns are used with highly aqueous mobile phases. *J. Chromatogr. A* 2005; **1075**: 177–183.

7. Zhang L, Sun L, Siepmann JI, Schure MR. Molecular simulation study of the bonded-phase structure in reversed-phase liquid chromatography with neat aqueous solvent. *J. Chromatogr. A* 2005; **1079**: 127–135.

8. Alpert AJ. Hydrophilic-interaction chromatography for the separation of peptides, nucleic acids and other polar compounds. *J. Chromatogr.* 1990; **499**: 177–196.

9. Liu M, Ostovic J, Chen EX, Cauchon N. Hydrophilic interaction liquid chromatography with alcohol as a weak eluent. *J. Chromatogr. A* 2009; **1216**: 2362–2370.

10. Karatapanis AE, Fiamegos YC, Sakkas VA, Stalikas CD. Effect of chromatographic parameters and detector settings on the response of HILIC–evaporative light-scattering detection system using experimental design approach and multi-criteria optimization methodology. *Talanta* 2011; **83**: 1126–1133.

11. Hatambeygi N, Abedi G, Talebi M. Method development and validation for optimised separation of salicylic, acetylsalicylic and ascorbic acid in pharmaceutical formulations by hydrophilic interaction chromatography and response surface methodology. *J. Chromatogr. A* 2011; **1218**: 5995–6003.

12. Risley D, Yang W, Peterson J. Analysis of mannitol in pharmaceutical formulations using hydrophilic interaction liquid chromatography with evaporative light-scattering detection. *J. Sep. Sci.* 2006; **29**: 256–264.

13. Hemström P, Irgum K. Hydrophilic interaction chromatography. *J. Sep. Sci.* 2006; **29**: 1784–1821.

14. Dejaegher B, Vander Heyden Y. HILIC methods in pharmaceutical analysis. *J. Sep. Sci.* 2010; **33**: 698–715.

15. Dejaegher B, Mangelings D, Vander Heyden Y. Method development for HILIC assays. *J. Sep. Sci.* 2008; **31**: 1438–1448.

16. Ma M, Liu M. HILIC retention behavior and method development for highly polar basic compounds used in pharmaceutical synthesis. In Wang PG, He W, eds., *Pharmaceutical Synthesis in Hydrophilic Interaction Liquid Chromatography (HILIC) and Advanced Applications*. Boca Raton, FL: CRC Press; 2011, pp. 345–372.

17. Sun M, Liu DQ. Analysis of pharmaceutical impurities using hydrophilic interaction liquid chromatography. In Wang PG, He W, eds., *Pharmaceutical Synthesis in Hydrophilic Interaction Liquid Chromatography (HILIC) and Advanced Applications*. Boca Raton, FL: CRC Press; 2011, pp. 259–290.

18. Wells JJ. *Pharmaceutical Preformulation—The Physicochemical Properties of Drug Substances*. Chichester, UK: Ellis Horwood Ltd.; 1998.

19. Peterson J, Risley D. Validation of an HPLC method for the determination of sodium in LY293111 sodium, a novel LTB4 receptor antagonist, using evaporative light scattering detection. *J. Liq. Chromatogr.* 1995; **18**: 331–338.

20. Risley D, Peterson J, Griffiths K, McCarthy S. An alternative method for the determination of chloride in pharmaceutical drug substances using HPLC and evaporative light-scattering detection. *LC-GC N. Am.* 1996; **14**: 1040–1047.

21. Pack B, Risley D. Evaluation of a monolithic silica column operated in the hydrophilic interaction chromatography mode with evaporative light scattering detection for the separation and detection of counter-ions. *J. Chromatogr. A* 2005; **1073**: 269–275.

22. McClintic C, Remick D, Peterson J, Risley D. Novel method for the determination of piperazine in pharmaceutical drug substances using hydrophilic interaction chromatography and evaporative light scattering detection. *J. Liq. Chromatogr. Relat. Technol.* 2003; **26**: 3093–3104.

23. Guo Y, Huang A. A HILIC method for the analysis of tromethamine as the counter ion in an investigational pharmaceutical salt. *J. Pharm. Biomed. Anal.* 2003; **31**: 1191–1201.

24. Guo Y, Srinivasan S, Gaiki S. Investigating the effect of chromatographic conditions on retention of organic acids in hydrophilic interaction chromatography using a design of experiment. *Chromatographia* 2007; **66**: 223–229.

25. Guo Y. Analysis of quaternary amine compounds by hydrophilic interaction chromatography/mass spectrometry (HILIC/MS). *J. Liq. Chromatogr. Relat. Technol.* 2005; **28**: 497–512.

26. Liu X, Pohl C. HILIC behavior of a reversed-phase/cation-exchange/anion-exchange trimode column. *J. Sep. Sci.* 2010; **33**: 779–786.

27. Huang Z, Richards M, Zha Y, Francis R, Lozano R, Ruan J. Determination of inorganic pharmaceutical counterions using hydrophilic interaction chromatography coupled with a Corona CAD detector. *J. Pharm. Biomed. Anal.* 2009; **50**: 809–814.

28. Cohen RD, Liu Y, Gong X. Analysis of volatile bases by high performance liquid chromatography with aerosol-based detection. *J. Chromatogr. A* 2012; **1229**: 172–179.

29. Risley D, Pack B. Simultaneous determination of positive and negative counterions using a hydrophilic interaction chromatography method. *LC-GC N. Am.* 2006; **24**: 776–785.

30. Crafts C, Bailey B, Plante M, Acworth I. Evaluation of methods for the simultaneous analysis of cations and anions using HPLC with charged aerosol detection and a zwitterionic stationary phase. *J. Chromatogr. Sci.* 2009; **47**: 534–539.

31. United States Pharmacopeia. USP 35-NF 30, General Chapter <905>, Uniformity of Dosage Units, 2012.

32. Matysová L, Havlíková L, Hájková R, Krivda A, Solich P. Application of HILIC stationary phase to determination of dimethindene maleate in topical gel. *J. Pharm. Biomed. Anal.* 2009; **50**: 23–26.

33. ICH Harmonised Tripartite Guideline. Validation of Analytical Procedures: Text And Methodology Q2(R1), November 2005.

34. Ali MS, Ghori M, Rafiuddin S, Khatri AR. A new hydrophilic interaction liquid chromatographic (HILIC) procedure for the simultaneous determination of pseudoephedrine hydrochloride (PSH), diphenhydramine hydrochloride (DPH) and dextromethorphan hydrobromide (DXH) in cough-cold formulations. *J. Pharm. Biomed. Anal.* 2007; **43**: 158–167.

35. Jia S, Park JH, Lee J, Kwon SW. Comparison of two aerosol-based detectors for the analysis of gabapentin in pharmaceutical formulations by hydrophilic interaction chromatography. *Talanta* 2011; **85**: 2301–2306.

36. Code of Federal Regulations. Title 21, Food and Drugs (General Services Administration, Washington DC, April 1973), Part 211.67.

37. ICH Harmonised Tripartite Guideline Q7A. Good Manufacturing Practice Guide for Active Pharmaceutical Ingredients, November 2000.

38. Lytle M. 2011. Unpublished results.

39. United States Pharmacopeia. USP 35-NF 30, General Chapter <711>, Dissolution, 2012.

40. FDA. FDA Alert for Healthcare Professionals. Hydromorphone hydrochloride extended release capsules. July 2005.

41. Traynor MJ, Brown MB, Pannala A, Beck P, Martin GP. Influence of alcohol on the release of tramadol from 24-h controlled release formulations during in vitro dissolution experiments. *Drug Dev. Ind. Pharm.* 2008; **34**: 885–889.

42. Olsen BA, Baertschi SW. Strategies for investigation and control of process and degradation-related impurities. In Ahuja S, Alsante KM, eds., *Handbook of Isolation and Characterization of Impurities in Pharmaceuticals*. San Diego, CA: Academic Press; 2003, pp. 89–117.

43. Olsen BA. Hydrophilic interaction chromatography using amino and silica columns for the determination of polar pharmaceuticals and impurities. *J. Chromatogr. A* 2001; **913**: 113–122.

44. Wang X, Li W, Rasmussen H. Orthogonal method development using hydrophilic interaction chromatography and reversed-phase high-performance liquid chromatography for the determination of pharmaceuticals and impurities. *J. Chromatogr. A* 2005; **1083**: 58–62.

45. Gavin PF, Olsen BA, Wirth DD, Lorenz KT. A quality evaluation strategy for multi-sourced active pharmaceutical ingredient (API) starting materials. *J. Pharm. Biomed. Anal.* 2006; **41**: 1251–1259.

46. Wang Y, Lu X, Xu G. Simultaneous separation of hydrophilic and hydrophobic compounds by using an online HILIC RPLC system with two detectors. *J. Sep. Sci.* 2008; **31**: 1564–1572.

47. Louw S, Pereira AS, Lynen F, Hanna-Brown M, Sandra P. Serial coupling of reversed-phase and hydrophilic interaction liquid chromatography to broaden the elution window for the analysis of pharmaceutical compounds. *J. Chromatogr. A* 2008; **1208**: 90–94.

48. Liu M, Ronk M, Ren D, Ostovic J, Cauchon N, Zhou ZS, Cheetham J. Structure elucidation of highly polar basic degradants by on-line hydrogen/deuterium exchange hydrophilic interaction chromatography coupled to tandem mass spectrometry. *J. Chromatogr. A* 2010; **1217**: 3596–3611.

49. Strege MA, Durant C, Boettinger J, Fogarty M. A hydrophilic interaction chromatography method for the purity analysis of cytosine. *LC-GC N. Am.* 2008; **26**(7): 632–642.

50. Guo Y, Gaiki S. Retention behavior of small polar compounds on polar stationary phases in hydrophilic interaction chromatography. *J. Chromatogr. A* 2005; **1074**: 71–80.

51. Marrubini G, Mendoza BEC, Massolini G. Separation of purine and pyrimidine bases and nucleosides by hydrophilic interaction chromatography. *J. Sep. Sci.* 2010; **33**: 803–816.

52. Yang L, Liu L, Olsen BA, Nussbaum MA. The determination of oxalic acid, oxamic acid, and oxamide in a drug substance by ion-exclusion chromatography. *J. Pharm. Biomed. Anal.* 2000; **22**: 487–493.

53. Ali MS, Rafiuddin S, Ghori M, Khatri AR. Simultaneous determination of metformin hydrochloride, cyanoguanidine and melamine in tablets by mixed-mode HILIC. *Chromatographia* 2008; **67**: 517–525.

54. Al-Rimawi F. Development and validation of an analytical method for metformin hydrochloride and its related compound (1-cyanoguanidine) in tablet formulations by HPLC-UV. *Talanta* 2009; **79**: 1368–1371.

55. Tittlemier SA. Methods for the analysis of melamine and related compounds in foods: A review. *Food Addit. Contam.* 2010; **27**: 129–145.

56. U.S. Department of Health and Human Services Food and Drug Administration Center for Drug Evaluation and Research (CDER) Center for Veterinary

Medicine (CVM). Guidance for Industry Pharmaceutical Components at Risk for Melamine Contamination, 2009.

57. Zheng X-L, Yu B-S, Li K-X, Dai Y-N. Determination of melamine in dairy products by HILIC-UV with NH_2 column. *Food Control* 2012; **23**: 245–250.

58. Turnipseed S, Casey C, Nochetto C, Heller DN. Determination of Melamine and Cyanuric Acid Residues in Infant Formula using LC-MS/MS. Laboratory Information Bulletin LIB 4421 2008; Volume 24.

59. Box GEP, Hunter JS, Hunter WG. *Statistics for Experimenters: Design, Innovation, and Discovery*, 2nd ed. Hoboken, NJ: Wiley; 2005.

60. Montgomery DC. *Design and Analysis of Experiments*, 7th ed. Hoboken, NJ: Wiley; 2008.

61. Webster G, Elliott A, Dahan A, Miller J. Analysis of PEG 400 in perfusate samples by aqueous normal phase (ANP) chromatography with evaporative light scattering detection. *Anal. Method* 2011; **3**: 742–744.

62. Beilmann B, Langguth P, Haeusler H, Grass P. High-performance liquid chromatography of lactose with evaporative light scattering detection applied to determine fine particle dose of carrier in dry powder inhalation products. *J. Chromatogr. A* 2006; **1107**: 204–207.

63. Risley D, Magnusson L, Morow P, Aburub A. Analysis of magnesium from magnesium stearate in pharmaceutical capsule formulations using hydrophilic interaction chromatography with nano quantity analyte detection, *J. Pharm. Biomed. Anal.*, submitted.

64. Liu Z, Yin G, Gong Y, Zhao N. Application of HPLC-ELSD in impurity test of 1-deoxynojirimycin, glucose and glucosamine. *Fenxi Shiyanshi* 2007; **26**: 83–86.

65. Karlsson G, Hinz A, Winge S. Determination of the stabilizer sucrose in a plasma-derived antithrombin process solution by hydrophilic interaction chromatography with evaporative light-scattering detection. *J. Chromatogr. Sci.* 2004; **42**: 361–365.

66. Karlsson G, Winge S, Sandberg H. Separation of monosaccharides by hydrophilic interaction chromatography with evaporative light scattering detection. *J. Chromatogr. A* 2005; **1092**: 246–249.

67. Hu F, Yuan H, Ying X, Du Y, Tian Z, Ye W. Manufacture of powder injection containing vancomycin hydrochloride, patent Application CN 2006-10049829 20060314, 2006.

68. Dousa M, Brichac J, Gibala P, Lehnert P. Rapid hydrophilic interaction chromatography determination of lysine in pharmaceutical preparations with fluorescence detection after postcolumn derivatization with *o*-phtaldialdehyde. *J. Pharm. Biomed. Anal.* 2011; **54**: 972–978.

69. Langrock T, Czihal P, Hoffmann R. Amino acid analysis by hydrophilic interaction chromatography coupled on-line to electrospray ionization mass spectrometry. *Amino Acids* 2006; **30**: 291–297.

70. Bhandare P, Madhavan P, Rao B, Rao N. Determination of arginine, lysine and histidine in drug substance and drug product without derivatisation by HILIC column LC technique. *J. Chem. Pharm. Res.* 2010; **2**(5): 580–586.

71. Hao Z, Lu C, Xiao B, Weng N, Parker B, Knapp M, Ho C. Separation of amino acids, peptides and corresponding amadori compounds on a silica column at elevated temperatures. *J. Chromatogr. A* 2007; **1147**: 165–171.

72. Hellmuth C, Koletzko B, Peissner W. Aqueous normal phase chromatography improves quantification and qualification of homocysteine, cysteine and methionine by liquid chromatography-tandem mass spectrometry. *J. Chromatogr. B* 2011; **879**: 83–89.

73. Ali I, Aboul-Enein Y. Role of polysaccharides in chiral separations by liquid chromatography and capillary electrophoresis. In Subramanian G, ed., *Chiral Separation Techniques: A Practical Approach*, 3rd ed. Weinheim, Germany: Wiley; 2007, pp. 29–97.

74. Risley DS, Strege MA. Chiral separations of polar compounds by hydrophilic interaction chromatography with evaporative light scattering detection. *Anal. Chem.* 2000; **72**: 1736–1739.

75. Guisbert AL, Sharp VS, Peterson JA, Risley DS. Enantiomeric separation of an AMPA antagonist using a Chirobiotic™ T column with HPLC and evaporative light-scattering detection. *J. Liq. Chromatogr. Relat. Technol.* 2000; **23**: 1019–1028.

76. Jin HL, Stalcup AM, Armstrong DW. Separation of cyclodextrins using cyclodextrin bonded phases. *J. Liq. Chromatogr* 1988; **11**: 3295–3304.

77. Aboul-Enein HY, Ali I. *Chiral Separations by Liquid Chromatography and Related Technologies*. New York: Marcel Dekker; 2003.

78. Wang C, Jiang C, Armstrong DW. Considerations on HILIC and polar organic solvent-based separations: Use of cyclodextrin and macrocyclic glycopeptide stationary phases. *J. Sep. Sci.* 2008; **31**: 1980–1990.

79. Chang SC, Reid GL III, Chen S, Chang CD, Armstrong DW. Evaluation of new polar-organic high-performance liquid chromatographic mobile phase for cyclodextrin-bonded chiral stationary phases. *Trends Anal. Chem.* 1993; **12**: 144–153.

80. Tang Y, Zukowski J, Armstrong DW. Investigation on enantiomeric separations of fluorenymethoxycarbonyl amino acids and peptides by high-performance liquid chromatography using native cyclodextrins as chiral stationary phases. *J. Chromatogr. A* 1996; **743**: 261–271.

81. Richards DS, Davidson SM, Holt RM. Detection of non-UV-absorbing chiral compounds by high performance liquid chromatography-mass spectrometry. *J. Chromatogr. A* 1996; **746**: 9–15.

82. Han SM, Armstrong DW. Use of microcolumn liquid chromatography with a chiral stationary phase for the separation of low-resolution enantiomers. *J. Chromatogr.* 1987; **389**: 256–260.

83. Seeman JI, Secor H, Armstrong DW, Timmons KD, Ward TJ. Enantiomeric resolution and chiral recognition of racemic nicotine and nicotine analogs by β-cyclodextrin complexation. Structure-enantiomeric resolution structure relationships in host-guest interactions. *Anal. Chem.* 1988; **60**: 2120–2127.

84. Armstrong DW, Tang Y, Chen S, Zhow Y, Bagwill C, Chen JR. Macrocyclic antibiotics as a new class of chiral selectors for liquid chromatography. *Anal. Chem.* 1994; **66**: 1473–1484.

85. Advanced Separation Technologies Inc. *Chirobiotic Handbook*, 5th ed. Whippany, NJ, USA 2004.

86. Xiao LX, Armstrong DW. Enantiomeric separations by HPLC using macrocyclic glycopeptides-based chiral stationary phases. In Gübitz G, Schmid MG, eds.,

Methods in Molecular Biology, Chiral Separations: Methods and Protocols. Totowa, NJ: Humana Press; 2004, pp. 113–171.

87. Cavazzini A, Nadalini G, Dondi F, Gasparrini F, Ciogli A, Villani C. Study of mechanisms of chiral discrimination of amino acids and their derivatives on a teicoplanin-based chiral stationary phase. *J. Chromatogr. A* 2004; **1031**: 143–158.

88. Peter A, Torok G, Armstrong DW. High-performance liquid chromatographic separation of enantiomers of unusual amino acids on a teicoplanin chiral stationary phase. *J. Chromatogr. A* 1998; **793**: 283–296.

89. Peter A, Lazar L, Fulop F, Armstrong DW. High-performance liquid chromatographic enantioseparation of β-amino acids. *J. Chromatogr. A* 2001; **926**: 229–238.

90. Beesley TE, Lee JT. Method development strategy and applications update for Chirobiotic chiral stationary phases. *J. Liq. Chromatogr. Relat. Technol.* 2009; **32**: 1733–1767.

91. Ekborg-Ott KH, Wang X, Armstrong DW. Effect of selector coverage and mobile phase composition on enantiomeric separations with ristocetin A chiral stationary phases. *Michrochem. J.* 1999; **62**: 26–49.

92. Ekborg-Ott KH, Liu Y, Armstrong DW. Highly enantioselective HPLC separations using the covalently bonded macrocyclic antibiotic, ristocetin A, chiral stationary phase. *Chirality* 1998; **10**: 434–483.

93. Peter A, Arki A, Tourwe D, Forro E, Fulop F, Armstrong DW. Comparison of separation efficiencies of Chirobiotic T and TAG columns in the separation of unusual amino acids. *J. Chromatogr. A* 2004; **1031**: 159–170.

94. Sztojkov-Ivanov A, Lazar L, Fulop F, Armstrong DW, Peter A. Comparison of separation efficiency of macrocyclic glycopeptide-based chiral stationary phases for the LC enantioseparation of β-amino acids. *Chromatographia* 2006; **64**: 89–94.

95. Peter A, Torok G, Armstrong DW, Toth G, Tourwe D. High-performance liquid chromatographic separation of enantiomers of unusual amino acids on a teicoplanin chiral stationary phase. *J. Chromatogr. A* 2000; **904**: 1–15.

96. Torok G, Peter A, Armstrong DW, Tourwe D, Toth G, Sapi J. Direct chiral separation of unnatural amino acids by high-performance liquid chromatography on a ristocetin A-bonded stationary phase. *Chirality* 2001; **13**: 648–656.

97. Chen S, Pawlowaka M, Armstrong DW. HPLC enantioseparation of di- and tripeptides on cyclodextrin bonded stationary phases after derivitization with 6-aminoquinolyl-*N*-hydroxysuccinimidyl carbamate (AQC). *J. Liq. Chromatogr.* 1994; **17**: 483–497.

98. Sharp VS. Solatol HPLC chiral separation comparison–Polar ionic and HILIC modes (chirobiotic T HPLC column), 2011. Unpublished results.

99. Hyun MH. Enantiomer separation by chiral crown ether stationary phases. In Subramanian, G, ed., *Chiral Separation Techniques: A Practical Approach*, 3rd ed. Weinheim, Germany: Wiley; 2007, pp. 275–299.

100. Hyun MH, Cho YJ, Kim JA, Jin JS. Preparation and application of a new modified liquid chromatographic chiral stationary phase based on (+)-(18-crown-6)-2,3,11,12-tetracarboxylic acid. *J. Chromatogr. A* 2003; **984**: 163–171.

101. Hyun MH, Jin JS, Lee W. Liquid chromatographic resolution of racemic amino acids and their derivatives on a new chiral stationary phase based on crown ether. *J. Chromatogr. A* 1998; **822**: 155–161.

102. Hyun MH, Min HJ, Cho YJ. Resolution of tocainide and its analogs on liquid chromatographic chiral stationary phases based on (+)-(18-crown-6)-2,3,11,12-tetracarboxylic acid. *Bull. Korean Chem. Soc.* 2003; **24**: 911–915.

103. Hyun MH, Cho YJ, Jin JS. Liquid chromatographic direct resolution of β-amino acids on a chiral crown ether stationary phase. *J. Sep. Sci.* 2002; **25**: 648–652.

104. Hyun MH, Han SC, Cho YJ, Jin JS, Lee W. Liquid chromatographic resolution of gemifloxacin mesylate on a chiral stationary phase derived from crown ether. *Biomed. Chromatogr.* 2002; **16**: 356–360.

105. Hyun MH, Jin JS, Han SC, Cho YJ. The effect of analyte lipophilicity on the resolution of α-amino acids on a HPLC chiral stationary phase based on crown ether. *Microchem. J.* 2001; **70**: 205–209.

106. Hyun MH, Jin JS, Koo HJ, Lee W. Liquid chromatographic resolution of racemic amines and amino alcohols on a chiral stationary phase derived from crown ether. *J. Chromatogr. A* 1999; **837**: 75–82.

107. Hyun MH. Characterization of liquid chromatographic chiral separation on chiral crown ether stationary phases. *J. Sep. Sci.* 2003; **26**: 242–250.

108. Sun P, Wang C, Breitbach ZS, Zhang Y, Armstrong DW. Development of new HPLC chiral stationary phases based on native and derivatized cyclofructans. *Anal. Chem.* 2009; **81**: 10215–10226.

109. Sigma-Aldrich Website. Cyclofructan-based chiral HPLC columns, 2012. http://www.sigmaaldrich.com/analytical-chromatography/hplc/columns/chiral/cyclofructan-csps.html (accessed October 5, 2011).

110. Sun P, Armstrong DW. Effective enantiomeric separations of racemic primary amines by the isopropyl carbamate-cyclofructan6 chiral stationary phase. *J. Chromatogr. A* 2010; **1217**: 4904–4918.

111. Sharp VS. 1-(2-Naphthyl)ethylamine chiral separation comparison–Polar ionic and HILIC modes (CF6 HPLC column), 2011. Unpublished results.

CHAPTER

5

HYDROPHILIC INTERACTION CHROMATOGRAPHY (HILIC) FOR DRUG DISCOVERY

ALFONSO ESPADA

Analytical Technologies Department,
Centro de Investigacion Lilly S.A., Madrid, Spain

MARK STREGE

Analytical Sciences Research and Development, Lilly Research Laboratories,
A Division of Eli Lilly and Company, Indianapolis, IN

5.1 DRUG DISCOVERY MODEL

By definition, drug discovery is a multistep and interactive process by which new drugs are designed and/or discovered. The key steps of the drug discovery process comprise target identification/validation, biological screening, lead identification, lead optimization, and candidate selection. The synergy between medicinal chemists, biologists, pharmacologists, bioanalysts (as part of drug disposition and absorption, distribution, metabolism, excretion [ADME]/Tox), and computational scientists has enabled great improvements in the drug discovery flow path over the last years, facilitating the introduction of new technologies with the main goal of delivering synthetic small molecules that have the potential to become beneficial therapeutic drugs [1]. Throughout the

Hydrophilic Interaction Chromatography: A Guide for Practitioners, First Edition.
Edited by Bernard A. Olsen and Brian W. Pack.
© 2013 John Wiley & Sons, Inc. Published 2013 by John Wiley & Sons, Inc.

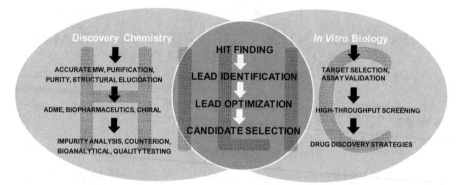

Figure 5.1. Schematic representation of a drug discovery model and the applications of HILIC across the chemistry and biology processes. MW, molecular weight.

long history of drug discovery, the overall efficiency of drug research has been, to a great extent, enabled by a single instrumental technique, liquid chromatography, and more recently by the interface of liquid chromatography with mass spectrometry (LC-MS) [2]. At present, high-performance liquid chromatography (HPLC) and LC-MS-based methodologies have become walk-up tools in all the key steps of drug discovery (including compound identification, quantitation, and purification), addressing needs for screening biology, structure–activity relationship (SAR), and physicochemical property profiling. Recent advances in the field of column chemistry technology are making possible the enhancement of the value of information derived from LC-MS assays for lead compound series characterization [3]. In this context, hydrophilic interaction chromatography (HILIC) is the latest mode of LC that is gaining widespread attention as an orthogonal separation technique to classical reversed-phase LC in the drug discovery arena. This chapter reviews the diverse applications of HILIC to meet the fundamental analytical requirements involved in drug discovery (see Fig. 5.1).

5.2 HILIC APPLICATIONS FOR *IN VITRO* BIOLOGY

5.2.1 Biological Screening and Hit Finding

The advent of applied genomic technologies has marked the beginning of an era in drug discovery biology with a plethora of new potential targets derived from the biological space or "target universe." The expected drug target pool from the target universe can be estimated by approximately 600 drug targets for small molecules and 1800 drug targets for protein therapeutics [4]. The size of the chemical universe is expected to be $>10^{200}$ compounds and the chemical space of low molecular mass is in the range of 10^{50}–10^{100} compounds. Therefore, the opportunity to find agents that are active against these targets is invaluable. Once a potential target is identified, its role in disease must be clearly defined throughout a validation process before it is used to screen large numbers of compounds in the search for new drug candidates [1,4,5].

5.2.1.1 Target Selection and Assay Validation

The comprehensive analysis of metabolic pathways in a biological system, well known as metabolomics, is becoming one of the most intense research areas in discovery biology. Metabolomics approaches have led to the wide scale measurement of vast numbers of endogenous metabolites in complex biological samples with the end focus on the identification of new potential biological targets. Although many of those metabolites are water-soluble compounds, the samples are typically analyzed by reversed-phase LC-MS, and consequently often with poor retention and poor MS ionization. In this scenario, the benefits of HILIC-MS are significant. As has been mentioned in previous chapters, the organic-rich mobile phases used in HILIC often result in increasing retention of extremely polar compounds and increased MS response, which can lead to lower limits of detection. For example, Lu et al. employed HILIC-MS and reversed-phase LC-MS to investigate the intracellular metabolic changes of >150 metabolites during tumorigenesis in metastasis cell lines [6]. The HILIC method was performed on an aminopropyl column at basic pH and coupled to an MS triple quadrupole (MS/MS) with positive mode electrospray ionization (ESI). The resolution power of the HILIC method and consequently the high sensitivity of the MS allowed studying the metabolic change accompanying tumor progression in pathways including glycolysis, the tricarboxylic acid cycle, and pentose phosphate, fatty acid, and nucleotide biosynthesis. The results of these studies suggest possible biomarkers of breast cancer and the opportunities to identify potential cancer-drug targets through the targeting of metabolic pathways (i.e., glycolysis and oxidative phosphorylation) [7,8]. Cases in point are pyruvate kinase (PK) and lactate dehydrogenase (LDH), which are important enzymes in glucose metabolism, and their inhibition seems to have antitumor effects [9,10]. A schematic representation of the screening assays used to measure PK activity is depicted in Figure 5.2.

PK inhibitors were identified by coupled PK and LDH assays employing luminescence and fluorescence detection methods [10]. However, the advent of new column technology and stationary phases has expanded the applications of HILIC for *in vitro* biology. Thus, we have developed HILIC-MS/MS-based methods to accurately detect pyruvate, lactate, and the cofactors NADH and NAD$^+$ in enzymatic and cell-based assays. The separation of the pair NADH/NAD$^+$ is illustrated in Figure 5.3. Although both analytes were baseline resolved in the XBridge Amide and ACE C18-PFP columns with ammonium formate buffer and acetonitrile (ACN; see Fig. 5.3A,B), superior chromatographic resolution was achieved on a perfluorinated column XSelect

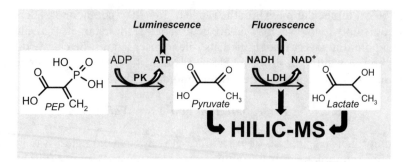

Figure 5.2. Schematic representation of the coupled reactions of pyruvate kinase (PK) and lactate dehydrogenase (LDH). Two biochemical techniques are used to measure PK activity, whereas HILIC-MS is employed to measure PK and LDH activities. ADP = adenosyl diphosphate; ATP = adenosyl triphosphate; NAD⁺ = nicotinamide adenine dinucleotide; NADH = NAD⁺ (reduced form); PEP = phosphoenolpyruvate. Reprinted and modified from Reference 10 with permission from Elsevier.

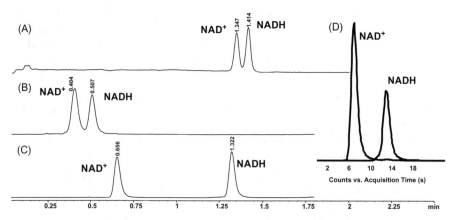

Figure 5.3. HILIC-ESI⁺-MS separation of the hydrophilic compounds NADH and NAD⁺. Conditions for (A): XBridge Amide 2.1 × 30 mm, 3.5 μm column. Solvent A: 10 mM ammonium formate + 5 mM formic acid in 50% ACN, solvent B: ACN. Gradient elution from 90% to 40% B in 1.5 min at a flow rate of 1.0 mL/min. Conditions for (B): ACE C18-PFP, 2.1 × 30 mm, 3 μm column. Solvent A: 10 mM ammonium formate + 5 mM formic acid, solvent B: 10 mM ammonium formate in MeOH, Gradient elution from 0% to 50% B in 1.5 min at a flow rate of 1.0 mL/min. Conditions for (C): XSelect PFP 2.1 × 30 mm, 3.5 μm column. Solvents as in B. Gradient elution from 5% to 70% B in 1.5 min at a flow rate of 1.0 mL/min. Conditions for (D): XSelect PFP 2.1 × 10 mm, 3.5 μm column. Solvents as in B. Gradient elution from 5% to 70% B in 18 s at a flow rate of 1.6 mL/min.

with mixtures of ammonium formate and methanol with fast gradients (see Fig. 5.3C). Remarkably, by switching from a 2.1 × 30 mm column size to 2.1 × 10 mm, an ultrafast gradient with analysis time below 25 s was achieved with excellent resolution and sensitivity (see Fig. 5.3D).

Chromatogram A in Figure 5.4 corresponds to the analysis of pyruvate, lactate, and ¹³C3-lactate internal standard (lactate-IS) on an XBridge Amide

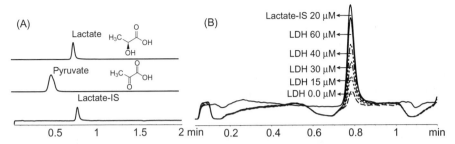

Figure 5.4. (A) HILIC-ESI¯MS separation of the water-soluble compounds lactate, pyruvate, and lactate-IS. (B) Measurement of lactate levels from biochemical reactions. Conditions: XBridge Amide 2.1 × 30 mm, 3.5 μm column. Solvent A: 10 mM ammonium formate + 5 mM formic acid in 50% ACN, solvent B: ACN. Gradient elution from 90% to 40% B in 1.5 min at a flow rate of 1.0 mL/min.

column. This HILIC method was employed to measure levels of lactate in biochemical reactions on an LC-MS/MS system where different concentrations of LDH were tested (see Fig. 5.4B). The examples shown here confirm the suitability of HILIC to deal with hydrophilic metabolites where reversed-phase LC fails. The aforementioned methods were found to be suited to validate LC-MS/MS assays to measure enzymatic activity with acceptable sample throughput [11].

5.2.1.2 High-Throughput Screening (HTS)

A typical approach for a synthetic organic therapeutic molecule begins with the screening of thousands of compounds by high-throughput assays. The objective of the HTS campaigns is to identify chemical hits from which lead series may be derived [12]. Compound bank collections (including synthetic organic compounds collections, combinatorial chemistry libraries, and isolated natural products) in the pharmaceutical industry often exceed 1 million compounds. In this situation, LC-MS-based methodologies represent the most valuable analytical technology for the accurate analysis of these collections of diverse compounds in terms of purity, identity, and quantity with high speed and resolution. However, LC-MS has also been proven to be an essential tool within screening campaigns [3]. Given that screening methods mainly rely on radiolabel or fluorescence assays, LC-MS (a label-free technology) has been recently used to interrogate enzymatic targets to detect active compounds (exhibiting acceptable potency and selectivity) without labeling of coupling reagents [13]. Zhang et al. has recently described an approach using an LC-MS/MS-based assay in a large screening campaign of approximately 800,000 compounds for inhibitors of human diacylglycerol acyltransferase [14]. The screening results showed good assay quality and hit-finding capacity. The LC method was on a reversed-phase Betabasic C4 2.1 × 20 mm column with ACN/water and ammonium formate as mobile phase. Li et al. described the use of a HILIC bridged ethyl-silica hybrid (BEH) amide 2.1 × 50 mm, 1.7 μm column with ammonium formate buffer for an ultra-performance liquid chromatography mass spectrometry (uPLC-MS)-based galactose-1-phosphate uridylyltransferase and galactokinase assays. The quantification of the enzymatic products $^{13}C_6$-uridine diphosphate galactose and $^{13}C_6$-galactose-1-phosphate was carried out in just 3 min with HILIC-MS/MS [15]. This HILIC-uPLC method eliminated the use of ion-pairing reagents to make the assay more robust and suitable for routine use. Although the method was developed as a diagnostic tool in biochemical genetics laboratories, this HILIC approach may be considered as an alternative to biochemical and/or reversed-phase LC-MS/MS screening assays for metabolic and signaling pathways [16].

5.2.2   New Drug Discovery Strategies

In addition to the "-omics" approaches and HTS, the pharmaceutical industry has room to improve the traditional target-directed drug discovery efforts. Fragment-based screening is one of the drug discovery strategies that has emerged as a promising source of new lead structures [17]. Fragments are compounds with a molecular weight of between 100 and 300 Daltons, with a limited number of functionalities in comparison with other small molecules. Therefore, they are considered substructures of drug-like compounds with certain affinity for individual protein epitopes (see Fig. 5.5 for examples). Many of these fragment substances are water-soluble compounds, often with weak chromophores. Therefore, their accurate qualitative/quantitative analysis by conventional reversed-phase LC columns is a difficult task. In this scenario, HILIC is becoming the method of choice.

Phenotypic lead generation is another strategy that is evolving in the pharmaceutical industry. The goal of this paradigm is to discover compounds that modulate complex, physiologically relevant biological processes that may lead to the discovery of new pathways of therapeutic value. From this perspective, Lilly launched the phenotypic drug discovery (PD^2) initiative where external research groups can submit compounds for testing in a panel of phenotypic screening assays in the therapeutic areas of metabolic and neurological disorders, as well as oncology [18]. Significantly, this initiative allows Lilly access to a vast number of compounds exhibiting novel structures with diverse physicochemical properties. In order to confirm the quality (compound purity, identity, quantity, and chemical stability) of diverse sets of compounds from academia and biotech companies worldwide, accurate reversed-phase LC-MS methodology is the right tool to deliver such relevant data in a timely fashion. Additionally, orthogonal methods to reversed-phase, such as HILIC, can be developed easily due to the utilization of a similar solvent system and applied to this strategy to minimize the risk of delivering inconsistent data. Chromatograms in Figure 5.6 illustrate the advantage of running orthogonal LC-MS methods to confirm compound purity and quality of a hydrophilic natural product sample. Analysis under reversed-phase conditions with a gradient designed for the retention of very polar compound yielded sufficient

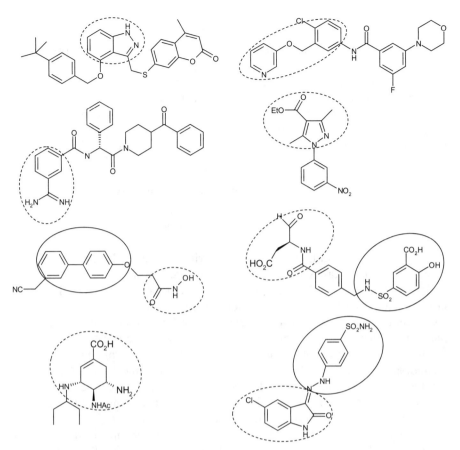

Figure 5.5. Examples of lead compounds derived from fragments approaches with the original starting fragments shown in solid and dashed circles. Reprinted and adapted from Reference 17 with permission from Elsevier.

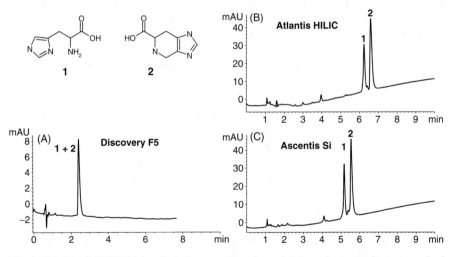

Figure 5.6. The LC/UV (214 nm) chromatograms of a natural product sample composed of spinacin (1) and histidine (2). Conditions A: Discovery F5 4.6 × 50 mm, 3.5 μm column. Solvent A: 0.05% trifluoroacetic acid in H_2O, solvent B: 0.05% trifluoroacetic acid in ACN. Gradient elution from 0 to 50% B in 6 min at a flow rate of 1.0 mL/min., Conditions B and C: Atlantis HILIC and Ascentis Si 4.6 × 100 mm, 5 μm columns, Solvent A: 0.05% trifluoroacetic acid in H_2O, solvent B: 0.05% trifluoroacetic acid in ACN. Gradient from 95% to 70% B in 10 min at a flow rate of 1.0 mL/min.

retention time of a single peak with a relative purity level of >97%. Conversely, two peaks were detected in a 4:6 ratio when the sample was analyzed under a HILIC LC-MS separation mode with an acetonitrile-rich mobile phase, highlighting the major powerful application of HILIC to the precise and accurate qualitative/quantitative measurement of water-soluble compounds.

5.3   HILIC APPLICATIONS FOR DISCOVERY CHEMISTRY

5.3.1   Lead Identification

As HTS hits are identified, a set of activities collectively known as lead identification (LI) takes place in an effort to confirm the hit and gain an understanding of the influence of structural changes on the biological activity. Dose–response curve generation is performed through the testing of several compound concentrations using the same assay. Orthogonal testing also takes place, as confirmed hits are analyzed using a different activity assay, which is usually closer to the target physiological condition, or by employing a different technology. Biophysical analysis techniques are commonly used to assess whether the compound binds effectively to the target, the stoichiometry of binding, the presence of any associated conformational change, and to identify nonspecific inhibitors. Through this work, a structure–activity relationship (SAR) is established, and LC-MS is used to support chemical synthesis and purification of the compound. During LI, biological models follow a progression from *in vitro* assays to cell culture screens (membrane permeability is usually a critical parameter) and ultimately to live animal models. Much effort is expended to understand these models prior to the next stage of drug discovery, lead optimization (LO). Assessing drug exposure in the *in vivo* pharmacology models also begins at this stage, based on the throughput provided by LC-MS methods. Initial investigations of the biotransformation of the lead molecules also occur during LI to identify metabolically labile positions in the molecules of interest and the potential for reactive (i.e., potentially toxic) metabolites.

The ruggedness and broad applicability of atmospheric pressure ionization (API) sources for mass spectrometry revolutionized multiple aspects of drug discovery during the 1990s. The concept of "open access" analysis, either through flow injection analysis or by LC-MS, provided a means for synthetic organic chemists to rapidly analyze their own samples without having to go through the work of submission to a separate analytical laboratory [2,19]. In the area of natural products, through LC-MS, chemists gained a powerful tool for the profiling of complex sample extracts and for rapidly determining a chemical "fingerprint" that could be associated with biological activity in HTS [20]. This approach has also been proven successful for evaluation of biologically active compounds in mixtures generated by synthetic combinatorial chemistry [21]. To accommodate wide structural diversity, gradient elution

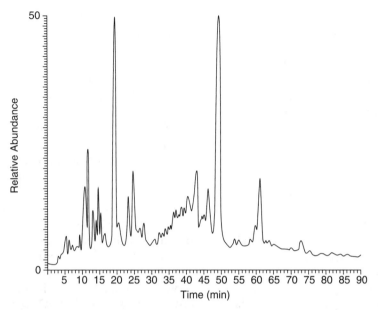

Figure 5.7. HILIC-ESI-MS separation of the polar components (unretained by reversed-phase solid-phase extraction pretreatment) of a fermentation extract, represented as a total ion chromatogram in positive ion ESI. The chromatography was obtained using a TSKgel Amide-80 packing, 6.5 mM ammonium acetate pH 5.5-buffered mobile phases, and a 90-min 10–40% aqueous gradient. Reprinted with permission from Strege MA. *Anal. Chem.* 1998; **70**(13): 2439. Copyright 1998 American Chemical Society.

became the default approach for use with reversed-phase HPLC, and HILIC has been demonstrated to be a powerful tool for LC-MS analyses of polar compounds that could not be analyzed by reversed phase [22]. Figure 5.7 shows a HILIC-MS separation of the components of a natural product extract that were not retained by a reversed-phase solid phase extraction cartridge.

Starting with HTS, as samples progress through the batteries of tests used in drug discovery, increasing amounts of material are needed for analysis. During the process of LI, milligram quantities of purified compounds are required, and these needs can typically be met through the use of semi-preparative chromatography columns. Over the years, purification systems consisting of columns integrated with a variety of detectors (including UV, MS, and evaporative light scattering detection [ELSD]) useful for selective triggering of fraction collection have been built and optimized, and are now commercially available [3,23,24]. Drug discovery purification support, however, remains a challenging task due to large sample loads and short turnaround time. The ability to rapidly scale up separations from analytical to preparative columns is critical for meeting timelines. Samples that present difficulty for purification include those that demonstrate minimal retention on reversed-phase packings due to high polarity, and those that demonstrate minimal separation by reversed phase due to structurally similar impurities such as diastereomers and regioisomers [24–26]. Among other approaches, Yan et al. have successfully employed HILIC to meet these specific purification challenges [26]. In this work, a prepacked HILIC preparative column was used with 0.1% formic acid mobile phases and an aqueous gradient in the presence of acetonitrile to isolate up to 450 mg of a polar compound from a synthetic mixture with an 82% yield at a purity of 98%. Boyman et al. found that the integration of a HILIC separation step within a purification procedure for a low molecular weight inhibitor of cardiac sodium–calcium exchanger (from calf ventricle tissue) was necessary for achieving acceptable purity of the isolated compound [27]. Along with HILIC, supercritical fluid chromatography (SFC) represents a newer technological advance in the field of analytical science with a significant impact in the analysis and purification of drug discovery compounds [28]. SFC uses carbon dioxide (CO_2) as the mobile phase, mixed with a solvent (known as a modifier) to facilitate the separation of compounds based on a normal-phase retention mechanism that occurs directly between the analytes and the hydrophilic stationary phase. Therefore, HILIC stationary phases can also typically be used for SFC, and the opposite is also true. Although several solvents have been evaluated in SFC and in some cases successfully, methanol is by far the most generic solvent in SFC. The use of polar stationary phases (i.e., silica, cyano, and diol phases) is favored by the nonpolar character of CO_2. SFC-MS technology has also been recently integrated in our purification platform as part of the Lilly green chemistry commitment. In this scenario, a new Luna HILIC column (cross-linked diol phase with ethylene bridges) has proven to be well suited to facilitate the small-scale purification of drug discovery compounds [29]. The elution gradient of polar modifier (from 5% to 50%) in CO_2 on this diol polar phase allows the elution of most moderate and polar compounds. A representative example is shown in Figure 5.8 in which good reproducibility was obtained from analytical to preparative SFC-MS on this packing. Within this application, the MS collection was triggered to the peak exhibiting a retention time of 2.2 min and an m/z 442 [M+H] in the analytical chromatogram. To increase mass loading (up to 100 mg/run) and to ensure good recovery and purity, a rapid gradient was applied to the preparative conditions.

The determination of compound purity is critical for successful LI, and in fact can often be the rate-limiting step during the early phases of drug discovery [30]. For example, during SAR elucidation, a detailed assessment of the influence of chemical structure upon biological activity occurs, and the presence of impure compounds can result in inaccurate conclusions. Because of the importance of compound purity, most pharmaceutical companies have adopted guidelines to govern the level of purity required as a function of the stage of drug discovery [31]. As a second example, natural product extract libraries for HTS may be composed of samples that are themselves complex mixtures (crude extracts containing 10–100 components or more), prefractionated extracts (targeted to be semi-pure mixtures of 5–10 compounds), or single purified biologically active compounds. Each library design presents

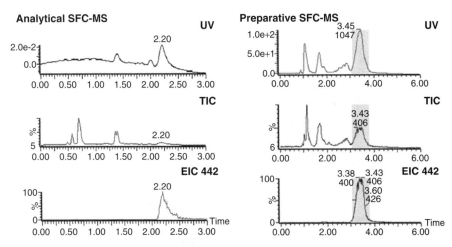

Figure 5.8. SFC-MS chromatograms of a crude mixture containing a lead compound. Analytical conditions: Luna HILIC-diol 2.1 × 150 mm, 5 μm column. Solvent A: CO_2, solvent B: 0.2% dimethylethylamine in MeOH, and gradient elution from 5% to 40% B in 2.5 min at a flow rate of 6 g/min. Preparative conditions: Luna HILIC-diol 30 × 150 mm, 5 μm column, solvent A and B as in the analytical conditions, and gradient elution from 25% to 40% B in 5.5 min at a flow rate of 100 g/min. TIC, total ion chromatogram; EIC = extracted ion chromatogram. Courtesy of C. Anta.

advantages and limitations in terms of analytical resources and requirements for the identification of bioactive constituent requirements [30]. At a minimum, a mass-based determination provided by "universal detection" instrumentation such as an ELSD, a chemiluminescence nitrogen detector (CLND), or a charged aerosol detector (CAD) is typically used in conjunction with LC-MS and ultraviolet (UV) absorbance detection for the estimation of purity [31–33]. In a specific example, Marin et al. examined optimization of the speed and effectiveness of the use of orthogonal methods for routine purity assessment and characterization [32]. In this study, chromatographic orthogonality was achieved through the use of different reversed-phase HPLC packings and mobile phase pH. To exploit the inherent orthogonality of the two chromatographic methods, reversed-phase and HILIC have also been coupled online to provide enhanced specificity for pharmaceutical analyses [34–36]. Figure 5.9 demonstrates the separation power of HILIC by providing reversed-phase and HILIC profiles of a microbial extract fraction enriched on a preparative column. Other multidimensional applications of HILIC are described in Chapter 9 of this book.

Regardless of the source of the chemical diversity that is subjected to HTS during drug discovery, the rapid and accurate elucidation of structure is another key factor of success. Samples generated by both combinatorial chemistry synthesis and natural product extraction often exist initially as complex mixtures from which individual chemical structures must be determined.

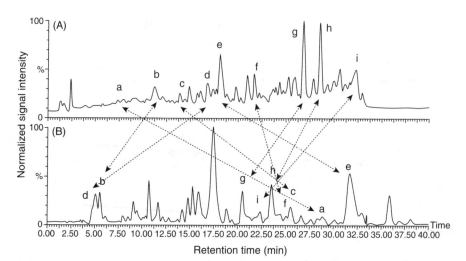

Figure 5.9. Separations of a semipurified microbial fermentation extract sample obtained using (A) RP-HPLC-ESI-MS and (B) ACE-HILIC-ESI-MS as represented by total negative ion mass spectrometry signal chromatograms. Nine individual sample components were tracked via their unique ESI-MS spectra as indicated by the dashed lines for the demonstration of separation orthogonality. Reprinted with permission from Strege MA et al. *Anal. Chem.* 2000; **72**: 4629–4633. Copyright 2000 American Chemical Society.

Powerful hyphenated LC, MS, and nuclear magnetic resonance (NMR) platforms are useful for online structural elucidation, as high-throughput alternatives to the traditional approach of compound isolation and characterization [30,37,38]. Automated database searching against libraries of UV, MS, NMR, and also Fourier transform infrared (FTIR) spectra is often used to facilitate rapid identification of compounds [39,40]. The integration of HILIC as part of these hyphenated systems has been demonstrated to provide full access of these chemical structure identification tools to polar compounds in mixtures that are not otherwise amenable to approaches based on reversed-phase chromatography [22,36].

5.3.2   Lead Optimization

In the drug discovery process, the final stage leading to the selection of a candidate is LO, which in many ways serves as an expanded version of LI in that it also follows a series of testing activities appropriate for the project. During LO, a greater degree of emphasis is placed on *in vivo* models for efficacy, including the determination of pharmacokinetic, pharmacodynamic, and toxicokinetic information for the lead compound. ADME characteristics of the molecule are exploited to gain an understanding of the factors that influence drug clearance and the potential for drug–drug interactions, and LC-MS is used to determine the major metabolic routes. Material supply is very important during LO, and traditional synthetic chemistry is used to produce the larger amounts of material required to perform toxicokinetic investigations and other compound-intensive studies. At this point in the process, increased attention is also given to the identification and qualification of biomarkers that can be used in the clinic for the prediction of biological outcomes (see also Section 5.3.3).

5.3.2.1 ADME Profile

The ADME drug metabolism profile demonstrates the process by which the body, primarily the liver, breaks down and converts a drug into other chemical substances known as metabolites. Metabolites are then bound to other substances for excretion from the body. Through a transition in strategy that took place during the 1990s, the fact that ADME properties are now investigated much earlier in the drug discovery process relative to what had been done in the past has resulted in a significant drop in clinical attrition of drug candidates. This process improvement has been largely attributed to the advent of high-throughput analytical chemistry techniques, specifically LC-MS, which have enabled routine method development for compounds of high structural diversity [2]. HILIC has played a significant role in this achievement, as is described in more detail in Chapter 7 of this book.

5.3.2.2 Biopharmaceutics

Biopharmaceutics is the study of the relationships between physical and chemical properties, dosage, and administration of a drug and its activity in humans and animals. In addition to the information generated through *in vitro* ADME profiling, the screening of key physiochemical properties such as solubility, lipophilicity, pK_a, and drug permeability provides important data needed for the evaluation of the compound of interest [41]. In many cases LC-UV-vis and LC-MS have served as important tools for support of these screening studies beyond simple offline quantitation of analytes [30].

Lipophilicity, expressed in various ways such as the logarithm of octanol–water partition coefficient (log P_{oct}) or distribution coefficient at a given pH (log D^{pH}), is one of the most important parameters to be determined at the early stages of drug discovery because it is a key descriptor of drug-like molecular properties that can influence other physiochemical and ADME properties. Chromatographic methods, which can facilitate the analysis of large numbers and low quantities of compounds with simplicity, high-throughput capacity, and excellent reproducibility, are now considered the preferred alternatives to the traditional shake-flask methodology [42–44]. Some limitations of this approach have been found to exist, however, including secondary interactions with the surface of the stationary phases and the incompatibility of the stationary phases with the high pH conditions necessary for the assessment of basic compounds. As an alternative, Kadar et al. demonstrated a linear relationship between log D^{pH} and the HILIC retention (log k) for a series of

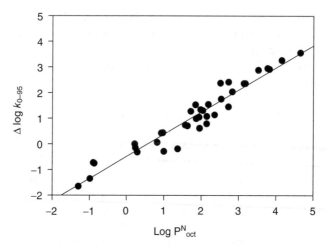

Figure 5.10. Relationship between Δlog $k_{0\text{-}95}$ (mobile phase = acetonitrile–buffer trifluoroacetic acid–ammonium pH 2 [I = 100 mM]) and log P_{oct}^N. Reprinted with permission from Bard B et al. *J. Med. Chem.* 2009; **52**: 3416–3419. Copyright 2009 American Chemical Society.

representative pharmaceutical compounds under generic HILIC conditions [45]. Bard et al. used a ZIC-HILIC stationary phase to successfully measure the log P_{oct} of the neutral form of basic drugs, including beta-blockers and local anesthetics, by measuring the difference between the retention (log k) of the cationic forms of the compounds under two dramatically different isocratic conditions. In order to do so, however, it was necessary to strictly control the composition of the mobile phase (pH, buffer nature, and ionic strength) [46]. At low concentrations of acetonitrile in the mobile phase (0% ACN), the log k measurement reflected reversed-phase-like hydrophobic retention, whereas at high acetonitrile concentration (95% ACN), log k was based on hydrophilic retention. Figure 5.10 shows the strong correlation between a measurement of lipophilicity in units of Δlog $k_{0\text{-}95}$, and the log $P_{octanol}$ values from the chemical literature.

The ionization state of a drug molecule, pK_a (dissociation constant), is a fundamental parameter for understanding and designing drug-like compounds with beneficial physicochemical properties and ADME profiles. Potentiometric titration has been the technique traditionally used within the pharmaceutical industry for the determination of pK_a, but it requires relatively large amounts (milligrams) of highly pure material of a compound possessing high aqueous solubility. Reversed-phase LC-MS has been used to determine the pK_a of a compound through the chromatography of the compound across a range of mobile phase pH [47]. Using this approach, plotting retention time versus pH enabled the identification of the pK_a. In theory, HILIC may also be used in this manner; however, one drawback associated with the use of any chromatographic technique for this application is the complexity of interpretation of the behavior of multifunctional compounds. Another methodology that has shown significant promise for the determination of pK_a is capillary electrophoresis (CE)-MS [48].

Although most drugs cross biological barriers by transcellular pathways where the main physical barrier is the lipid matrix of the membranes, octanol–water lipophilicity measurements have not proven sufficient for prediction of drug permeation. Therefore, several liquid chromatography approaches have been proposed to predict permeability through the three main biological membranes (intestinal, blood–brain, and percutaneous barriers) as an alternative to traditional cell-based assays, based on the addition of amphiphilic components in either the stationary phase or the mobile phase. These models have included the use of immobilized artificial membranes (IAM) [49], immobilized liposome chromatography (ILC) [48], and biopartitioning micellar chromatography (BMC) [50,51]. These methodologies generate reasonably accurate predictions, but suffer from long retention time, specifically for lipophilic compounds. Because HILIC retention is based on hydrophilicity and not lipophilicity, the technique may demonstrate future application in this area.

5.3.2.3 Chiral Purity

Chiral analysis is an essential undertaking throughout drug discovery and development since the enantiomeric forms of a chiral drug substance often possess completely different pharmacological effects, with one enantiomer producing a potent, desired effect and the other being inactive or even toxic. The achievement and measurement of chiral purity becomes very significant during LO, and both analytical and preparative chromatography techniques, including HILIC, are key tools. As an example, the separation of pharmaceutical compound enantiomers has been demonstrated through the use of stationary phases derivatized with chiral selectors and operated in HILIC mode [52]. For a detailed summary of the use of HILIC for chiral purification and analysis of chiral purity, see Chapter 4 of this book.

5.3.3 Candidate Selection

Candidate selection (CS) is the event of formal approval for the allocation of resources to gather the necessary data to allow Phase 1 testing in human subjects for the lead candidate after sufficient merit has been demonstrated through LO. After a new drug candidate passes CS, it formally enters the clinical development pipeline. During the work process from CS to Product Decision (investment in resources for Phase 3 clinical testing), it is important to have phase-appropriate formulation and analytical strategies to balance technical challenges, aggressive development timelines, and limited resources. At CS, the goal is to develop a dosage form for the candidate with sufficient physical and chemical stability for the first human dose (FHD) studies with the shortest development time and limited quantities of drug substance. The time from CS to FHD is dependent on several factors, including bulk drug availability, resource allocations, and the degree of development difficulty for a specific compound. Within this broad area of focus, HILIC has demonstrated great application and impact.

For HPLC methods used in pharmaceutical analysis, the specificity (or "selectivity" or "stability indicating ability") is a critical attribute that must be thoroughly investigated and demonstrated through method development and validation. Specificity is the ability of the method to separate and detect known, unknown, and potential impurities (including degradation products), and is especially important for methods intended for early-phase drug development following CS when the chemical and physical properties of the drug substance are not fully understood and the synthetic processes are not fully in control. Therefore, the assurance of patient safety in clinical trials of a candidate relies heavily on the ability of analytical methods for the detection and quantitation of unknown impurities that may pose safety concerns.

Investigations have demonstrated that HILIC is a practical alternative to reversed-phase and ion-pair chromatography for impurities determination. Hmelnickis et al. developed a method using a zwitterionic sulfobetaine stationary phase to determine six impurities of mildronate drug substance [53]. Strege et al. developed a purity method for cytosine (a starting material for drug substance synthesis) using an Amide 80 column [54], and an impurity method using a diol packing was developed to serve as a complement to reversed-phase for the impurities analysis of a proprietary compound in pharmaceutical development by Wang et al. [55]. All three methods were validated in terms of specificity, limit of detection and quantitation, linearity, accuracy, and precision, as per the International Conference on Harmonization (ICH) guidelines [56]. HILIC-ESI-MS positive ion total ion chromatograms of cytosine and two impurities are shown in Figure 5.11, demonstrating the specificity provided by the method. A stability-indicating HILIC method was also developed and validated for the quantitative determination of brimonidine tartrate formulated as an ophthalmic solution [57]. In this study, the drug was subjected to oxidative, hydrolytic, photolytic, and thermal stress conditions, and complete separation was achieved for the parent compound and degradation products. There is currently much focus in the pharmaceutical industry on the control of genotoxic impurities in drug substances and drug products. As an example of genotoxic impurities analysis, alkyl esters resulting from the use of alchohols as solvents for crystallization and sulfonic acid salt forms of drug substances were determined at trace levels by An et al. using HILIC interfaced with LC-MS [58]. For a more detailed description of the applications of HILIC in the area of pharmaceutical impurities analysis, see Chapter 4 of this book.

During the development activities taking place after CS, selection of an appropriate and stable salt form of a new drug substance serves to improve the physicochemical properties and water solubility that will favor drug bioavailability and assist in the formulation of a suitable dosage form. Therefore, inorganic ions and organic acids and bases are often employed as counterions to prepare salts of ionizable drug substances. Analytical methods are required

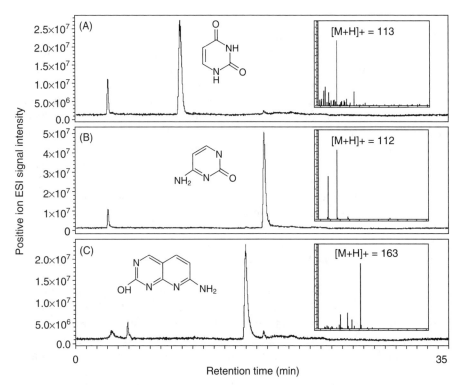

Figure 5.11. HILIC-ESI-MS positive ion total ion chromatograms (TICs) of (A) uracil (MW 112), (B) cytosine (MW 111), and (C) 7-ADOP (MW 162), with structures and extracted mass spectra taken from the center of each peak. The separation was obtained using an Amide 80 packing and mobile phases containing 0.2% formic acid pH 3.5 (MPA = water, MPB = acetonitrile). The gradient conditions consisted of a 5-min hold at 5% aqueous, followed by a 10-min linear gradient to 25% aqueous, followed by a 5-min hold at 25% aqueous.

to confirm production of the correct salt form and stoichiometry of each batch of drug substance manufacture. Among other methodologies [59], the application of HILIC for the determination of counterions of pharmaceuticals has been reported in the literature using an aminopropyl stationary phase and refractive index detection [60], a zwitterionic stationary phase and evaporative light scattering detection (ELSD) [61] or charged aerosol detection (CAD) [62], and a monolithic stationary phase with ELSD [63]. For a more detailed description of the applications of HILIC in the area of drug substance counterion analysis, see Chapter 4 of this book.

The term "bioanalysis" pertains to the quantitative analysis of drugs and their metabolites in biological matrices such as plasma, and is a foundational aspect supporting the *in vivo* studies that occur throughout the drug discovery process, especially following CS. Examples of key *in vivo* studies include the determination of an ADME profile, metabolomics (as described in Section 5.2.1), and determination of biomarkers (substances that are evaluated as indicators of normal biological processes, pathogenic processes, or pharmaco-logic responses to therapeutic interventions) [64–68]. HILIC, specifically in LC-MS mode, has made a great impact in these areas of focus, and a number of comprehensive reviews describing the HILIC applications have been published [69–72]. For a more detailed description of the applications of HILIC in the area of bioanalysis, the reader is also directed to Chapter 7 of this book.

Near the conclusion of drug discovery and development, a manufacturing process for the drug product is established and analytical methods are incorporated into quality control (QC) laboratories for support. A number of HILIC methods have been developed and validated for QC testing purposes for a variety of specific applications, as reviewed by Dejaegher et al. [73,74]. For a more detailed description of the applications of HILIC for QC testing, the reader is also directed to Chapter 4 of this book.

5.4   PRACTICAL CONSIDERATIONS

The analysis of polar and/or hydrophilic small molecules from hit finding to CS depends heavily on the resolving power of HPLC and LC-MS methodologies to efficiently identify characterize and quantify a vast numbers of samples and compounds. An essential part in this process is the development of orthogonal methods where the correct selection of the stationary phase, organic solvent/water content, buffer type, mobile phase pH, and other chromatographic parameters have important effects on the successful applications of HILIC to support the increasing analytical needs for *in vitro* biology and discovery chemistry. From a practical perspective, HILIC retention can be achieved using virtually any stationary phase demonstrating polar character, ranging from simple underivatized silica to complex functionalities such as cyclodextrins and macrocyclic compounds anchored to silica particles [52]. It is important to keep in mind that depending on the nature of the stationary phase used for HILIC, multimodal separations can occur (including HILIC combined with ion exchange or chiral separations, for example, as discussed in Chapter 1).

For most drug discovery applications, neutral and charged stationary phases (e.g., amide, diol, amino, etc.) are a good choice to enable the analysis and measurement of hydrophilic compounds from biological samples as well as enzymatic and cell-based assays. These phases have received much attention from pharmaceutical researchers since many new and robust stationary phases and column formats (diverse particle size and dimensions) have become commercially available in the last few years. The availability of robust stationary phases and a growing knowledge base facilitates method development and the selection of the optimal column chemistry in a systematic manner in a short time [75]. For example, an LC-MS method with a BEH amide column with acetonitrile as a weak solvent, 10 mM of ammonium acetate buffer at the desired pH value (depending on the nature of the analytes) and fast generic gradients is a general good starting point. In order to attain different retention and selectivity, Luna NH2 and/or Luna HILIC (diol) can be used as a reasonable substitute. It is worth mentioning that BEH amide and Luna NH2 columns can be operated in a wide pH range (pH 2–10). Upon selection of the column chemistry, by systematically varying acetonitrile/water content, mobile phase pH, and buffer type (ionic strength), successful separations are achieved. To ensure retention and reproducibility, the use of acetonitrile concentrations greater than 70% of the total mobile phase composition is recommended. Additionally, highly orthogonal phases to those conventionally used for HILIC, such as fluorinated stationary phases, should be taken into consideration, particularly when analyzing basic polar compounds [76]. An example is shown in Figure 5.3 where the HILIC method worked well, but the reversed-phase method with a fluorinated column was still superior for the pair NADH/NAD$^+$.

Compound purity assessment typically demands more sophisticated separations to cover the wide diversity of hydrophilic compounds encountered in the drug discovery arena. To achieve this goal, chromatographers have focused on HILIC as an alternative to reversed phase, where the combination of neutral, charged, and zwitterionic phases have proven to be quite successful in method development [77]. As discussed in Section 5.3, several papers have been published describing the use of underivatized silica, diol, Amide 80, ZIC-HILIC, and zwitterionic sulfobetaine packings in various discovery applications. Among these phases, bare silica and Amide 80 columns remain popular for HILIC, particularly in the QC and purification areas. Combining these stationary phases with a scouting gradient of decreasing acetonitrile from 95% to 50% is a recommended strategy that allows chromatographers to determine quickly the optimal HPLC and/or LC-MS conditions for a desired HILIC application. At this point, if adequate separation is achieved but superior resolution between peaks is still needed, it is advisable to reduce the gradient slope by changing the ending gradient concentration (e.g., 95% to 70% or 95% to 90% acetonitrile) or switching in a straightforward manner to isocratic conditions. Adjusting buffer or additive concentration and ionic strength can alter retention, selectivity, and sensitivity, and must therefore be selected with caution. It is also important to note that previous reports concluded that when analytes are in their ionized form they exhibit the greatest retention as well as higher MS signal response [75]. Therefore, maintaining an effective buffer concentration and ionic strength in the mixed mobile phase is of paramount importance for maximum reproducibility and ruggedness.

5.5 CONCLUSIONS

In this chapter, the current impact and future potential of HILIC for pharmaceutical drug discovery has been described. The subjects considered in this report have included both *in vitro* biological applications and applications supporting discovery chemistry with an emphasis on the stages of lead identification, lead optimization, and candidate selectionCS. Practical considerations from the author's own laboratory on the key chromatographic parameters to be successful on the implementation and application of HILIC for various discovery purposes have been briefly discussed.

REFERENCES

1. Bunnage ME. Getting pharmaceutical R&D back on target. *Nat. Chem. Biol.* 2010; **7**: 335–339.
2. Ackermann BL, Berna M, Eckstein JE, Ott LW, Chauddhary AK. Current applications of liquid chromatography/mass spectrometry in pharmaceutical discovery after a decade of innovation. *Annu. Rev. Anal. Chem. (Palo Alto Calif.)* 2008; **1**: 357–396.
3. Espada A, Molina-Martin M, Dage J, Kuo M-S. Application of LC/MS and related techniques to high-throughput drud discovery. *Drug Discov. Today* 2008; **13**: 417–423.
4. Betz UAK. How many genomic targets can a portfolio afford? *Drug Discov. Today* 2005; **10**: 1057–1063.
5. Russ AP, Lampel S. The druggable genome: An update. *Drug Discov. Today* 2005; **10**: 1607–1610.
6. Lu X, Bennet B, Mu E, Rabinowitz J, Kang Y. Metabolic changes accompanying transformation and acquisition of metastatic potential in a syngeneic mouse mammary tumor model. *J. Biol. Chem.* 2010; **285**: 9317–9321.
7. Kaelin WG, Thompson CB. Clues from cell metabolism. *Nature* 2010; **465**: 562–564.
8. Cairn RA, Harris IS, Mak TW. Regulation of cancer cell metabolism. *Nat. Rev. Cancer* 2011; **11**: 85–95.
9. Christofk HR, Vander Heiden MG, Harris MH, Ramanathan A, Gerszten RE, Wei R, et al. The M2 splice isoform of pyruvate kinase is important for cancer metabolism and tumor growth. *Nature* 2008; **452**: 230–233.
10. Vander Heiden MG, Christofk HR, Schuman E, Subtelny AO, Sharfi H, Harlow EE, Xian J, Cantley LC. Identification of small molecule inhibitors of pyruvate kinase M2. *Biochem. Pharmacol.* 2010; **79**: 1118–1124.
11. Espada A., unpublished results.
12. Macarron R, Banks MN, Bojanic D, Burns DJ, Cirovic DA, Garyantes T, Green DV, Hertzberg RP, Janzen WP, Paslay JW, Schopfer U, Sittampalam GS. Impact of high-throughput screening in biomedical research. *Nat. Rev. Drug Discov.* 2011; **10**: 188–195.
13. Roddy TP, Horvath CR, Stout ST, Kenney KL, Ho P-I ZJ-H, Vickers C, Kaushik V, Hubbard B, Wang YK. Mass spectrometry techniques for label-free high-throughput screening in drug discovery. *Anal. Chem.* 2007; **79**: 8207–8213.
14. Zhang J-H, Roddy TP, Ho P-I, Horvath CR, Vickers C, Stout S, Hubbard B, Wang YK, Hill WA, Bojanic D. Assay development and screening of human DGAT1inhibitors with an LC/MS-based assay: Application of mass spectrometry for large-scale primary screening. *J. Biomol. Screen.* 2010; **15**: 695–702.

15. Li Y, Ptolemy AS, Harmonay L, Kellogg M, Berry GT. Ultra fast and sensitive liquid chromatography tandem mass spectrometry based assay for galactose-1-phosphate uridyltransferase and galactokinase deficiencies. *Mol. Genet. Metab.* 2011; **102**: 33–40.

16. Dang L, White DW, Gross S, Bennet BD, Bittinger MA, Driggers EM, Fantin VR, Jang HG, Jin S, Keenan MC, Marks KM, Prins RM, Ward PS, Yen KE, Liau LM, Rabinowitz JD, Cantley LC, Thompson CB, Vander Heiden MG, Su SM. Cancer-associated IDH1 mutations produce 2-hydroxyglutarate. *Nature* 2009; **462**: 739–744.

17. Carr RAE, Congreve M, Murray CW, Rees DC. Fragment-based lead discovery: Leads by design. *Drug Discov. Today* 2005; **10**: 987–992.

18. Lee JA, Chu S, Willard FS, Cox KL, Galvin RJS, Peery RB, Oliver SE, Oler J, Meredith TD, Heidler SA, Gough WH, Husain S, Palkowitz AD, Moxham CM. Open innovation for phenotypic drug discovery: The PD^2 assay panel. *J. Biomol. Screen.* 2011; **16**: 588–602.

19. Mallis LM, Sarkahian AB, Kulishoff JM, Watts WL. Open access liquid chromatography-mass spectrometry in a drug discovery environment. *J. Mass Spectrom.* 2002; **37**: 889–896.

20. Strege MA. High performance liquid chromatographic-electrospray ionization mass spectrometric analyses for the integration of natural products with modern high-throughput screening. *J. Chromatogr. B* 1999; **725**: 67–78.

21. Peake DA, Duckworth DC, Perun TJ, Scott WL, Kulanthaivel P, Strege MA. Analytical and biological evaluation of high throughput screen actives using evaporative light scattering, chemiluminescent nitrogen detection, and accurate mass HPLC-MS-MS. *Comb. Chem. High Throughput Screen.* 2005; **8**: 477–487.

22. Strege MA. Hydrophilic interaction chromatography-electrospray mass spectrometry analysis of polar compounds for natural product drug discovery. *Anal. Chem.* 1998; **70**(13): 2439–2445.

23. Edwards C, Liu J, Smith TJ, Brooke D, Hunter DJ. Parallel preparative high performance liquid chromatography with on-line molecular mass characterization. *Rapid Commun. Mass Spectrom.* 2003; **17**: 2027–2033.

24. Fitzgibbons J, Op S, Hobson A, Schaffter L. Novel approach to optimization of a high-throughput semipreparative LC/MS system. *J. Comb. Chem.* 2009; **11**: 592–597.

25. Espada A, Marin A, Anta C. Optimization strategies for the analysis and purification of drug discovery compounds by reversed-phase high-performance liquid chromatography with high-pH mobile phases. *J. Chromatogr. A* 2004; **1030**(1–2): 43–51.

26. Yan TQ, Bradow J, Chang S, Depianta R, Philippe L. Approaches to singleton achiral purification of difficult samples for discovery research support. *LC-GC N. Am.* 2009; **27**(4): 340–345.

27. Boyman L, Hiller R, Shpak B, Yomtov E, Sphak C, Khananshvili D. Advanced procedure for separation and analysis of low molecular weight inhibitor (NCX_{IF}) of the cardiac sodium-calcium exchanger. *Biochem. Biophys. Res. Commun.* 2005; **337**: 936–943.

28. Weller HN, Ebinger K, Bullock W, Edinger KJ, Hermsmeier MA, Hofmman SL, Nirschl DS, Swann T, Zhao J, Kiplinger J, Lefebvre P. Orthogonality of supercritical

fluid chromatography versus HPLC for small molecule library separation. *J. Comb. Chem.* 2010; **12**: 877–882.

29. Lopez-Soto P, Anta C, de la Puente ML. HILIC-Diol as a generic stationary phase for achiral SFC, Abstract of papers. SFC2010-the 4th International Conference on Packed Column SFC 2010; September 15–16, Stockholm, Sweden.

30. Nicoli R, Martel S, Rudaz S, Wolfender J, Veuthey J, Carrupt P, Guillarme D. Advances in LC platforms for drug discovery. *Expert Opin. Drug Discov.* 2010; **5**(5): 475–489.

31. Lawrence K. The plight of purity. *Drug Discov. Dev.* 2009; **12**(7): 30–34.

32. Marin A, Burton K, Rivera-Sagredo A, Espada A, Byrne C, White C, Sharman G, Goodwin L. Optimization and standardization of liquid chromatography-mass spectrometry systems for the analysis of drug discovery compounds. *J. Liq. Chromatogr. Relat. Technol.* 2008; **31**: 2–22.

33. Riley J, Everatt B, Aldcroft C. Implementation of charged aerosol detection in routine reversed phase liquid chromatography methods. *J. Liq. Chromatogr. Relat. Technol.* 2008; **31**: 3132–3142.

34. Liu Y, Xue X, Guo Z, Xu Q, Zhang F, Liang X. Novel two-dimensional reversed-phase liquid chromatography/hydrophilic interaction chromatography, an excellent orthogonal system for practical analysis. *J. Chromatogr. A* 2008; **1208**: 133–140.

35. Louw S, Pereira A, Lynen F, Hanna-Brown M, Sandra P. Serial coupling of reversed-phase and hydrophilic interaction liquid chromatography to broaden the elution window for the analysis of pharmaceutical compounds. *J. Chromatogr. A* 2008; **1208**: 90–94.

36. Strege MA, Stevenson S, Lawrence SM. Mixed-mode anion-cation exchange/hydrophilic interaction chromatography-electrospray mass spectrometry as an alternative to reversed phase for small molecule drug discovery. *Anal. Chem.* 2000; **72**: 4629–4633.

37. Wolfender J, Queiroz E, Hostettmann K. The importance of hyphenated techniques in the discovery of new lead compounds from nature. *Expert Opin. Drug Discov.* 2006; **1**(3): 237–260.

38. Chen G, Pramanik BN, Liu Y, Mirza UA. Applications of LC/MS in structure identifications of small molecules and proteins in drug discovery. *J. Mass Spectrom.* 2007; **42**(3): 279–287.

39. Urban S, Separovic F. Developments in hyphenated spectroscopic methods in natural product profiling. *Front. Drug Des. Discov.* 2005; **1**: 113–166.

40. Sashidhara KV, Rosaiah JN. Various dereplication strategies using LC-MS for rapid natural product lead identification and drug discovery. *Nat. Prod. Commun.* 2007; **2**(2): 193–202.

41. Wan J, Holmen AG. High throughput screening of physicochemical properties and *in vitro* ADME profiling in drug discovery. *Comb. Chem. High Throughput Screen.* 2009; **12**: 315–329.

42. Poole SK, Poole CF. Separation methods for estimating octanol-water partition coefficients. *J. Chromatogr. B* 2003; **797**: 3–19.

43. Nasal A, Kaliszan R. Progress in the use of HPLC for evaluation of lipophilicity. *Curr. Comput. Aided Drug Des.* 2006; **2**: 327–340.

44. Benhaim D, Grushka E. Effect of *n*-octanol in the mobile phase on lipophilicity determination by reversed phase high performance liquid chromatography on a modified silica column. *J. Chromatogr. A* 2008; **1209**: 111–119.

45. Kadar EP, Wujcik CE, Wolford DP, Kavetskaia O. Rapid determination of the applicability of hydrophilic interaction chromatography utilizing ACD Labs Log D Suite: A bioanalytical application. *J. Chromatogr. B* 2008; **863**: 1–8.

46. Bard B, Carrupt P, Martel S. Lipophilicity of basic drugs measured by hydrophilic interaction chromatography. *J. Med. Chem.* 2009; **52**: 3416–3419.

47. Wan H, Ulander J. High throughput pK_a screening and prediction amenable for ADME profiling. *Expert Opin. Drug Metab. Toxicol.* 2006; **2**: 139–155.

48. Wan H, Holmen AG, Wang YD, Lindbergh W, Englund M, Nagard MB, Thompson RA. High throughput screening of pKa values of pharmaceuticals by pressure-assisted capillary electrophoresis and mass spectrometry. *Rapid Commun. Mass Spectrom.* 2003; **17**: 2639–2648.

49. Taillardat-Bertschinger A, Carrupt PA, Barbato F, Testa B. Immobilized artificial membrane HPLC in drug research. *J. Med. Chem.* 2003; **46**: 655–665.

50. Escuder-Gilabert L, Martinez-Pla JJ, Sagrado S, Villanueva-Camañas RM, Medina-Hernandez MJ. Biopartitioning micellar separation methods: Modeling drug absorption. *J. Chromatogr. B* 2003; **797**: 21–35.

51. Molero-Monfort M, Escuder-Gilabert L, Villanueva-Camanas RM, Sagrado S, Medina-Hernandez MJ. Biopartitioning micellar chromatography: An *in vitro* technique for predicting human drug absorption. *J. Chromatogr. B* 2001; **753**: 225–236.

52. Risley DS, Strege MA. Chiral separations of polar compounds by hydrophilic interaction chromatography with evaporative light scattering detection. *Anal. Chem.* 2000; **72**(8): 1736–1739.

53. Hmelnickis J, Pugovics O, Kazoka H, Visna A, Susinskis I, Kokums K. Application of hydrophilic interaction chromatography for simultaneous separation of six impurities of mildronate substance. *J. Pharm. Biomed. Anal.* 2008; **48**: 649–656.

54. Strege MA, Durant C, Boettinger J, Fogarty M. A hydrophilic interaction chromatography method for the purity analysis of cytosine. *LC-GC N. Am.* 2008; **26**(7): 632–642.

55. Wang X, Li W, Rasmussen HT. Orthogonal method development using hydrophilic interaction chromatography and reversed-phase high-performance liquid chromatography for the determination of pharmaceuticals and impurities. *J. Chromatogr. A* 2005; **1083**: 58–62.

56. ICH Quality Guidelines. Topic Q2 (R1): Validation of Analytical Procedures: Text and Methodology, 2005. Available from: http//www.ich.org/products/guidelines/quality/article/quality-guidelines.html.

57. Ali MS, Khatri AR, Munir MI, Ghori M. A stability-indicating assay of brimonidine tartrate ophthalmic solution and stress testing using HILIC. *Chromatographia* 2009; **70**(3–4): 539–544.

58. An J, Sun M, Bai L, Chen T, Liu DQ, Kord A. A practical derivatization LC/MS approach for determination of trace level alkyl sulfonates and dialkyl sulfates genotoxic impurities in drug substances. *J. Pharm. Biomed. Anal.* 2008; **48**: 1006–1010.

59. Rocheleau M. Analytical methods for determination of counter-ions in pharmaceutical salts. *Curr. Pharm. Anal.* 2008; **4**: 25–32.

60. Guo Y, Huang A. A HILIC method for the analysis of tromethamine as the counterion in an investigational pharmaceutical salt. *J. Pharm. Biomed. Anal.* 2003; **31**: 1191–1201.

61. Risley DS, Pack BW. Simultaneous determination of positive and negative counterions using a hydrophilic interaction chromatography method. *LCGC N. Am.* 2006; **24**: 776–785.

62. Huang Z, Richards MA, Zha Y, Francis R, Lozano R, Ruan J. Determination of inorganic pharmaceutical counterions using hydrophilic interaction chromatography coupled with a Corona® CAD detector. *J. Pharm. Biomed. Anal.* 2009; **50**: 809–814.

63. Pack BW, Risley DS. Evaluation of a monolithic silica column operated in the hydrophilic interaction chromatography mode with evaporative light scattering detection for the separation and detection of counterions. *J. Chromatogr. A* 2005; **1073**: 269–275.

64. Arnold JN, Saldova R, Galligan MC, Murphy TB, Mimura-Kimura Y, Telford JE, Godwin AK, Rudd PM. Novel glycan biomarkers for the detection of lung cancer. *J. Proteome Res.* 2011; **10**(4): 1755–1764.

65. Shen G, Chen Y, Sun J, Zhang R, Zhang Y, He J, Tian Y, Song Y, Chen X, Abliz Z. Time-course changes in potential biomarkers detected using a metabonomic approach in Walker 256 tumor-bearing rats. *J. Proteome Res.* 2011; **10**(4): 1953–1961.

66. Lin L, Huang Z, Gao Y, Yan X, Xing J, Hang W. LC-MS based serum metabonomic analysis for renal cell carcinoma diagnosis, staging, and biomarker discovery. *J. Proteome Res.* 2011; **10**(3): 1396–1405.

67. Scherer M, Schmitz G, Liebisch G. Simultaneous quantification of cardiolipin, bis(monoacylglycero)phosphate and their precursors by hydrophilic interaction LC-MS/MS including correction of isotopic overlap. *Anal. Chem.* 2010; **82**(21): 8794–8799.

68. Mohamed R, Varesio E, Ivosev G, Burton L, Bonner R, Hopfgartner G. Comprehensive analytical strategy for biomarker identification based on liquid chromatography coupled to mass spectrometry and new candidate confirmation tools. *Anal. Chem.* 2009; **81**(18): 7677–7694.

69. Hsieh Y. Hydrophilic interaction liquid chromatography-tandem mass spectrometry for drug development. *Curr. Drug Discov. Technol.* 2010; **7**: 223–231.

70. Onorato JM, Langish R, Bellamine A, Shipkova P. Applications of HILIC for targeted and non-targeted LC/MS analyses in drug discovery. *J. Sep. Sci.* 2010; **33**: 923–929.

71. Jian W, Edom RW, Xu Y, Weng N. Recent advances in application of hydrophilic interaction chromatography for quantitative bioanalysis. *J. Sep. Sci.* 2010; **33**: 681–697.

72. Spagou K, Tsoukali H, Raikos N, Gika H, Wilson ID, Theodoridis G. Hydrophilic interaction chromatography coupled to MS for metabonomic/metabolomic studies. *J. Sep. Sci.* 2010; **33**: 716–727.

73. Dejaegher B, Vander Heyden Y. HILIC methods in pharmaceutical analysis. *J. Sep. Sci.* 2010; **33**: 698–715.

74. Dejaegher B, Pieters S, Vander Heyden Y. Emerging analytical separation techniques with high throughput potential for pharmaceutical analysis, part II: Novel chromatographic modes. *Comb. Chem. High Throughput Screen.* 2010; **13**: 530–547.

75. Fountain KJ, Xu J, Diehl DM, Morrison D. Influence of stationary phase chemistry and mobile-phase composition on retention, selectivity, and MS response in hydrophilic interaction chromatography. *J. Sep. Sci.* 2010; **33**: 740–751.

76. Bicking MKL, Henry RA. A global approach to HPLC column selection using reversed phase and HILIC modes: What to try when C18 doesn't work. *LCGC N. Am.* 2010; **28**: 234–244.

77. Guo Y, Gaiki S. Retention and selectivity of stationary phases for hydrophilic interaction chromatography. *J. Chromatogr. A* 2011; **1218**: 5920–5938.

CHAPTER

6

ADVANCES IN HYDROPHILIC INTERACTION CHROMATOGRAPHY (HILIC) FOR BIOCHEMICAL APPLICATIONS

FRED RABEL

ChromHELP, LLC, Woodbury, NJ

BERNARD A. OLSEN

Olsen Pharmaceutical Consulting, LLC, West Lafayette, IN

6.1 INTRODUCTION

The polar nature of many biochemicals has made hydrophilic interaction chromatography (HILIC) a valuable analytical tool for biochemical research. In fact, much of the development of HILIC has been driven by biochemical applications. Early work on carbohydrate analysis employed what are now considered to be HILIC conditions [1]. Alpert's landmark paper introducing the term HILIC described the separation of amino acids, peptides, carbohydrates, and oligonucleotides [2]. Much work has been performed since Alpert's original paper aimed at understanding HILIC retention mechanisms, developing new or improved HILIC stationary phases, and using HILIC to solve analytical problems. As is apparent from the other chapters in this book, compounds of

Hydrophilic Interaction Chromatography: A Guide for Practitioners, First Edition.
Edited by Bernard A. Olsen and Brian W. Pack.
© 2013 John Wiley & Sons, Inc. Published 2013 by John Wiley & Sons, Inc.

biochemical interest have been involved in all three of the above activities. In this chapter we will discuss the use of HILIC for analysis of biochemical compounds. Some compounds such as amino acids, carbohydrates, nucleobases, vitamins, peptides, and proteins are of interest as components in pharmaceutical or food products, and applications in those areas are described in Chapter 4 and Chapter 8. The complex nature of many biological samples, such as protein digests, has drawn some investigators to two-dimensional (2D) separations where HILIC is used as one of the separation modes. This topic is discussed in depth in Chapter 9. We will focus on recent examples of HILIC for biochemical applications and illustrate the general approach to successful HILIC separations. Leading references will be provided for further consultation. For example, the excellent review by Hemström and Irgum provides many references for biochemical and other applications up to 2006 [3]. The review organized by HILIC stationary phase by Jandera also describes separations of several biomolecules [4]. Additional specific reviews are cited in the following sections.

6.2 CARBOHYDRATES

6.2.1 Mono- and Disaccharides

One of the oldest determinations of mono- and disaccharides using the HILIC mode was cited in 1976 [1]. The authors described analysis on a Partisil PAC (polar amino cyano) column. The acetonitrile:water (80:20) mobile phase might lead one to think the separation mode was reversed-phase. However, increasing the water content of the mobile phase decreased the elution times of the carbohydrates, the opposite of reversed-phase behavior. The separation was at that time described as a normal phase partition mode, similar to the operation of an adsorption separation but with a polar mobile phase. Today this mechanism is recognized as a classic HILIC separation.

Since the early work described above, HILIC has grown in popularity for carbohydrate separations [5]. HILIC has been used to separate monosaccharides that are common in N-linked oligosaccharides in glycoproteins and other compounds [6]. A TSKgel Amide-80 column was eluted with 82% acetonitrile and 18% 5 mM ammonium formate (pH 5.5). Evaporative light scattering (ELSD) was used for detection. An elevated column temperature (60°C) was used to prevent split peaks due to mutarotation (anomer interconversion). In this phenomenon, the interconversion of α and β anomeric forms of reducing sugars is slow enough compared with the chromatographic timescale that the individual forms can be separated. The conversion results in doublet peaks in the chromatograms. The split peaks can be avoided by increasing the rate of anomer interconversion by increasing the column temperature during separation. With this method, L-fucose, D-galactose, D-mannose, N-acetyl-D-glucosamine, N-acetylneuraminic acid, and D-glucuronic acid were separated.

Ikegami et al. described the separation of mono-, di-, and trisaccharides using a polyacrylamide-modified monolithic silica capillary column with mass spectrometric detection [7]. Ricochon et al. abandoned amino stationary phases because of loss of separation and sensitivity attributed to Schiff base formation [8]. They also found that a ZIC-HILIC zwitterionic phase did not separate the large number of monosaccharides they studied. They chose a Polyamine II column (silica with a polymer layer containing secondary and tertiary amine groups) to prevent Schiff base formation while still providing good selectivity. They used a small amount of chloroform added to the acetonitrile portion of the mobile phase (70%) and monitored chloride adducts of the sugars in negative ion atmospheric pressure chemical ionization (APCI) mode.

6.2.2 Oligosaccharides and Polysaccharides

HILIC has also been applied to the separation of oligosaccharides (3–10 saccharide units) and polysaccharides (>10 saccharide units). Brokl et al. investigated a number of HPLC methods for analyzing mixtures of neutral oligosaccharides including various tetra- and pentasaccharides, and maltose saccharides (-triose to -heptose) [9]. One system used an anion exchange column with pulsed amperometric detection (PAD). The other HPLC systems used either reversed-phase, graphitized carbon, or a HILIC (Xbridge BEH [bridged ethyl silica hybrid] Amide column) mode with electrospray ionization mass spectrometry (ESI-MS) detection. They found that both the HILIC and carbon columns gave the best overall separations of these mixtures. HILIC was superior for separation by degree of polymerization while the graphitized carbon columns performed well for isomeric oligosaccharides (see Fig. 6.1). These authors also encountered mutarotation of some oligosaccharides during separation which resulted in split peaks. Instead of increasing the column temperature as mentioned above, these authors used a basic mobile phase to increase the rate of anomer interconversion, thereby avoiding split peaks. Leijdekkers et al. also described the benefits of HILIC compared with other techniques such as RP-HPLC, graphitized carbon chromatography, and capillary electrophoresis [10]. Separation of acidic oligosaccharides with different degrees of polymerization but with the same charge was achieved using a 1.7-μm particle size BEH amide column. Mass spectrometric detection also facilitated structure elucidation, particularly with multiple MS stages (MS^n).

In other work, zwitterionic, polyhydroxyethyl aspartamide, and BEH amide phases were compared for the separation of galactooligosaccharides with MS detection. The amide column with a mobile phase of acetonitrile:water with 0.1% ammonium hydroxide provided the best results [11]. An example of the separation of various oligosaccharides is shown in Figure 6.2.

6.2.3 Glycans

Glycosylation is a common post-translational modification of proteins, and variations of glycosylation can have a significant impact on biological activity. N-glycosylation is the attachment of a glycan (oligo or polysaccharide) through

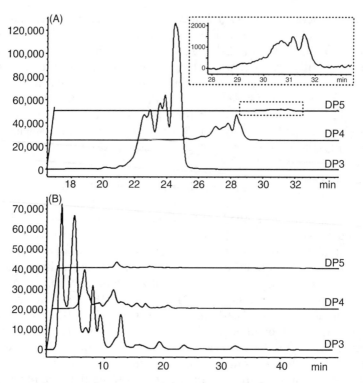

Figure 6.1. Analysis of gentiooligosaccharides (glucose polymers from starch) by (A) HILIC and (B) graphitized carbon chromatography under optimized conditions. Mass spectrometry with selected ion monitoring acquisition of characteristic m/z ratios of tri-, tetra-, and pentasaccharides. DP, degree of polymerization. Reprinted from Reference 9 with permission from Elsevier.

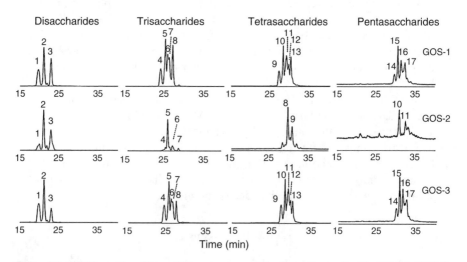

Figure 6.2. HILIC profiles of three different commercial galactooligosaccharide (GOS-1, GOS-2, GOS-3) prebiotic supplement samples separated on a BEH amide column. Reprinted from Reference 11 with permission from Elsevier.

the amide nitrogen of asparagine while *O*-glycosylation takes place through the hydroxyl group of serine or threonine. Characterization of protein glycosylation in biological systems is important in understanding protein function and role in various disease states. The composition, length, and branching of the glycan chain as well as the glycosylation site can influence the overall structure and function of the protein. These glycosylation parameters are also important to understand in proteins used as therapeutic agents [12].

The analysis and characterization of glycans from glycoprotein, proteoglycan, and glycolipid conjugates using a variety of separation techniques has received much attention [13–15]. An overview of protein glycosylation analysis is provided by Mariño et al. [16], and Zauner reviewed advances in HILIC for structural glycomics [17]. Various approaches have been applied for the analysis of glycoproteins depending on the type of information desired. The polar nature of oligosaccharides and glycopeptides produces selective interactions with the polar stationary phases providing application opportunities for HILIC in both sample enrichment/purification and analysis.

6.2.3.1 Glycan and Glycopeptide Analysis

HILIC has been applied to several aspects of glycosylation analysis. In some cases glycans released from glycoproteins are targeted and in other situations separation of intact glycopeptides is of interest. Fluorescence detection of labeled glycans and mass spectrometric analysis after HILIC separation are both widely used. Techniques for release of glycans from glycoconjugates, labeling for detection, and interpretation of results are beyond the scope of this chapter but are obviously critical aspects of the analytical methodology.

Melmer et al. described HILIC as the method of choice for characterizing protein glycans because of the separating power and the straightforward method development HILIC offers [18]. HILIC phases with amide functionality are often used for the separation of glycans that have been released from glycoproteins and subsequently derivatized with 2-aminobenzamide to allow fluorescence detection. Acetonitrile with a gradient of increasing ammonium formate aqueous buffer is typically used as a mobile phase. In general, the oligosaccharides are eluted in order of size. Comparison of retention time with a dextran ladder standard containing linear chains of 2–18 glucose units can be used to estimate the identity of separated glycans. A database has also been constructed to aid in interpretation of HILIC glycan profiles [19,20]. Reversed-phase and porous graphitic carbon separations were compared with HILIC for glycan analysis [21]. HILIC was judged to be the superior general choice for glycans. The different chromatographic modes provided orthogonal selectivity and a combination of techniques with MS detection was recommended for complex samples. Melmer et al. validated a method using a TSKgel Amide-80 column with an ammonium formate/acetonitrile mobile phase to profile 2-aminobenzamide-derivatized glycans from a monoclonal

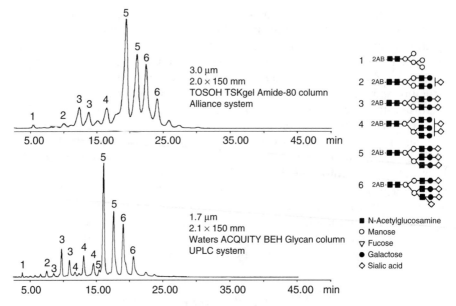

Figure 6.3. Separtion of 2-aminobenzamide (2-AB)-labeled glycans from fetuin using 3.0-μm TSKgel Amide-80 column and 1.7-μm BEH Glycan column. Peaks labeled with the same number indicate isomers. Reprinted from Reference 22 with permission from Elsevier.

antibody [18]. The method was optimized to distinguish pharmacologically relevant glycan variants.

Ahn et al. used a 1.7-μm BEH Glycan column (amide functionality) for analysis of 2-aminobenzamide-labeled glycans with fluorescence detection [22]. Information on use of this column is also available from the supplier [23,24]. The authors compared the performance of the BEH column with that of a 3.0-μm particle TSKgel Amide-80 column (see Fig. 6.3). As expected, the smaller 1.7-μm particle column provided greater efficiency and therefore greater peak capacity. The authors also noted that separation selectivity for glycans changes with initial gradient conditions and gradient slope. Gilar et al. also utilized a 1.7-μm BEH Glycan column for the separation of glycoforms of glycopeptides using MS detection [25]. For complex samples the glycopeptides were initially separated by reversed-phase high-performance liquid chromatography (RP-HPLC).

Bones et al. conducted further work using a BEH Glycan sub-2 μm column for profiling serum N-glycans [26]. They noted decreased analysis time with improved selectivity compared with previous methods. In another paper, Bones et al. found the orthogonal selectivity of HILIC compared with capillary electrophoresis beneficial in characterizing components of a complex oligosaccharide profile from recombinant β-glucuronidase [27].

Takegawa et al. discussed the electrostatic and hydrophilic retention aspects of 2-aminopyridine derivatives of sialic acid-containing N-glycans on

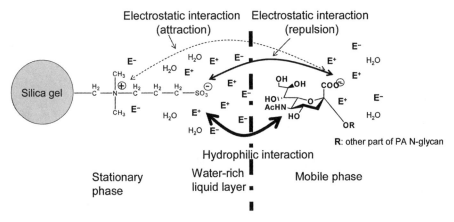

Figure 6.4. Schematic diagram of ZIC-HILIC interactions of sialylated 2-aminopyridine *N*-glycan. Only the sialic acid residue of PA *N*-glycan is represented. Electrostatic (attraction and repulsion) and hydrophilic interactions between the sialic acid and the sulfobetaine group on the surface of the ZIC column are schematically shown. E^+ and E^- are positive- and negative-electrolyte ions (i.e., ammonium and acetate ions) in the eluent, respectively. Reprinted from Reference 28 with permission from Wiley.

a sulfobetaine ZIC-HILIC column [28]. Increased retention of *N*-glycans with negatively charged sialic acid groups with increased electrolyte concentration was attributed to decreased electrostatic repulsion from the negatively charged sulfonate groups on the stationary phase (see Fig. 6.4). A slight increase in retention of neutral *N*-glyans as electrolyte concentration increased could be caused by an increase in the water layer on the stationary phase and greater HILIC partitioning. The authors showed that the retention of sialated *N*-glycans could be tuned by varying the electrolyte concentration.

HILIC with electrospray ionization (ESI)-MS or matrix-assisted laser desorption/ionization time-of-flight mass spectrometry (MALDI-TOF-MS) is a powerful tool for characterization of released glycans and glyopeptides. Wuhrer et al. listed a variety of HILIC stationary phases used with MS analysis including silica-based ion exchange, zwitterionic, and nonionic phases [29]. Silica and amino columns suffer from irreversible adsorption and reactivity for glycan analytes. Amide columns have been used most frequently among the nonionic phases. Most applications used ammonium formate or ammonium acetate buffers and acetonitrile for compatibility with MS detection. MALDI-TOF-MS analysis is often performed on collected fractions from the HILIC separation while ESI-MS is performed in an online mode. Analysis of glycopeptides from glycoprotein digests is conducted to gain information about sites of glycosylation. Mauko et al. described glycan profiling of monoclonal antibodies using a ZIC-HILIC column with ESI-MS detection. The method offered speed and simplicity advantages compared with the more standard amide column separation with fluorescence detection [30].

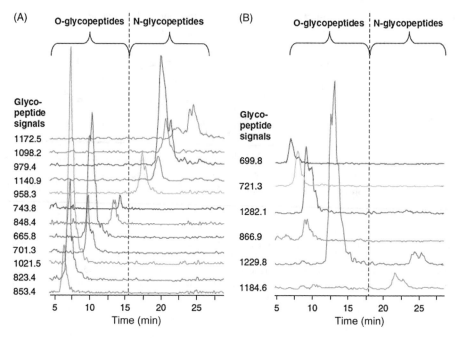

Figure 6.5. (A) HILIC extracted ion chromatograms for seven *O*-glycosylated peptides and five *N*-glycosylated peptides from an asialofetuin Proteinase K digest. (B) HILIC extracted ion chromatograms for five *O*-glycopeptides and two *N*-glycopeptides from a fetuin Proteinase K digest. Reprinted from Reference 31 with permission from Wiley.

Zauner et al. separated N-linked and *O*-linked glycopeptides using a nanoscale Amide-80 column [31]. Retention was controlled primarily by the glycan portion of the glycopeptides and was used to differentiate *N*- or O-linked species as well as glycan size (Fig. 6.5). The retention time information simplified glycopeptide structural assignments from the ESI-MS data.

6.2.3.2   HILIC for Sample Enrichment

Biological samples are often very complex, and it can be difficult to overcome interference or detection sensitivity challenges with direct analysis. Sample pretreatment to enrich glycopeptides of interest has been facilitated with HILIC approaches [32]. An and Cipollo described HILIC as the method of choice for glycopeptide enrichment with advantages over methods such as hydrazide capture, size exclusion chromatography, or lectin affinity chromatography [33]. They demonstrated that enrichment using a TSKgel Amide-80 phase provided signal increases up to 40-fold compared to unenriched samples. Furthermore, the glycan profiles of five model proteins were not biased by the enrichment procedure. Hägglund et al. used ZIC-HILIC particles packed in pipet tip microcolumns to retain hydrophilic glycopeptides while allowing hydrophobic peptides to pass through [34]. The retained species were eluted with 0.5% aqueous formic acid for further analysis. Nettleship et al. used a similar approach for glycopeptide enrichment in the study of N-glycosylation site occupancy in glycoproteins [35]. Yu et al. used amino propyl silica in a 96-well microelution format to purify glycans released from glycoproteins prior to analysis by MALDI-MS [36]. Kondo et al. used a combined hydrophobic–hydrophilic workflow for analysis of glycopeptides derived from β2-glycoprotein samples. A ZIC-HILIC phase was used for the capture of glycopeptides after removal of other components with an RP microcolumn [37]. A ZIC-HILIC microcolumn was also used for enrichment of glycosylphosphatidylinositol-anchored peptides, which were further characterized by MALDI-TOF-MS [38]. Calvano et al. also used a ZIC-HILIC phase in combination with a lectin microcolumn for enrichment of glycoproteins before further MS analysis [39].

6.3   NUCLEOBASES AND NUCLEOSIDES

Pyrimidine and purine nucleobases and their nucleosides have been widely studied using HILIC. Some investigations have used nucleobases and/or nucleotides to investigate retention behavior and mechanisms on HILIC columns (see Chapter 1 and Chapter 2). Analysis of these types of compounds used as drugs is described in Chapter 4. Here we will focus on application of HILIC to separations of these types of compounds in problems of biochemical and related interests.

Chen et al. showed different retention behavior of 14 nucleosides and nucleobases under RP (C18) and HILIC (TSKgel Amide-80) conditions [40]. Although the run time was quite long (110 min), the HILIC method separated guanine and hypoxanthine which were not separated by RP. The method was applied to extracts of two traditional Chinese medicines.

Rodríguez-Gonzalo et al. compared the separation of several methylated and hydroxylated nucleobases and nucleosides on pentafluorophenylpropyl (RP), cross-linked diol (HILIC), and zwitterionic (HILIC) phases [41]. They concluded that the ZIC-HILIC zwitterionic phase using 80% acetonitrile:20% formic acid (2.6 mM) provided the best results, with both hydrophilic partitioning and weak electrostatic interactions contributing to retention and selectivity. The HILIC mode also provided better MS detection sensitivity than RP. The same authors coupled similar HILIC conditions for analysis of nucleosides and nucleobases extracted from urine with online restricted-access material (RAM) [42]. The method was used to differentiate analyte profiles between cancer patients and healthy volunteers.

Johnsen et al. found that a polymeric zwitterionic phase (ZIC-pHILIC) gave better separation of nucleoside triphosphates than the silica-based phase [43]. They also noted that decreasing the column temperature resulted in

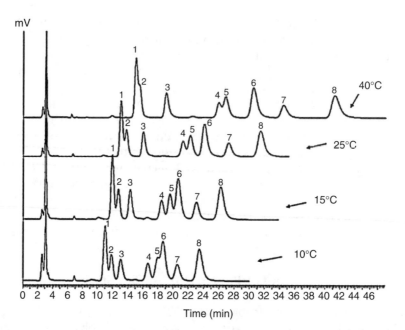

Figure 6.6. Separation of the eight nucleotides in an *Escherichia coli* cell sample (wild-type strain SMG3) on two ZIC-pHILIC columns at 15°C at 100 μL/min. Chromatogram recording began 15.0 min after injection. Isocratic elution was performed with 70/30 (v/v) ACN/(NH$_4$)$_2$CO$_3$ (pH 8.9, 100 mM) at 200 μL/min. The injection volume was 5 μL and the UV absorbance was set to 254 nm. Reprinted from Reference 43 with permission from Elsevier.

decreased retention of the nucleotides (see Fig. 6.6). This retention phenomenon was attributed to a decrease in hydrophilic partitioning brought about by increased clustering of water molecules and buffer ions around the polar triphosphate functional groups as the temperature decreased.

In the development of a method to determine a single nucleotide in liver extracts, Pucci et al. investigated ion-pairing, RP, and HILIC (silica, zwitterionic, and aminopropyl phases) separations [44]. Aminopropyl columns separated the 2′-C-methylcytidine-triphosphate analyte from an endogenous interference, and the acetonitrile/aqueous ammonium acetate mobile phase facilitated sensitive ESI-MS-MS detection. Zhou and Lucy found that phosphate-containing mobile phases along with high acetonitrile concentrations reduced ligand-exchange interactions and promoted HILIC retention for nucleotides on a bare titania (TiO$_2$) column [45]. With a gradient of increasing aqueous content, phosphate concentration, and pH, 15 nucleotides and intermediates could be separated in 26 min. Enhanced fluidity mobile phases prepared by addition of carbon dioxide have also been described for separations of nucleobases and nucleotides [46,47] with amide columns. Methanol was used instead of acetonitrile, and sodium chloride was added to increase retention and selectivity [47].

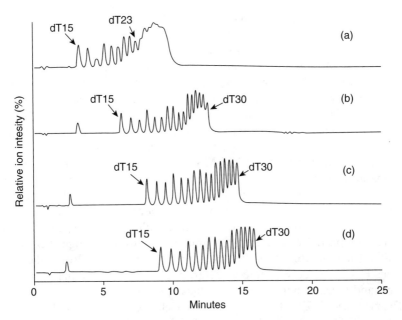

Figure 6.7. Separation of oligonucleotides dT15-30. PEEK ZIC®-HILIC, 100 × 2.1 mm, 3.5 μm column. Mobile phase A: Milli-Q H2O; B: acetonitrile; C: 100 mM ammonium acetate, pH 5.8. Gradient from 70% to 60% B in 15 min, with constant C 5% (a), 10% (b), 15% (c) and 20% (d), flow rate, 0.6 mL/min but only 0.2 mL/min split to MS, temperature, 50°C; 10 picomole each injected. Reprinted from Reference 48 with permission from Elsevier.

6.4 OLIGONUCLEOTIDES

Oligonucleotides can also be separated using HILIC methods. Gong and McCullagh separated 20-mer oligonucleotides with single nucleotide resolution using a ZIC-HILIC column [48]. Different length nucleotides could also be separated as shown in Figure 6.7. The aqueous ammonium acetate/acetonitrile mobile phase facilitated negative ion ESI-MS detection. The authors found that column performance degraded after about 200 injections. They incorporated a column regeneration procedure involving washing with water and buffer to remove highly retained materials and restore initial column performance.

A few more specialized methods have been reported for oligonucleotide determination. Inductively coupled plasma MS detection has been coupled with HILIC for analysis of nucleotides [49]. The signal at m/z 47 corresponding to PO⁺ was monitored. Lengthy re-equilibration times were reported for a Luna cross-linked diol phase so the authors chose a TSKgel Amide-80 for the method with an ammonium acetate buffer in the mobile phase. Other HILIC separation conditions such as a hydroxymethyl methacrylate polymerized monolithic capillary column [50] have been reported.

6.5   AMINO ACIDS AND PEPTIDES

The analysis of amino acids has long been of interest in biochemical, food, and drug research. Most amino acid separations are routinely done on hydrolysates of peptides and proteins for biological research. Because of the wide interest in this area for many years, these separations were the earliest to use the HILIC mode. HILIC conditions for some individual amino acids used in pharmaceutical applications have been described in Chapter 4. Recent general applications of HILIC for amino acids are described here.

Langrock et al. used HILIC ESI-MS for the detection of amino acids with the separation performed on a TSKgel Amide-80 column [51]. The mobile phase was acetonitrile with aqueous ammonium acetate where both the aqueous content and buffer concentration were increased in a gradient. Retention times for amino acids are shown in Table 6.1. Where peaks co-eluted or were only partially separated, the amino acids were distinguished by their m/z ratio. The ultimate aim of this research was to classify hydrolyzed collagen, and this HILIC mode was able to determine the hydoxyproline-isomers, isoleucine, and leucine that differentiate these proteins.

The zwitterionic ZIC-HILIC is a common choice for amino acid separations. For example, Dell'mour et al. used a ZIC-HILIC column with an acetonitrile-

**Table 6.1.   Retention Times for a Test Mixture
Containing Amino and Imino Acids Using a TSKgel
Amide-80 Column (150 × 2 mm, 5 μm)**

Amino Acid	Retention Time (min)
Trp	10.9
Phe	11.5
Leu	12.4
Ile	13.6
Met	14.4
Tyr	15.4
Val	16.4
Pro	18.7
Ala	21.1
Thr	22.1
Gly	22.9
Glu	23.3
Asp	23.9
Ser	24.4
Gln	24.5
Asn	25.0

Eluents were: (A) 90% aqueous acetonitrile containing 0.5 mM ammonium acetate, pH 5.5, and (B) 60% aqueous acetonitrile containing 2.5 mM ammonium acetate, pH 5.5. The amino acids were eluted by a water gradient using two linear increments first from 5% to 60% eluent B for 28.5 min and then to 95% eluent B in 5 min with a flow rate of 150 μL/min.
Source: Adapted from Reference 51.

aqueous formic acid gradient [52] ranging from about 10% to 90% aqueous content. This wide gradient range involves HILIC characteristics at the beginning but transitions to other mechanisms at high aqueous content. Sixteen underivatized amino acids could be determined using MS-MS in multiple reaction monitoring mode for detection. Amino acids eluted in the general order of nonpolar side chains, polar side chains, and basic amino acids (histidine, lysine, arginine). Tyrosine was an exception because it has a polar side chain but eluted earlier than the less polar proline and alanine. The 19-min screening method was applied to samples from rhizosphere studies. Kato et al. used a ZIC-HILIC method with isotope dilution MS for the determination of underivatized amino acids in protein hydrolysates. They also employed a wide gradient from 25% to 90% aqueous acetic acid [53]. They obtained good quantitative results for protein reference materials, with better recovery and precision than a derivatization-based method (aminoquinolylhydroxysuccinimidyl carbamate). Other work with ZIC-HILIC columns for amino acids has also been reported [54–56].

Yoshida reviewed several aspects of HILIC for peptide separations [57]. The use of a poly(2-hydroxyethyl aspartamide)-silica column (commercially available as PolyHydroxyethyl A) with mobile phases containing salts such as triethylamine phosphate or sodium perchlorate is described [57]. Salts, often with a concentration gradient, are necessary to provide acceptable peak shape and consistent retention times due to mixed mode HILIC and weak ion-exchange interactions. Variability of the ion-exchange contribution to retention can be reduced by control of salt content in the mobile phase. Hydrophilic retention coefficients for amino acids were described for prediction of peptide retention on a TSKgel Amide-80 column based on amino acid composition of the peptide. Gilar and Jaworski studied the retention of peptides on bare silica, bridge-ethyl hybrid (BEH) silica, and a BEH amide phase [58]. They developed retention prediction models based on amino acid composition and could rationalize the significant pH effects that were observed. On an Atlantis silica column, peptides with different isoelectric points (pI < 5, 5 < pI < 9, pI > 9) showed different selectivity as the mobile phase pH was varied from 3.5 to 10. The orthogonality of retention versus pH could be exploited in a 2D separation with HILIC as the mode in both dimensions. The largest changes in selectivity were for peptides containing acidic (aspartic acid, glutamic acid) and basic (histidine, lysine, arginine) residues. The charge of both the amino acid and the silica were considered in explaining the retention behavior. Acidic residues might be expected to have greater retention coefficients at high pH where they are charged and more hydrophilic. However, the silica is also charged at a high pH, and electrostatic repulsion caused a decrease in retention coefficient. Basic residues lysine and arginine are protonated at pH 10 and interact more strongly with deprotonated silica, which led to greater retention coefficients than at pH 3.0 or 4.5. The increased retention coefficient was not observed for histidine, which is uncharged at pH 10.

Mant and coworkers have published several papers related to various aspects of peptide separations using HILIC [59–62]. They used a poly(2-sulphoethyl aspartamide)-silica (commercially available as polysulphoethyl

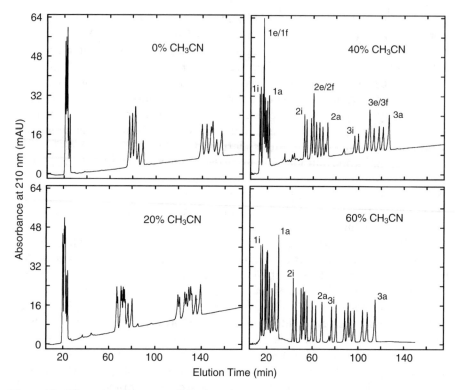

Figure 6.8. Effect of varying acetonitrile (ACN) concentration, at fixed salt gradient rate, on HILIC/CEX profile of 27 peptide standards. Conditions: linear AB gradient (1 mM NaClO$_4$/min) at a flow rate of 0.3 mL/min, where eluent A is 5 mM aq.triethylammonium phosphate, pH 4.5, and eluent B is eluent A plus 500 mM NaClO$_4$ (0, 20, 40% ACN runs) or 250 mM NaClO$_4$ (60% ACN run), both eluents also containing the corresponding 0%, 20%, 40%, or 60% v/v ACN; temperature 25°C. Reprinted from Reference 62 with permission from Wiley.

A) strong cation exchange (CEX) column in a HILIC/CEX mode using a gradient with increasing sodium perchlorate concentration [62]. Figure 6.8 shows the increased retention and separation of peptides when higher acetonitrile concentrations were utilized. Separations of peptides with the same composition but different sequence could be achieved.

Yang et al. studied the separation of protein digests using narrow bore or capillary columns packed with an amide- or an amino-modified silica-based phase [63]. The TSKgel Amide-80 column gave good separations of peptides over a range of low to neutral pH. Lower resolution was obtained with the amino phase. Singer et al. achieved resolution of mono- and di-phosphorylated peptide isomers with an amino phase with a combined salt and water gradient [64]. Di Palma et al. investigated the separation of peptide mixtures on two types of ZIC-HILIC zwitterionic columns with different positively and negatively charged groups in opposite positions on the bonded phase chain [65]. The HILIC separations were similar on the two phases and showed promise when coupled to RP in 2D system for proteome analysis. Boersema et al. also reviewed the potential for HILIC as one mode in 2D systems for proteomics research [66]. The orthogonality of retention compared with RP separations makes HILIC an attractive alternative to strong cation exchange for coupling in 2D systems. Mihailova et al. showed that HILIC using a ZIC-HILIC phase provided superior fractionation of rat brain neuropeptides compared with strong cation exchange as the first dimension in a 2D system [67]. A ZIC-HILIC separation was used between size exclusion fractionation and RP-nanoLC-MS-MS in a multidimensional approach to proteomics [68]. The method was used to identify 1955 serum proteins and 375 phosphoproteins. Loftheim et al. have also used the ZIC-HILIC column coupled with RP-LC-MS-MS for urinary proteomics studies [69].

The more recently developed electrostatic repulsion-hydrophilic interaction chromatography (ERLIC) technique (see Chapter 1) has been applied to peptide separations [70]. In another 2D system for proteome research, Hao et al. coupled ERLIC with RP modes [71]. The approach was to use an increasing water gradient with decreasing pH to take advantage of ionic and hydrophilic mixed-mode interactions in the HILIC dimension. Peptide mixtures were fractionated by HILIC and collected for further analysis by RP-LC-MS-MS analysis. Compared with strong cation exchange-RP, ERLIC-RP identified a larger number of proteins and unique peptides, particularly basic and hydrophobic peptides. Other ERLIC peptide separations have also been reported [72–74].

6.6 PROTEINS

HILIC has also been applied to the analysis of intact proteins. Tetaz et al. analyzed lipophilic and membrane-associated proteins using different HILIC phases: PolyHydroxyethyl A. ZIC-HILIC, silica, TSKgel Amide-80 [75]. Shifts in retention with continued use of the PolyHydroxyethyl A and TSKgel Amide-80 columns were observed but were not investigated for the other columns. All the columns showed separation of multiple isoforms of apoM protein. Solubility of water-soluble proteins in mobile phases with high acetonitrile content could be a limiting factor for more general use in intact protein separations. Recovery of activity when HILIC is used for protein isolation/purification is also a concern.

6.7 PHOSPHOLIPIDS

The hydrophobic nature of lipids normally makes them not amenable to HILIC separations. However, the polar end group of phospholipids provides opportunities for interactions with polar stationary phases. Phospholipids have classically been separated with a combination of normal phase HPLC followed by a reversed-phase separation. Characterization by mass spectrometry was complicated by the buffers and solvents used in these steps. Amphiphilic compounds such as phospholipids can, however, lend themselves well to HILIC separations. This phospholipid separation was first investigated by Schwalbe-Herman et al. [76]. They first extracted lipids from blood and then performed a step-wise fractionation of lipid classes on an aminopropyl solid-phase extraction column using different elution solvent combinations. The phospholipid fraction was collected after elution with methanol. For the separation of phospholipids, a HILIC silica column was used with an optimized mobile phase of acetonitrile:methanol:10 mM ammonium acetate (55:35:10) with a flow rate of 0.6 mL/min. Detection was by ESI-MS. All five phospholipids classes were separated as well as some of their subclasses as shown in Figure 6.9 for a standard mixture and blood plasma sample.

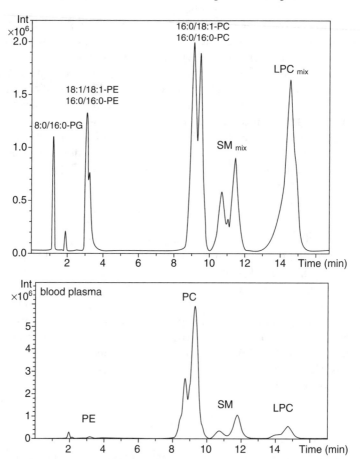

Figure 6.9. Chromatograms of the HILIC separation of the five phopholipid fractions (PG = phosphatidylglycerol; PE = phosphatidylethanolamine; PC = phosphatidylcholine; SM = sphingomyelin; LPC = lysophosphatidylcholine) with a mobile phase of acetonitrile/methanol/10 mM ammonium acetate (5:35:10). Peak detection was carried out by ESI-MS in the positive ion mode. Top chromatogram, phospholipids standard mixture sample; bottom chromatogram, blood plasma sample. Reprinted from Reference 76 with permission from Elsevier.

Donato et al. analyzed phospholipids in milk using HILIC with a fused core silica packing and ELSD or MS detection [77]. Phospholipid extracts from milk samples were injected onto a Ascentis Express column (150 × 2.1 mm, 2.7 μm). Gradient elution with acetonitrile:water mobile phase components was used with several steps covering a range of 0–33% water. Although a fused core column is often used to obtain higher efficiency, faster separations and lower back pressure for simpler analyses, this separation is not taking advantage of these properties. The fused core particle, smaller column diameter and the gradient do, however, allow for a small sample amount to be used and for increased detection limits. The fused core packing also enabled equilibration to starting conditions more rapidly than with a totally porous particle packing. This analysis scheme allowed the researchers to separate compounds within phospholipid classes including up to 10 different molecular species within the PC class alone.

Lísa et al. employed multidimensional chromatography combining HILIC and reversed-phase for the analysis of lipids [78]. The orthogonality of these modes with MS detection excels in separation of complex lipid mixtures. In the HILIC mode the lipids were separated into classes according to their polarity and electrostatic interactions. In the reversed-phase mode the retention of the lipids was controlled by their acyl chain length and the number of double bonds. In the first dimension, HILIC was used for the fractionation of total lipid extracts into lipid classes using a silica column and a mobile phase of acetonitrile with an increasing gradient of 5 mM ammonium acetate. Detection was with ESI-MS. In the second dimension, collected fractions of lipid classes were analyzed using two RP-HPLC methods according to their polarity. Additional discussion of 2D separations is given in Chapter 9.

6.8 CONCLUSIONS

The number of HILIC applications for analysis of biomolecules has increased rapidly in the last few years. Many recent glycan analysis applications have employed zwitterionic ZIC-HILIC or neutral amide stationary phases. Other types of HILIC columns have been used for certain applications, but it seems that the field is moving toward the former two column types as first-line choices. Fluorescence detection of derivatized glycans or mass spectrometric detection is clearly the mode of choice, with MS providing superior data for structural characterization.

Amide or zwitterionic phases have also become popular for other biomolecules such as nucleobases, nucleosides, nucleotides, oligonucleotides, amino acids, and peptides. Further work will determine whether methods using these phases will become more standardized for common applications. Also, the development of new HILIC phases will certainly offer additional opportunities for researchers interested in the analysis of polar biomolecules.

REFERENCES

1. Rabel F, Caputo A, Butts E. Separation of carbohydrates on a new polar bonded phase material. *J. Chromatogr.* 1976; **126**: 731–740.

2. Alpert A. Hydrophilic-interaction chromatography for the separation of peptides, nucleic acids and other polar compounds. *J. Chromatogr.* 1990; **499**: 177–196.

3. Hemström P, Irgum K. Hydrophilic interaction chromatography. *J. Sep. Sci.* 2006; **29**: 1784–1821.

4. Jandera P. Stationary and mobile phases in hydrophilic interaction chromatography: A review. *Anal. Chim. Acta* 2011; **692**: 1–25.

5. Karlsson G. Development and application of methods for separation of carbohydrates by hydrophilic interaction liquid chromatography. In Wang PG, He W, eds., *Hydrophilic Interaction Liquid Chromatography (HILIC) and Advanced Applications*. Boca Raton, FL: CRC Press; 2011, pp. 491–522.

6. Karlsson G, Winge S, Sandberg H. Separation of monosaccharides by hydrophilic interaction chromatography with evaporative light scattering detection. *J. Chromatogr. A* 2005; **1092**: 246–249.

7. Ikegami T, Horie K, Saad N, Hosoya K, Fiehn O, Tanaka N. Highly efficient analysis of underivatized carbohydrates using monolithic-silica-based capillary hydrophilic interaction (HILIC) HPLC. *Anal. Bioanal. Chem.* 2008; **391**: 2533–2542.

8. Ricochon G, Paris C, Girardin M, Muniglia L. Highly sensitive, quick and simple quantification method for mono and disaccharides in aqueous media using liquid chromatography–atmospheric pressure chemical ionization–mass spectrometry (LC–APCI–MS). *J. Chromatogr. B* 2011; **879**: 1529–1536.

9. Brokl M, Hernandex-Hernances M, Soria AC, Sanz ML. Evaluation of different operation modes of HPLC for the analysis of complex mixtures of neutral oligosaccharides. *J. Chromatogr. A* 2011; **1218**: 7697–7703.

10. Leijdekkers AGM, Sanders MG, Schols HA, Gruppen H. Characterizing plant cell wall derived oligosaccharides using hydrophilic interaction chromatography with mass spectrometry detection. *J. Chromatogr. A* 2011; **1218**: 9227–9235.

11. Hernández-Hernández O, Calvillo I, Lebron-Aguilarb R, Morenoc FJ, Sanza ML. Hydrophilic interaction liquid chromatography coupled to mass spectrometry for the characterization of prebiotic galactooligosaccharides. *J. Chromatogr. A* 2012; **1220**: 57–67.

12. ICH Harmonised Tripartite Guideline. Specifications: Test Procedures and Acceptance Criteria for Biotechnological/Biological Products Q6B, March 1999.

13. Yamada K, Kakehi K. Recent advances in the analysis of carbohydrates for biomedical use. *J. Pharm. Biomed. Anal.* 2011; **55**: 702–727.

14. Nettleship JN. Hydrophilic interaction liquid chromatography in the characterization of glycoproteins. In Wang PG, He W, eds., *Hydrophilic Interaction Liquid Chromatography (HILIC) and Advanced Applications*. Boca Raton, FL: CRC Press; 2011, pp. 523–550.

15. Thaysen-Anderson M, Engholm-Keller K, Roepstorff P. Analysis of protein glycosylation and phosphorylation using HILIC-MS. In Wang PG, He W, eds., *Hydrophilic Interaction Liquid Chromatography (HILIC) and Advanced Applications*. Boca Raton, FL: CRC Press; 2011, pp. 551–575.

16. Mariño K, Bones J, Kattla JJ, Rudd PM. A systematic approach to protein glycosylation analysis: A path through the maze. *Nat. Chem. Biol.* 2010; **6**: 713–723.

17. Zauner G, Deelder AM, Wuhrer M. Recent advances in hydrophilic interaction liquid chromatography (HILIC) for structural glycomics. *Electrophoresis* 2011; **32**: 3456–3466.

18. Melmer M, Stangler T, Schiefermeier M, Brunner W, Toll H, Rupprechter A, Lindner W, Premstaller A. HILIC analysis of fluorescence-labeled *N*-glycans from recombinant biopharmaceuticals. *Anal. Bioanal. Chem.* 2010; **398**: 905–914.

19. Campbell MP, Royle L, Radcliffe CM, Dwek RA, Glyco RPM. Base and autoGU: Tools for HPLC-based glycan analysis. *Bioinformatics* 2008; **24**: 1214–1216.

20. National Institute for Bioprocessing Research and Training. NIBRT Database Collection. 2012. http://glycobase.nibrt.ie/glycobase/show_nibrt.action, accessed April 6, 2012.

21. Melmer M, Stangler T, Premstaller A, Lindner W. Comparison of hydrophilic-interaction, reversed-phase and porous graphitic carbon chromatography for glycan analysis. *J. Chromatogr. A* 2011; **1218**: 118–123.

22. Ahn J, Bones J, Yua YQ, Rudd PM, Gilar M. Separation of 2-aminobenzamide labeled glycans using hydrophilic interaction chromatography columns packed with 1.7 μm sorbent. *J. Chromatogr. B* 2010; **878**: 403–408.

23. Waters. Acquity UPLC BEH Glyan columns, Care and Use Manual. 2012. http://www.waters.com/webassets/cms/support/docs/720003042en.pdf, accessed Feb. 1, 2012.

24. Waters. Glycan Separation Technology. 2009. http://www.waters.com/webassets/cms/library/docs/720002981en.pdf, accessed April 6, 2012.

25. Gilar M, Yu Y-Q, Ahn J, Xie H, Han H, Ying W, Qian X. Characterization of glycoprotein digests with hydrophilic interaction chromatography and mass spectrometry. *Anal. Biochem.* 2011; **417**: 80–88.

26. Bones J, Mittermayr S, O'Donoghue N, Guttman A, Rudd PM. Ultra performance liquid chromatographic profiling of serum *N*-glycans for fast and efficient identification of cancer associated alterations in glycosylation. *Anal. Chem.* 2010; **82**: 10208–10215.

27. Bones J, Mittermayr S, McLoughlin N, Hilliard M, Wynne K, Johnson GR, Grubb JH, Sly WS, Rudd PM. Identification of *N*-glycans displaying mannose-6-phosphate and their site of attachment on therapeutic enzymes for lysosomal storage disorder treatment. *Anal. Chem.* 2011; **83**: 5344–5352.

28. Takegawa Y, Deguchi K, Ito H, Keira T, Nakagawa H, Nishimura S-I. Simple separation of isomeric sialylated *N*-glycopeptides by a zwitterionic type of hydrophilic interaction chromatography. *J. Sep. Sci.* 2006; **29**: 2533–2540.

29. Wuhrer M, de Boer AR, Deelder AM. Structural glycomics using hydrophilic interaction chromatography (HILIC) with mass spectrometry. *Mass Spectrom. Rev.* 2009; **28**: 192–206.

30. Mauko L, Nordborg A, Hutchinson JP, Lacher NA, Hilder EF, Haddad PR. Glycan profiling of monoclonal antibodies using zwitterionic-type hydrophilic interaction chromatography coupled with electrospray ionization mass spectrometry detection. *Anal. Biochem.* 2011; **408**: 235–241.

31. Zauner G, Koeleman CAM, Deelder AM, Wuhrer M. Protein glycosylation analysis by HILIC-LCMS of proteinase K-generated *N*- and *O*-glycopeptides. *J. Sep. Sci.* 2010; **33**: 903–910.

32. Calvano DC. Hydrophilic interaction chromatography-based enrichment protocol coupled to mass spectrometry for glycoproteome analysis. In Wang PG, He W, eds., *Hydrophilic Interaction Liquid Chromatography (HILIC) and Advanced Applications.* Boca Raton, FL: CRC Press; 2011, pp. 469–489.

33. An Y, Cipollo JF. An unbiased approach for analysis of protein glycosylation and application to influenza vaccine hemagglutinin. *Anal. Biochem.* 2011; **415**: 67–80.

34. Hägglund P, Bunkenborg J, Elortza F, Jensen ON, Roepstorff P. A new strategy for identification of *N*-glycosylated proteins and unambiguous assignment of their glycosylation sites using HILIC enrichment and partial deglycosylation. *J. Proteome Res.* 2004; **3**: 556–566.

35. Nettleship JE, Aplin R, Aricescu AR, Evans EJ, Davis SJ, Crispin M, Owens RJ. Analysis of variable *N*-glycosylation site occupancy in glycoproteins by liquid chromatography electrospray ionization mass spectrometry. *Anal. Biochem.* 2007; **361**: 149–151.

36. Yu YQ, Gilar M, Kaska J, Gebler JC. A rapid sample preparation method for mass spectrometric characterization of N-linked glycans. *Rapid Commun. Mass Spectrom.* 2005; **19**: 2331–2336.

37. Kondo A, Thaysen-Andersen M, Hjernø K, Jensen ON. Characterization of sialylated and fucosylated glycopeptides of β2-glycoprotein I by a combination of HILIC LC and MALDI MS/MS. *J. Sep. Sci.* 2010; **33**: 891–902.

38. Omaetxebarria MJ, Hägglund P, Elortza F, Hooper NM, Arizmendi JM, Jensen ON. Isolation and characterization of glycosylphosphatidylinositol-anchored peptides by hydrophilic interaction chromatography and MALDI tandem mass spectrometry. *Anal. Chem.* 2006; **78**: 3335–3341.

39. Calvano CD, Zambonin CG, Jensen ON. Assessment of lectin and HILIC based enrichment protocols for characterization of serum glycoproteins by mass spectrometry. *J. Proteomics.* 2008; **71**: 304–317.

40. Chen P, Li W, Li Q, Wang Y, Li Z, Ni Y, Koike K. Identification and quantification of nucleosides and nucleobases in Geosaurus and Leech by hydrophilic-interaction chromatography. *Talanta* 2011; **85**: 1634–1641.

41. Rodríguez-Gonzalo E, García-Gómez D, Carabias-Martínez R. Study of retention behaviour and mass spectrometry compatibility in zwitterionic hydrophilic interaction chromatography for the separation of modified nucleosides and nucleobases. *J. Chromatogr. A* 2011; **1218**: 3994–4001.

42. Rodríguez-Gonzalo E, García-Gómez D, Carabias-Martínez R. Development and validation of a hydrophilic interaction chromatography–tandem mass spectrometry method with on-line polar extraction for the analysis of urinary nucleosides. Potential application in clinical diagnosis. *J. Chromatogr. A* 2011; **1218**: 9055–9063.

43. Johnsen E, Wilson SR, Odsbu I, Krappa A, Malerod H, Skarstad K, Lundanes E. Hydrophilic interaction chromatography of nucleoside triphosphates with temperature as a separation parameter. *J. Chromatogr. A* 2011; **1218**: 5981–5986.

44. Pucci V, Giuliano C, Zhang R, Koeplinger KA, LeoneJF ME, Bonelli F. HILIC LC-MS for the determination of 29-C-methylcytidine-triphosphate in rat liver. *J. Sep. Sci.* 2009; **32**: 1275–1283.

45. Zhou T, Lucy CA. Hydrophilic interaction chromatography of nucleotides and their pathway intermediates on titania. *J. Chromatogr. A* 2008; **1187**: 87–93.

46. Treadway JW, Philibert GS, Olesik SV. Enhanced fluidity liquid chromatography for hydrophilic interaction separation of nucleosides. *J. Chromatogr. A* 2011; **1218**: 5897–5902.

47. Philibert GS, Olesik SV. Characterization of enhanced-fluidity liquid hydrophilic interaction chromatography for the separation of nucleosides and nucleotides. *J. Chromatogr. A* 2011; **1218**: 8222–8230.

48. Gong L, McCullagh JSO. Analysis of oligonucleotides by hydrophilic interaction liquid chromatography coupled to negative ion electrospray ionization mass spectrometry. *J. Chromatogr. A* 2011; **1218**: 5480–5486.

49. Easter RN, Kröning KK, Caruso JA, Limbach PA. Separation and identification of oligonucleotides by hydrophilic interaction liquid chromatography (HILIC)—inductively coupled plasma mass spectrometry (ICPMS). *Analyst* 2010; **135**: 2560–2565.

50. Holdšvendová P, Suchánková J, Bunček M, Bačkovská V, Coufal P. Hydroxymethyl methacrylate-based monolithic columns designed for separation of oligonucleotides in hydrophilic-interaction capillary liquid chromatography. *J. Biochem. Biophys. Methods* 2007; **70**: 23–29.

51. Langrock T, Czihal P, Hoffmann R. Amino acid analysis by hydrophilic interaction chromatography coupled on-line to electrospray ionization mass spectrometry. *Amino Acids* 2006; **30**: 291–297.

52. Dell'mour M, Jaitz L, Oburger E, Puschenreiter M, Koellensperger G, Hann S. Hydrophilic interaction LC combined with electrospray MS for highly sensitive analysis of underivatized amino acids in rhizosphere research. *J. Sep. Sci.* 2010; **33**: 911–922.

53. Kato M, Kato H, Eyama S, Takatsu A. Application of amino acid analysis using hydrophilic interaction liquid chromatography coupled with isotope dilution mass spectrometry for peptide and protein quantification. *J. Chromatogr. B* 2009; **877**: 3059–3064.

54. Schettgen T, Tings A, Brodowsky C, Muller-Lux A, Musiol A, Kraus T. Simultaneous determination of the advanced glycation end product N^ε-carboxymethyllysine and its precursor, lysine, in exhaled breath condensate using isotope-dilution-hydrophilic-interaction liquid chromatography coupled to tandem mass spectrometry. *Anal. Bioanal. Chem.* 2007; **387**: 2783–2791.

55. Conventz A, Musiol A, Brodowsky C, Mueller-Lux A, Dewes P, Kraus T, Schettgen T. Simultaneous determination of 3-nitrotyrosine, tyrosine, hydroxyproline and proline in exhaled breath condensate by hydrophilic interaction liquid chromatography/electrospray ionization tandem mass spectrometry. *J. Chromatogr. B* 2007; **860**: 78–85.

56. Preinerstorfer B, Schiesel S, Lämmerhofer M, Lindner W. Metabolic profiling of intracellular metabolites in fermentation broths from β-lactam antibiotics production by liquid chromatography–tandem mass spectrometry methods. *J. Chromatogr. A* 2010; **1217**: 312–328.

57. Yoshida T. Peptide separation by hydrophilic-interaction chromatography: A review. *J. Biochem. Biophys. Methods* 2004; **60**: 265–280.

58. Gilar M, Jaworski A. Retention behavior of peptides in hydrophilic-interaction chromatography. *J. Chromatogr. A* 2011; **1218**: 8890–8896.

59. Mant CT, Litowski JR, Hodges RS. Hydrophilic interaction/cation-exchange chromatography for separation of amphipathic α-helical peptides. *J. Chromatogr. A* 1998; **816**: 65–78.

60. Mant CT, Kondejewski LH, Hodges RS. Hydrophilic interaction/cation-exchange chromatography for separation of cyclic peptides. *J. Chromatogr. A* 1998; **816**: 79–88.

61. Hartmann E, Chen Y, Mant CT, Jungbauer A, Hodges RS. Comparison of reversed-phase liquid chromatography and hydrophilic interaction/cation-exchange chromatography for the separation of amphipathic α-helical peptides with L- and D-amino acid substitutions in the hydrophilic face. *J. Chromatogr. A* 2003; **1009**: 61–71.

62. Mant CT, Hodges RS. Mixed-mode hydrophilic interaction/cationexchange chromatography: Separation of complex mixtures of peptides of varying charge and hydrophobicity. *J. Sep. Sci.* 2008; **31**: 1573–1584.

63. Yang Y, Boysen RI, Hearn MTW. Hydrophilic interaction chromatography coupled to electrospray mass spectrometry for the separation of peptides and protein digests. *J. Chromatogr. A* 2009; **1216**: 5518–5524.

64. Singer D, Kuhlmann J, Muschket M, Hoffmann R. Separation of multiphosphorylated peptide isomers by hydrophilic interaction chromatography on an aminopropyl phase. *Anal. Chem.* 2010; **82**: 6409–6414.

65. Di Palma S, Boersema PJ, Heck AJR, Mohammed S. Zwitterionic hydrophilic interaction liquid chromatography (ZIC-HILIC and ZIC-cHILIC) provide high resolution separation and increase sensitivity in proteome analysis. *Anal. Chem.* 2011; **83**: 3440–3447.

66. Boersema PJ, Mohammed S, Heck AJR. Hydrophilic interaction liquid chromatography (HILIC) in proteomics. *Anal. Bioanal. Chem.* 2008; **391**: 151–159.

67. Mihailova A, Malerød H, Wilson SR, Karaszewski B, Hauser R, Lundanes E, Greibrokk T. Improving the resolution of neuropeptides in rat brain with on-line HILIC-RP compared to on-line SCX-RP. *J. Sep. Sci.* 2008; **31**: 459–467.

68. Garbis SD, Roumeliotis TI, Tyritzis SI, Zorpas KM, Pavlakis K, Constantinides CA. A novel multidimensional protein identification technology approach combining protein size exclusion prefractionation, peptide zwitterion-ion hydrophilic interaction chromatography, and nano-ultraperformance RP chromatography/NESI-MS2 for the in-depth analysis of the serum proteome and phosphoproteome: Application to clinical sera derived from humans with benign prostate hyperplasia. *Anal. Chem.* 2011; **83**: 708–718.

69. Loftheim H, Nguyen TD, Malerød H, Lundanes E, Åsberg A, Reubsaet L. 2-D hydrophilic interaction liquid chromatography-RP separation in urinary proteomics—Minimizing variability through improved downstream workflow compatibility. *J. Sep. Sci.* 2010; **33**: 864–872.

70. Alpert AJ. Electrostatic repulsion hydrophilic interaction chromatography for isocratic separation of charged solutes and selective isolation of phosphopeptides. *Anal. Chem.* 2008; **80**: 62–76.

71. Hao P, Guo T, Li X, Adav SS, Yang J, Wei M, Sze SK. Novel application of electrostatic repulsion-hydrophilic interaction chromatography (ERLIC) in shotgun proteomics: Comprehensive profiling of rat kidney proteome. *J. Proteome Res.* 2010; **9**: 3520–3526.

72. Gan CS, Guo T, Zhang H, Lim SK, Sze SK. A comparative study of electrostatic repulsion-hydrophilic interaction chromatography (ERLIC) versus SCX-IMAC-based methods for phosphopeptide isolation/enrichment. *J. Proteome Res.* 2008; **7**: 4869–4877.

73. Lewandrowski U, Lohrig K, Zahedi RP, Walter D, Sickmann A. Glycosylation site analysis of human platelets by electrostatic repulsion hydrophilic interaction chromatography. *Clin. Proteomics* 2008; **4**: 25–36.

74. Zhang H, Guo T, Li X, Datta A, Park JE, Yang J, Lim SK, Tam JP, Sze SK. Simultaneous characterization of glyco- and phosphoproteomes of mouse brain membrane proteome with electrostatic repulsion hydrophilic interaction chromatography. *Mol. Cell Proteomics* 2010; **9**: 635–647.

75. Tetaz T, Detzner S, Friedlein A, Molitor B Mary J-L. Hydrophilic interaction chromatography of intact, soluble proteins. *J. Chromatogr. A* 2011; **1218**: 5892–5896.

76. Schwalbe-Hermann M, Willmann J, Leibritz D. Separation of phospholipid classes by hydrophilic interaction chromatography detected by electrospray ionization mass spectrometry. *J. Chromatogr. A* 2010; **1217**: 5179–5183.

77. Donato P, Cacciola F, Cichello F, Russo M, Dugo P, Mondello L. Determination of phospholipids in milk sample by means of hydrophilic interaction liquid chromatography coupled to evaporative light scattering and mass spectrometry detection. *J. Chromatogr. A* 2011; **1218**: 6476–6482.

78. Lísa M, Cífková E, Holčapek M. Lipidomic profiling of biological tissues using off-line two-dimensional high-performance liquid chromatography–mass spectrometry. *J. Chromatogr. A* 2011; **1218**: 5146–5156.

CHAPTER
7

HILIC-MS FOR TARGETED METABOLOMICS AND SMALL MOLECULE BIOANALYSIS

HIEN P. NGUYEN, HEATHER D. TIPPENS, and KEVIN A. SCHUG

*Department of Chemistry and Biochemistry,
The University of Texas at Arlington, Arlington, TX*

7.1   INTRODUCTION

It is clear that the shear breadth of bioanalytical applications of HILIC is ever-expanding. While we focus primarily here on metabolite analysis as a representative area, HILIC clearly applies to other areas of pharmacokinetic, stability, drug monitoring, enzyme activity, and biotransformation analysis. Some excellent reviews on various aspects of these topics are available in the literature [1–4]. As cited by Hsieh, one of the major causes of failure for drug candidates has been attributed to drug metabolism and pharmokinetic issues; thus, the driver for methods that can probe metabolites and drug compounds in complex matrix. From an analytical perspective it is desirable to have a generic RP-HPLC method; however, as the polarity of the analyte increases this desire is to difficult to achieve. In addition, the metabolites of these polar analytes are often even more polar than the parent itself making retention and separation even more difficult. To complicate matters further, the matrices themselves can be quite complex and the analyte concentrations low (e.g., ng/mL sensitivity may be required). Common matrices are human urine, human plasma, human blood, cerebrospinal fluid, liver tissues etc. It is important to

Hydrophilic Interaction Chromatography: A Guide for Practitioners, First Edition.
Edited by Bernard A. Olsen and Brian W. Pack.
© 2013 John Wiley & Sons, Inc. Published 2013 by John Wiley & Sons, Inc.

demonstrate during analytical method validation that the analyte is retained in a region of minimal matrix interferences where ion suppression may cause quantitation errors in LCMS analyses. As will be discussed later in this chapter, sample preparation is an extremely important aspect of bioanalytical applications and should be thoroughly investigated during characterization of the method. Novakova presents a nice review of the most common sample extraction procedures used for bioanalytical application [5]. In relation to the separation technique, liquid-liquid extraction (LLE) is commonly carried out with organic solvents and is naturally compatible with HILIC mobile phase conditions. One further advantage of using HILIC which is discussed thoroughly below, but applies to all bioanalytical applications, is the prevalent use of mass spectrometry due to the sensitivity and selectivity advantages that are commonly reported.

Metabolomics, along with other technologies such as transcriptomics and proteomics, can give us a more complete picture of biological systems. Metabolomics involves the analysis of all known and unknown metabolites in biological samples, and is a critical part of pharmacokinetic studies on adsorption, distribution, metabolism, and excretion (ADME) of drug compounds. Targeted metabolomics refers specifically to the quantitation of selected metabolites from dozens to hundreds of known compounds [6]. Biomonitoring of these compounds, along with their parent compound, reveals the scientific basis for understanding metabolic pathways, to predict the incidence or outcome of diseases [7,8], as well as molecular mechanisms of chemically-induced beneficial or toxicological effects [9]. Due to the wide range of metabolites, which exhibit different polarities and ionizabilities, arising from various functional groups such as amino, carboxyl, hydroxyl, thiol, sulfate, and phosphate motifs, chromatographic retention to achieve comprehensive quantitative analysis is a major challenge.

Many analytical methods have been applied to differentiate and identify metabolites, such as nuclear magnetic resonance (NMR) spectroscopy [10,11] or mass spectrometry (MS) coupled to separation techniques [12–15]. Among those techniques, reversed-phase (RP) chromatography coupled with MS has been widely used in targeted metabolomics because of its robust, well-understood, and well-established separation mechanism. However, metabolites are typically moderately to highly polar/hydrophilic molecules and, therefore, they can be poorly retained on RP columns. Although this can be resolved by either derivatization of the analytes or addition of ion pair reagents into mobile phases to improve the retention and selectivity of polar/hydrophilic compounds, derivatization requires additional steps for sample preparation while the use of ion pair reagents can affect the peak shape, shorten column life time, and foul mass spectrometer inlets. Hydrophilic interaction chromatography (HILIC), using polar stationary phases and aqueous mobile phases, can provide a complementary separation for RPLC to not only effectively retain and separate hydrophilic compounds but also enhance the sensitivity when coupled with MS [16].

In this chapter, and in the context of targeted metabolomics and other bioanalytical studies, we first discuss the potential role of HILIC, compared to other separation modes, when coupled to mass spectrometric detection. Throughout the chapter we refer to various literature studies, but this presentation is not an exhaustive review of the tremendous amount of work that has been done in the area. We next turn to specific method development strategies for targeted metabolomics experiments. The choice of stationary phase, and its ability to match and retain the classes of metabolites of interest, is of paramount importance. In this context, we also include some data from our own lab to demonstrate some of the differences encountered in practice, both for comparing of RP to HILIC separation modes, and to take into consideration the prevalence of matrix effects, which must be addressed for biomonitoring applications from complex biological matrices when using MS detection.

7.2 THE ROLE OF HILIC-MS IN TARGETED METABOLOMICS VERSUS OTHER LC MODES

Targeted metabolomics has been investigated with a variety of different separation modes, most often in conjunction with MS detection. Capillary electrophoresis (CE)-MS has been reported to comprehensively analyze many charged species in biological samples with fast analysis times, high sensitivity, and high efficiency [13,17–21]. However, in order to be sufficiently separated, the analytes have to be chargeable in liquid phases. Additionally, technical difficulties, including reproducibility of migration times, make this method less amenable for routine analyses. Currently, the most common chromatographic method in MS metabolomics relies on RP separation, an approach that has shown coverage across a wide range of various metabolites [22,23]. Weak and moderate polar/hydrophilic compounds can be directly separated by RP [6,24], while the analysis of strongly polar or highly hydrophilic compounds can be carried out with prior derivatization [25,26] or using ion-pair (IP) reagents [27–31].

Derivatization can serve multiple purposes, including increasing sensitivity, especially in the negative electrospray ionization (ESI-MS) detection mode, or improving retention in the RP separation mode. Functional groups that exhibit poor ionization efficiency in ESI-MS, such as thiols or phenols, can be derivatized to significantly improve signal response [6,25,32–37]. However, when coupled to common MS quantitation methods, such as multiple reaction monitoring (MRM), neutral loss of analyte-specific fragments often occurs, reducing the specificity of the method. For example, derivatization of phenolic estrogen compounds by a common derivatization reagent, dansyl chloride [38–44] can increase sensitivity by several orders of magnitude, but the predominant fragment observed by MRM is m/z 171, which corresponds to the appended aminonaphthyl sulfonate moiety from the reagent. Consequently, a lack of analyte-specific structural information and, additionally, cross-talk can occur in this case. In metabolomics, isotope enrichment of stable isotope-labeled metabolites is regularly used during metabolomic pathway studies or flux analysis [45–47]; therefore, the neutral loss of an isotope-labeled analyte can also be disadvantageous for this technique [48].

Instead of derivatization, another option for the analysis of highly polar/hydrophilic compounds is through the addition of IP reagents in the mobile phase. This technique does not require extra steps during sample preparation. However, choosing IP reagents and subsequent chromatographic method development can be a challenge. First, these reagents must be volatile to be compatible with MS detection. If the reagents are nonvolatile, then they will foul the inlet to the mass spectrometer. Second, reagents need to be amphiphilic so that they can interact with both the analyte and the stationary phase to assist analyte retention [49]. Since formation of an electrical charge double layer is key, the technique can also require a long time to equilibrate the stationary phase, and it is very sensitive to changes in the mobile phase, such as pH and ionic strength. Third, the amphiphilic nature of the IP reagents make them particularly detectable by ESI-MS. As such, some consideration has to be given to their potential contributions to ion suppression or adduct formation.

In a complementary role, or sometimes as a viable replacement, HILIC has gradually found increased utility in the analysis of various small hydrophilic compounds to avoid extra steps of derivatization and the use of IP reagents. Its application in targeted metabolomics has dramatically increased in the last few years, especially for the analysis of metabolites in cell extracts and biological samples such as biofluids and tissues [17]. HILIC-MS has been shown to enhance sensitivity of hydrophilic metabolite analysis for two reasons: (1) it provides a better separation selectivity and efficiency; and (2) it uses a higher organic content mobile phase [16] which facilitates solvent evaporation and analyte ionization in the mass spectrometer.

A plethora of HILIC stationary phases have been reported to efficiently separate diverse groups of hydrophilic metabolites. These metabolites include amino acids [50–55], carbohydrates [17,50,52–56], nucleosides/nucleotides [17,52,55,56], organic acids [6,53–56], vitamins/drugs [52,55,57–61], and their precursors, as well as derivatives. Some of these hydrophilic compounds, such as organic acids, can be measured by RPLC-MS, but the total number of such analytes which can be reliably determined can be limited in some cases to only one or two compounds in a given run [62,63]. The mobile phase must be highly aqueous (up to 100% water) to retain the analytes in the RP mode, but they are still often eluted near the void volume. This lack of retention provides a fundamental limit to the separation power available for addressing complex mixtures of a large number of analytes at one time. Although hydrophilic compounds can be retained on HILIC stationary phases, different groups of metabolites require different HILIC conditions, including the selection of stationary phase chemistries and mobile phase compositions. Amine compounds have been demonstrated to have better separation on a cyano column at acidic pH mobile phase [52] while the amide stationary phase, on which a wider range of metabolites can be separated, seems to be among the most popular phases [6,50–52]. Recently, a new silica hydride-based stationary phase has been introduced into HILIC analysis with an ability to separate many hydrophilic metabolites, boasting highly reproducible retention and fast equilibration times [53].

It is generally well-known that HILIC mobile phases are more amenable to increased ionization efficiencies for analytes when coupled to MS detection, compared to RP mode [16,64,65]. To clearly demonstrate this effect, we designed a preliminary experiment to compare the sensitivity of HILIC-MS and RP-MS for analysis of estrogen glucuronides as model compounds. These are good test probes because they can be retained in both separation modes. The sensitivity was evaluated based on the slopes of calibration curves, showing that HILIC-MS provided a better sensitivity than RP-MS by at least one order of magnitude (Fig. 7.1). In another similar study in the literature, several organic compounds such as citric and oxaloacetic acids were not detectable at parts-per-million level using IP-RP-MS [27], while they were easily detectable by HILIC-MS at the same level [53]. That said, it should be noted that detection limits were improved down to the parts-per-billion level when derivatization was used [27].

Although HILIC-MS is a very promising method for the analysis of hydrophilic metabolites in terms of chromatography and sensitivity, it may still not be the best in all cases. For example, a study by Yoshida et al. definitively demonstrated the analysis of 137 different metabolites on a pentafluorophenylpropyl-bonded silica column [66]. The RP method was able to separate different groups of metabolites including amino acids, amines, organic acids, nucleosides, and nucleotides. Some of them (~20%) could be detected in the range of nM concentrations.

7.3   STRATEGIES FOR METHOD DEVELOPMENT BASED ON RETENTION BEHAVIOR OF TARGETED METABOLITES ON HILIC STATIONARY PHASES

It is not easy to predict the retention behavior of hydrophilic metabolites on HILIC stationary phases because of the diversity of analyte functional units as well as the wide variety of stationary phases available. Each particular group of metabolites (amino acid, carbohydrate, organic acids, etc.) behaves differently on individual HILIC columns, which greatly vary in functional groups bonded to stationary particles. The variety ranges from negatively charged stationary phases (silica) to positively charged ones (amino), or they can possess both charges (zwitterions). The uncharged stationary phases can present hydrogen bonding acceptor (cyano) or both acceptor and donor (amide) groups. In RP, hydrophobic interactions are predominant whereas mixed modes often occur during HILIC analysis, including hydrogen bonding,

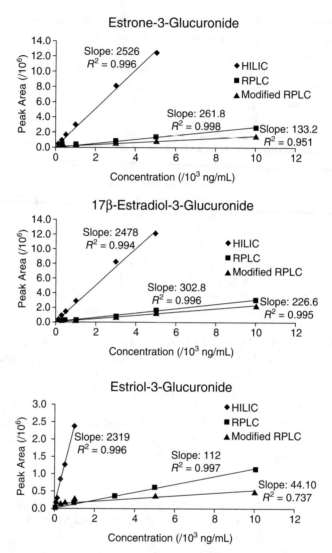

Figure 7.1. Calibration curves for three different estrogen glucuronides evaluated based on detection sensitivity following HILIC (Tosoh Amide-80 column, 100 × 2 mm), RP (Varian Pursuit C18, 100 × 2 mm), and modified RP (polar-embedded group; Varian Polaris Amide-C18, 100 × 2 mm) separations. Separation and detection was performed on a Shimadzu LCMS-2010 single quadrupole instrument using acetonitrile/water mobile phases modified with 5 mM ammonium acetate. Approximately one order of magnitude higher detection sensitivity (compare slopes of the plots) was observed for the HILIC mode separation.

electrostatic, dipole-dipole, hydrophilic, and hydrophobic partitioning interactions (see Chapter 1). Among HILIC interactions, hydrogen bonding appears to be the most important [67,68], compared to other interactions. Extensive electrostatic interactions are less desirable because they often require mobile phases with higher ionic strength, and hence more salts are introduced into a mass spectrometer detector. Significant electrostatic interactions have also been shown to be the cause of broad, asymmetric peak shapes, and prolonged separation times [68–71]. The use and properties of different HILIC stationary phases have been discussed in Chapter 2 and in several excellent reviews [72,73]. We focus our discussion on only the HILIC stationary phases and associated method development strategies commonly used in targeted metabolomic analysis.

There are two major aspects that need to be considered during method development: (1) retention behavior of hydrophilic metabolites on the HILIC phase; and (2) robustness of the method in terms of mobile phase compositions and matrix effects. These are addressed in two separate sections below.

7.3.1 Retention Behavior of Metabolites on HILIC Stationary Phases

The selection of the stationary phase is first determined based on the nature of analytes. For specific groups of metabolites, the stationary phase is selected to be complementary with the hydrophilicity, charge, and ionizability of the metabolites in a given mobile phase. For ionizable compounds it is most desirable to operate under pH conditions that favor a predominant ionization form to minimize the multitude of different interactions that can occur if multiple forms of functional groups for a particular analyte are present. A representation of some of the more common HILIC stationary phases is given in Figure 7.2.

Amine compounds such as amino acids, nucleosides/nucleotides, and redox-electron carriers, contain a basic nitrogen atom and can be both hydrogen bond acceptors and donors. Thus, they are well separated on a wide range of HILIC stationary phases such as neutral (cyano [52], amide [6,50–52]), negatively-charged (silica [72,74], diol [6,54,72]), and positively-charged (amino [52,54]) stationary phases. Interactions between the analyte and the stationary phase are generally attributed to hydrogen bonding, although hydrophilic partitioning into an adsorbed water layer and/or electrostatic interactions can also contribute to retention. Typically, the application of a cyano stationary phase is limited because it lacks hydrogen bond donor sites, yielding low retention in HILIC analysis. However, it was found to provide outstanding chromatography for many amines in acidic mobile phase [52]. Hydrogen bonding was surmised to be the predominant interaction in this case, assisted by the presence of protons in an acidic mobile phase. Other secondary interactions were deemed to be minimal; tailing was eliminated and good peak shape was observed. The analysis of amine metabolites should be performed in slightly basic mobile phase for the rest of the stationary phases. At a pH > pK_a of the

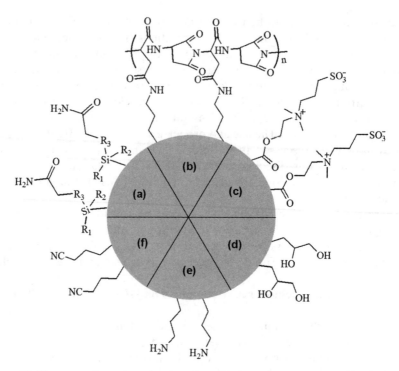

Figure 7.2. Representation of some common HILIC stationary phases: (a) amide; (b) poly(succinimide); (c) sulfoalkylbetaine (zwitterionic); (d) diol; (e) amine; and (f) cyano. Reprinted from Reference 16 with permission from Wiley.

amines, ionization is suppressed, and thus, electrostatic interactions with chargeable stationary phases are minimized [74] to favor retention by hydrogen bonding interactions. It is noted that irreversible adsorption of basic compounds may occur on silica stationary phase when using neutral or weakly acidic mobile phases. This effect depends strongly on the pH but not the ionic strength of the mobile phase [64].

In contrast, negatively-charged metabolites, such as phosphorylated and sulfated compounds or organic acids, are not detectable or are eluted with a poor peak shape when negatively-charged (silica) or nonhydrogen bonding donor (cyano) stationary phases are used [52]. This is mainly due to the repulsion interaction and/or the lack of hydrogen bonding. These metabolites can be retained and separated on other neutral (amide, diol) [6,52,64,72], positively charged (amino) [52,64,72], or zwitterionic stationary phases [64,73]. Similarly to the interaction between basic compounds and silica stationary phase, acidic compounds such as organic acids [64] or glucuronides [68] exhibit a high affinity with amino stationary phases. To temper the interaction, a slightly basic mobile phase can be used to facilitate better separation [52].

Acidic or basic mobile phases are more likely to provide better chromatography, in general, for metabolites with multiple ionizable groups, because the number of opposing ionic forms (e.g., in an analyte that has both basic and acidic functional units) in equilibrium are reduced relative to a neutral environment. Although split peaks have been observed for slightly basic/acidic or neutral mobile phases [52,68], these conditions are still used in many cases, given the good compatibility between silica-supported phases and mild pH environments.

The more diverse the metabolites of interest in a mixture are, the harder it is to choose an appropriate stationary phase. In such cases, neutral or zwitterionic stationary phases are often favored because they can behave as hydrogen bond acceptors and donors or possess both negative and positive charges. The most popular neutral stationary phases used in targeted metabolomics are amide- and diol-bonded columns. Several different groups of metabolites which can be chromatographically separated on these columns include organic acids, carbohydrates, and amine compounds such as amino acids and nucleosides/nucleotides [50–52,54]. However, fine tuning of the pH and ionic strength of the mobile phase for such a wide range of metabolites is not easy. For metabolites with multiple pK_a's, such as amino acids or multifunctional acids, the mobile phase pH plays a key role in terms of peak shape and separation selectivity. The addition of buffer salts in the mobile phase promotes interactions between the analytes and stationary phase [68] but it should be also limited in order to provide a compromise with the ionic strength and potential signal suppression during MS detection. In our experience, the use of 5–20 mM ammonium acetate as a mobile phase buffer provides the best performance for coupling HILIC separations with MS detection.

To analyze as many groups of metabolites as possible, HILIC has been used in combination with other separation modes, since no single chromatographic method is ideal for all classes of metabolites, which differ greatly in physical and chemical properties. Many studies have been reported using multiple chromatographic modes in parallel either on-line or off-line [6,75,76]. Furthermore, multiple HILIC stationary phases have also been coupled to analyze a wider variety of hydrophilic metabolites [50,52,54]. For example, a silica hydride-based stationary phase was used to separate multiple classes of hydrophilic metabolites using different mobile phase conditions (pH, additives, gradient, etc.) and ESI modes (negative and positive).

7.3.2 Robustness, Mobile Phase Compositions, and Matrix Effects

Although HILIC-MS has been demonstrated to provide a better sensitivity relative to RP mode [16], there are several issues associated with its robustness during quantitation. As a rule of thumb, HILIC separations are often subject to poor reproducibility and efficiency. HILIC typically requires a long equilibration to stabilize the immobilized water-enriched layer on the stationary phase. Any change in this hydrated layer, for example, through a change in salt concentration or pH during a gradient separation program, can affect the analyte retention and peak shape. In general, HILIC is operated at high organic solvent content, and that has a significant effect on effective buffer and analyte pK_a values, especially when the organic solvent content is more than 50% [77,78]. Although deleterious effects of lengthy equilibration times and pK_a shifts can be diminished using isocratic elution, this elution mode can lead to broad peaks and limits the complexity of analytes which can be simultaneously separated in a given run under reasonable time constraints. Therefore, all aspects should be taken into account, including analyte, buffer, and stationary phase properties to fine tune the buffer concentration and pH of the mobile phase as discussed in the method development chapter of this book (Chapter 3). The salt concentration should be adjusted to promote hydrophilic interaction and improve the analyte retention without affecting ESI signals. However, buffer salt concentrations that are too high increase the ionic strength of the mobile phase, and can cause early elution of the analytes. The mobile phase pH should be carefully chosen, based on the buffer and analyte pK_a at the specific composition of the mobile phase. An appropriate degree of analyte protonation or dissociation possibly prevents secondary interactions, such as electrostatic or ion exchange, which can lead to poor chromatographic performance. If a gradient separation is to be performed, care should be taken to try to maintain a constant ionic strength and pH throughout the separation process by incorporating consistent mobile phase additives in all of the solvent reservoirs.

While robustness in HILIC can be achieved by judiciously optimizing the separation conditions for a given set of analytes, applications of HILIC-MS for targeted metabolomics still faces a significant challenge in dealing with matrix effects, which may arise from components in complex biological samples. Ion enhancement or suppression, caused by co-elution of interfering species in the biological samples, is a problem with ESI-MS detection. Depending on characteristics of biological samples, matrix effects can be more or less significant for HILIC-MS compared with RPLC-MS. For example, phospholipids, considered as the major cause of matrix effects for RPLC-MS [79,80] analysis of plasma, are prone to have less effect on HILIC-MS analysis [81,82]. The retention behavior of phospholipids highly varies with different HILIC stationary phases. They can be retained on silica and diol stationary phase due to hydrophilic interactions between their polar heads with the stationary phase. However, less retentive behavior was observed on amino and cyano-bonded stationary phases, resulting in reduced matrix effects for later eluting metabolites. In contrast, matrix effects become more troublesome in HILIC-MS for urine analysis. Urine contains a large amount of metabolic waste, biological salts, and polar organic compounds. These hydrophilic interferences can be well retained and co-eluted with the analytes, affecting the analyte signal response and overloading the columns.

To demonstrate the problem of matrix interferences from urine samples, we investigated the change in sensitivity associated with different solid phase

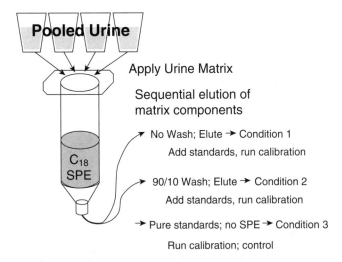

Figure 7.3. Schematic of an experimental design to investigate the matrix effects contributed to estrogen metabolite analysis from different SPE preparations of urine. Pooled urine was loaded onto a C18 SPE phase, and then was either directly eluted with 75:25 acetonitrile/methanol (Condition 1), or eluted following an initial wash 90:10 water/acetonitrile (Condition 2). Each matrix extract was spiked with standards after collection, and the LC-MS response was evaluated relative to that obtained for standards in pure solvent (Condition 3) [83].

extraction (SPE) preparations of estrogen glucuronide metabolites (Fig. 7.3). Our in-house studies showed that the tendency for retaining interfering hydrophilic species in urine samples during HILIC mode was a potential problem for ESI-MS detection. The matrix effect was evaluated based on comparing the slopes of calibration curves obtained in conjunction with different experimental designs. We tested two HILIC columns, amide- and diol-bonded stationary phases, for the analysis of estrogen glucuronides. Matrix effects arising from the constituents present in different SPE fractions of urine samples could be studied by spiking the analytes at different concentrations after extraction of the matrix. Interestingly, while ion suppression was observed on the amide-bonded column, ion enhancement was encountered on diol-bonded column (Fig. 7.4). There also appears to have been extensive changes to the linear ranges of the analyses. The differences in the matrix effect could be attributed to the presence of different interferences co-eluting with the compounds of interest and their retention shift on different stationary phases. Even so, it is difficult to attribute specific effects to particular coeluting interferences, many of which would be present in much higher concentration than the analytes themselves. Even if only in a cursory manner, this data indicates that very significant effects can arise and matrix effects should be carefully studied to select the appropriate stationary phase to limit the co-elution of the analytes and other interfering species when ESI-MS detection is used in conjunction with HILIC separations. That said, an important advantage of coupling RP-SPE

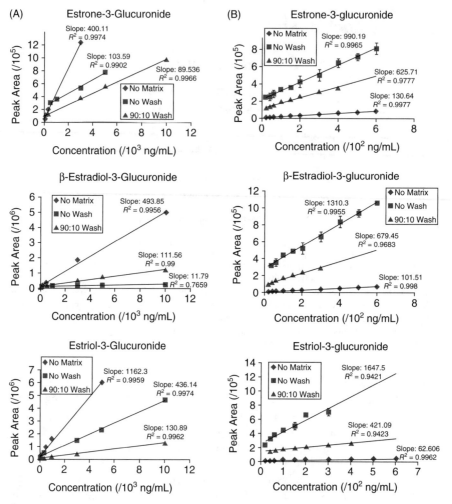

Figure 7.4. Variations in matrix effects observed for different sample preparations (see Fig. 7.3) prior to HILIC LC-MS analysis of estrogen metabolites on (A) amide (Tosoh Amide-80, 100×2 mm) and (B) diol (Phenomenex Luna HILIC, 100×2 mm) stationary phases.

sample preparation with HILIC separations is also illustrated by this work-flow. Analytes can be eluted from the SPE cartridge with a high ACN content, and they are then ready for immediate injection into the HILIC separation system, without need for evaporation/reconstitution or dilution [16,84]. Song determined that methyl-*t*-butyl ether (MTBE), ethyl acetate, and diethyl ether could be injected directly onto a silica-based HILIC column with no deterioration in peak shape [84]. As a result of these efforts, it was demonstrated that the time-consuming steps involved in LLE (i.e., evaporation and reconstitution) could be eliminated because of the compatibility of the extraction solvent (weak HILIC solvent) with the HILIC mobile phase.

To reduce the matrix effect of complex biological samples in HILIC-ESI-MS analysis, various clean-up procedures can be used during sample preparation, such as SPE [85–87], liquid-liquid extraction (LLE) [81,88,89], organic dilution, or on-line trap column [90]. However, the most effective method to address matrix effects during LC-ESI-MS is the isotope dilution technique, where a stable isotopically-labeled version of the analyte(s) of interest is incorporated as an internal standard (IS) [91,92]. It has been widely applied in many fields because of its simplicity and reliability to compensate for variations in instrumental response. Even so, the availability and cost of stable isotope-labeled IS variants can be a major problem for targeted metabolomics, especially intracellular metabolomics or flux analysis. As a new means to address this challenge, Heijnen's group has developed a method in which quantitation of metabolites of interest in cell extracts was performed using mass isotopomer ratio analysis of U-13C labeled extracts (MIRACLE) [93]. The method involves providing a C-13 carbon source during cell culture and co-extracting a known amount of U-13C labeled cells with the unlabeled samples. An improved precision was obtained compared with conventional analysis without IS but only relative ratios between unlabeled and labeled metabolites were determined. The method did not allow analysis of the absolute metabolite concentrations in samples. They improved the method using the labeled cell extracts as an IS [94]. The use of an IS effectively reduced errors in instrumental response and increased the robustness as well as precision in metabolome quantitative analyses.

7.4 SUMMARY

For both novice and advanced practitioners, HILIC-MS for targeted metabolomics and bioanalysis remains a challenge. There are many significant benefits of combining a separation mode that provides retention of highly polar/hydrophilic analytes with a detection mode, such as ESI-MS, that can offer excellent sensitivity and specificity. However, the challenges include choosing an appropriate stationary phase and analytical conditions to attain reproducible and reliable results. It is certain that more than one phase can possibly work for a given application, but some will be better than others. It is the view of the authors that general selection strategies to choose a phase for a given analyte target are still being investigated. Rather, the exclusion of some phases based on prior work may be more reliable. Thus, a trial and error process must still be invoked to some extent, and this is made more difficult by the plethora of commercial products offered by manufacturers. Once a phase is selected, method development to attain reproducible separations with symmetrical peaks will still require some time. Luckily, the choices of appropriate mobile phase compositions for HILIC-MS [acetonitrile/water with some volatile buffer (e.g., ammonium acetate or ammonium formate) and pH modifiers] limits trial and error in this regard, to some extent. Even so, the prevalence of mixed mode interactions will convolute efforts, and even slight changes in buffer content, pH, or even sample injection solvent, can have marked effects. A systematic design of experiments approach, as described in Chapter 3, can be helpful in investigating the effects of parameters that may have a complex impact on results.

The goal of targeted metabolomics is trace quantitative analysis of defined metabolites, most often from complex matrices. While MS, specifically ESI-MS combined with MS/MS techniques, represents the most attractive approach with regard to sensitivity and specificity, special considerations have to also be given here, in a system-dependent fashion. More specifically, different analytes to be determined in different matrices will have different propensities for being subject to matrix effects that can change analyte response. Method validation to establish the precision, accuracy, limits of detection, and linear ranges under strict guidelines for robustness are recommended. Even small changes in sample, sample preparation, and operating conditions, partially demonstrated in this work by our studies on estrogen metabolites, can have significant effects on these important analytical figures of merit, and such changes should be carefully characterized. Nevertheless, there is little doubt that the use of HILIC-MS is here to stay, and that every time a new application is reported, new information can be gathered to add to our base of knowledge and enable informed decisions. As metabolomics continues to grow in importance as a field, researchers will need to continuously review the literature in order to make the most effective use of the growing technology base in the area of HILIC-MS analysis.

ACKNOWLEDGMENTS

The authors wish to acknowledge support for this work from the National Science Foundation (CHE-0846310), Shimadzu Scientific Instruments, Varian (Agilent), and Eli Lilly and Company.

REFERENCES

1. Xu RN, Fan L, Rieser MJ, El-Shourbagy TA. Recent advances in high-throughput quantitative bioanalysis by LC-MS/MS. *J. Pharm. Biomed. Anal.* 2007; **44**: 342–355.

2. Hsieh Y. Potential of HILIC-MS in quantitative bioanalysis of drugs and drug metabolites. *J. Sep. Sci.* 2008; **31**: 1481–1491.

3. Xu RN, Rieser MJ, El-Shourbagy TA. Bioanalytical hydrophilic interaction chromatography: Recent challenges, solutions and applications. *Bioanalysis* 2009; **1**: 239–253.

4. Hebb ALO, Bhan V, Wishart AD, Moore CS, Robertson GS, Hsieh Y. Hydrophilic interaction liquid chromatography-tandem mass spectrometry for drug development. *Curr. Drug Discov. Technol.* 2010; **7**: 223–231.

5. Novakova L, Vickova H. A review of current trends and advances in modern bio-analytical methods: Chromatography and sample preparation. *Anal. Chim. Acta* 2009; **656**: 8–35.

6. Lu W, Bennett BD, Rabinowitz JD. Analytical strategies for LC-MS-based targeted metabolomics. *J. Chromatogr. B* 2008; **871**: 236–242.

7. Toniolo P, Boffetta P, Shuker DEG, Rothman N, Hulka B, Pearce N, eds. Application of biomarkers in cancer epidemiology, workshop report. *ARC Sci. Publ.* 1997; **142**: 1–335.

8. Albertini R, Bird M, Doerrer N, Needham L, Robison S, Sheldon L, Zenick H. The use of biomonitoring data in exposure and human health risk assessments. *Environ. Health Persp.* 2006; **114**: 1755–1762.

9. Chen C, Gonzalez FJ. LC-MS-based metabolomics in drug metabolism. *Drug Metab. Rev.* 2007; **39**: 581–597.

10. Nicholson JK, Lindon JC, Homes E. "Metabonomics": Understanding the metabolic responses of living systems to pathophysiological stimuli via multivariate statistical analysis of biological NMR data. *Xenobiotica* 1999; **29**: 1181–1189.

11. Robertson DG, Reily MD, Sigler RE, Wells DF, Paterson DA, Braden TK. Metabonomics: Evaluation of nuclear magnetic resonance (NMR) and pattern recognition technology for rapid in vivo screening of liver and kidney toxicants. *Toxicol. Sci.* 2000; **57**: 326–337.

12. Gullberg J, Jonsson P, Nordstrom A, Sjostrom M, Moritz T. Design of experiments: An efficient strategy to identify factors influencing extraction and derivatization of *Arabidopsis thaliana* samples in metabolomic studies with gas chromatography/mass spectrometry. *Anal. Biochem.* 2004; **331**: 283–295.

13. Monton MRN, Soga T. Metabolome analysis by capillary electrophoresis–mass spectrometry. *J. Chromatogr. A* 2007; **1168**: 237–246.

14. Sumner LW. Current status and forward looking thoughts on LC/MS metabolomics. *Biotechnol. Agric. Forest.* 2006; **57**: 21–32.

15. Spagou K, Tsoukali H, Raikos N, Gika H, Wilson ID, Theodoridis G. Hydrophilic interaction chromatography coupled to MS for metabonomic/metabolomic studies. *J. Sep. Sci.* 2010; **33**: 716–727.

16. Nguyen H, Schug KA. The advantages of ESI-MS detection in conjunction with HILIC mode separations: Fundamentals and applications. *J. Sep. Sci.* 2008; **31**: 1465–1480.

17. Cubbon S, Antonio C, Wilson J, Thomas-Oasters J. Metabolomic applications of HILIC–LC–MS. *Mass Spectrom. Rev.* 2010; **29**: 671–684.

18. Hollywood K, Brison DR, Goodacre R. Metabolomics: Current technologies and future trends. *Proteomics* 2006; **6**: 4716–4723.

19. Soga T, Ueno Y, Naraoka H, Ohashi Y, Tomita M, Nishioka T. Simultaneous determination of anionic intermediates for *Bacillus subtilis* metabolic pathways by capillary electrophoresis electrospray ionization mass spectrometry. *Anal. Chem.* 2002; **74**: 2233–2239.

20. Soga T, Ohashi Y, Ueno Y, Naraoka H, Tomita M, Nishioka T. Quantitative metabolome analysis using capillary electrophoresis mass spectrometry. *J. Proteome Res.* 2003; **2**: 488–494.

21. Ullsten S, Danielsson R, Backstrom D, Sjoberg P, Bergquist J. Urine profiling using capillary electrophoresis-mass spectrometry and multivariate data analysis. *J. Chromatogr. A* 2006; **1117**: 87–93.

22. Lenz EM, Wilson ID. Analytical strategies in metabonomics. *J. Proteome Res.* 2007; **6**: 443–458.

23. Wilson ID, Plum R, Granger J, Major H, Williams R, Lenz EM. HPLC-MS-based methods for the study of metabonomics. *J. Chromatogr. B* 2005; **817**: 67–76.

24. Lu WY, Kimball E, Rabinowitz JD. A high-performance liquid chromatography-tandem mass spectrometry method for quantitation of nitrogen-containing intracellular metabolites. *J. Am. Soc. Mass Spectrom.* 2006; **17**: 37–50.

25. Jaitz L, Mueller B, Koellensperger G, Huber D, Oburger E, Puschenreiter M, Hann S. LC–MS analysis of low molecular weight organic acids derived from root exudation. *Anal. Bioanal. Chem.* 2011; **400**: 2587–2596.

26. Tsukamoto Y, Santa T, Saimaru H, Imai K, Funaysu T. Synthesis of benzofurazan derivatization reagents for carboxylic acids and its application to analysis of fatty acids in rat plasma by high-performance liquid chromatography–electrospray ionization mass spectrometry. *Biomed. Chromatogr.* 2005; **19**: 802–808.

27. Luo B, Groenke K, Takors R, Wandrey C, Oldiges M. Simultaneous determination of multiple intracellular metabolites in glycolysis, pentose phosphate pathway and tricarboxylic acid cycle by liquid chromatography–mass spectrometry. *J. Chromatogr. A* 2007; **1147**: 153–164.

28. Kiefer P, Delmotte N, Vorholt JA. Nanoscale ion-pair reversed-phase HPLC–MS for sensitive metabolome analysis. *Anal. Chem.* 2011; **83**: 850–855.

29. Coulier L, Bas R, Jespersen S, Verheij E, Van Der Werf MJ, Hankemeier T. Simultaneous quantitative analysis of metabolites using ion-pair liquid chromatography – electrospray ionization mass spectrometry. *Anal. Chem.* 2006; **78**: 6573–6582.

30. Zoppa M, Gallo L, Zacchello F, Giordano G. Method for the quantification of underivatized amino acids on dry blood spots from newborn screening by HPLC–ESI–MS/MS. *J. Chromatogr. B* 2006; **831**: 267–273.

31. Armstrong M, Jonscher K, Reisdorph NA. Analysis of 25 underivatized amino acids in human plasma using ion-pairing reversed-phase liquid chromatography/time-of-flight mass spectrometry. *Rapid Commun. Mass Spectrom.* 2007; **21**: 2717–2726.

32. Dettmer K, Aronov PA, Hammock BD. Mass spectrometry-based metabolomics. *Mass Spectrom. Rev.* 2007; **26**: 51–78.

33. Van Der Werf MJ, Overkamp KM, Muilwijk B, Coulier L, Hankemeier T. Microbial metabolomics: Toward a platform with full metabolome coverage. *Anal. Biochem.* 2007; **370**: 17–25.

34. Karala AR, Ruddock LW. Does S-methyl methanethiosulfonate trap the thiol–disulfide state of proteins? *Antioxid. Redox Signal.* 2007; **9**: 527–531.

35. Shortreed MR, Lamos SM, Frey BL, Phillips MF, Patel M, Belshaw PJ, Smith LM. Ionizable isotopic labeling reagent for relative quantification of amine metabolites by mass spectrometry. *Anal. Chem.* 2006; **78**: 6398–6403.

36. Lamos SM, Shortreed MR, Frey BL, Belshaw PJ, Smith LM. Relative quantification of carboxylic acid metabolites by liquid chromatography–mass spectrometry using isotopic variants of cholamine. *Anal.Chem.* 2007; **79**: 5143–5149.

37. Tsukamoto Y, Santa T, Saimaru H, Imai K, Funatsu T. Synthesis of benzofurazan derivatization reagents for carboxylic acids and its application to analysis of fatty acids in rat plasma by high-performance liquid chromatography–electrospray ionization mass spectrometry. *Biomed. Chromatogr.* 2005; **19**: 802–808.

38. Xia YQ, Chang SW, Patel S, Bakhtiar R, Karanam B, Evans DC. Trace level quantification of deuterated 17β-estradiol and estrone in ovariectomized mouse plasma and brain using liquid chromatography/tandem mass spectrometry following dansylation reaction. *Rapid Commun. Mass Spectrom.* 2004; **18**: 1621–1628.

39. Nelson RE, Grebe SK, O'Kane DJ, Singh RJ. Liquid Chromatography–tandem mass spectrometry assay for simultaneous measurement of estradiol and estrone in human plasma. *Clin. Chem.* 2004; **50**: 373–384.

40. Toran-Allerand CD, Tinnikov AA, Singh RJ, Nethrapalli IS. 17α-estradiol: A brain-active estrogen? *Endocrinology* 2005; **146**: 3843–3850.

41. Anari MR, Bakhtiar R, Zhu B, Huskey S, Franklin RB, Evans DC. Derivatization of ethinylestradiol with dansyl chloride to enhance electrospray ionization:? Application in trace analysis of ethinylestradiol in rhesus monkey plasma. *Anal. Chem.* 2002; **74**: 4136–4144.

42. Kushnir MM, Rockwood AL, Bergquist J. Liquid chromatography–tandem mass spectrometry applications in endocrinology. *Mass Spectrom. Rev.* 2010; **29**: 480–502.

43. Nguyen HP, Li L, Gatson JW, Maass D, Wigginton JG, Simpkins JW, Schug KA. Simultaneous quantification of four native estrogen hormones at trace levels in human cerebrospinal fluid using liquid chromatography–tandem mass spectrometry. *J. Pharm. Biomed. Anal.* 2011; **54**: 830–837.

44. Nguyen HP, Li L, Nethrapalli IS, Guo N, Toran-Allerand CD, Harrison DE, Astle CM, Schug KA. Evaluation of matrix effects in analysis of estrogen using liquid chromatography–tandem mass spectrometry. *J. Sep. Sci.* 2011; **34**: 1781–1787.

45. Furch T, Preusse M, Tomasch J, Zech H, Wagner-Dobler I, Rabus R, Wittmann C. Metabolic fluxes in the central carbon metabolism of *Dinoroseobacter shibae* and *Phaeobacter gallaeciensis*, two members of the marine *Roseobacter* clade. *BMC Microbiol.* 2009; **9**: 209.

46. Rhee KY, Sorio De Carvalho LP, Bryk R, Ehrt S, Marrero J, Park SW, Schnappinger D, Venugopal A, Nathan C. Central carbon metabolism in *Mycobacterium tuberculosis*: An unexpected frontier. *Trends Microbiol.* 2011; **19**: 307–314.

47. Yang L, Kombu RS, Kasumov T, Zhu SH, Candrowski AV, David F, Anderson V, Kelleher JK, Brunengraber H. Metabolomic and mass isotopomer analysis of liver gluconeogenesis and citric acid cycle. *J. Biol. Chem.* 2008; **283**: 21978–21987.

48. Ma S, Chowdhury SK. Analytical strategies for assessment of human metabolites in preclinical safety testing. *Anal. Chem.* 2011; **83**: 5028–5036.

49. Stahlberg J. Retention models for ions in chromatography. *J. Chromatogr. A* 1999; **855**: 3–55.

50. Tolstikov VV, Fiehn O. Analysis of highly polar compounds of plant origin: Combination of hydrophilic interaction chromatography and electrospray ion trap mass spectrometry. *Anal. Biochem.* 2002; **301**: 298–307.

51. Langrock T, Czihal P, Hoffmann R. Amino acid analysis by hydrophilic interaction chromatography coupled on-line to electrospray ionization mass spectrometry. *Amino Acids* 2006; **30**: 291–297.

52. Bajad SU, Lu W, Kimball EH, Yuan J, Peterson C, Rabinowitz JD. Separation and quantitation of water soluble cellular metabolites by hydrophilic interaction chromatography-tandem mass spectrometry. *J. Chromatogr. A* 2006; **1125**: 76–88.

53. Pesek JJ, Matyska MT, Fischer SM, Sana TR. Analysis of hydrophilic metabolites by high-performance liquid chromatography–mass spectrometry using a silica hydride-based stationary phase. *J. Chromatogr. A* 2008; **1204**: 48–55.

54. Onorato JM, Langish R, Bellamine A, Shipkova P. Applications of HILIC for targeted and non-targeted LC/MS analyses in drug discovery. *J. Sep. Sci.* 2010; **33**: 923–929.

55. Padivitage NLT, Armstrong DW. Sulfonated cyclofructan 6 based stationary phase for hydrophilic interaction chromatography. *J. Sep. Sci.* 2011; **34**: 1636–1647.

56. Qiu H, Loukotkova L, Sun P, Tesarova E, Bosakova Z, Armstrong DW. Cyclofructan 6 based stationary phases for hydrophilic interaction liquid chromatography. *J. Chromatogr. A* 2011; **1218**: 270–279.

57. Alvarez-Sanchez B, Priego-Capote F, Mata-Granados JM, Luque De Castro MD. Automated determination of folate catabolites in human biofluids (urine, breast milk and serum) by on-line SPE–HILIC–MS/MS. *J. Chromatogr. A* 2010; **1217**: 4688–4695.

58. Georgakakou S, Kazanis M, Panderi I. Hydrophilic interaction liquid chromatography/positive ion electrospray ionization mass spectrometry method for the quantification of perindopril and its main metabolite in human plasma. *Anal. Bioanal. Chem.* 2010; **397**: 2161–2170.

59. Kolmonen M, Leinonen A, Kuuranne T, Pelander A, Ojanpera L. Hydrophilic interaction liquid chromatography and accurate mass measurement for quantification and confirmation of morphine, codeine and their glucuronide conjugates in human urine. *J. Chromatogr. B* 2010; **878**: 2959–2966.

60. Eckert E, Drexler H, Goen T. Determination of six hydroxyalkyl mercapturic acids in human urine using hydrophilic interaction liquid chromatography with tandem mass spectrometry (HILIC–ESI-MS/MS). *J. Chromatogr. B* 2010; **878**: 2506–2514.

61. Jian W, Edom RQ, Xu Y, Gallagher J, Weng N. Potential bias and mitigations when using stable isotope labeled parent drug as internal standard for LC–MS/MS quantitation of metabolites. *J. Chromatogr. B* 2010; **878**: 3267–3276.

62. Keevil BG, Owen L, Thornton S, Kavanagh J. Measurement of citrate in urine using liquid chromatography tandem mass spectrometry: Comparison with an enzymatic method. *Ann. Clin. Biochem.* 2005; **42**: 357–363.

63. Fernandez-Fernandez R, Lopez-Martinez JC, Romero-Gonzalez R, Martinez-Vidal JL, Flores MIA, Frenich AG. Simple LC–MS determination of citric and malic acids in fruits and vegetables. *Chromatographia* 2010; **72**: 55–61.

64. Naidong W. Bioanalytical liquid chromatography tandem mass spectrometry methods on underivatized silica columns with aqueous/organic mobile phases. *J. Chromatogr. B* 2003; **796**: 209–224.

65. Dejaegher B, Heyden YV. HILIC methods in pharmaceutical analysis. *J. Sep. Sci.* 2010; **33**: 698–715.

66. Yoshida H, Yamazaki J, Ozawa S, Mizukoshi T, Miyano H. Advantage of LC-MS metabolomics methodology targeting hydrophilic compounds in the studies of fermented food samples. *J. Agric. Food Chem.* 2009; **57**: 1119–1126.

67. Berthod A, Chang SSC, Kullman JPS, Armstrong DW. Practice and mechanism of HPLC oligosaccharide separation with a cyclodextrin bonded phase. *Talanta* 1998; **47**: 1001–1012.

68. Nguyen HP, Yang SH, Wigginton JG, Simpkins JW, Schug KA. Retention behavior of estrogen metabolites on hydrophilic interaction chromatography stationary phases. *J. Sep. Sci.* 2010; **33**: 793–802.

69. Liu M, Chen EX, Ji R, Semin D. Stability-indicating hydrophilic interaction liquid chromatography method for highly polar and basic compounds. *J. Chromatogr. A* 2008; **1188**: 255–263.

70. Guo Y, Gaiki S. Retention behavior of small polar compounds on polar stationary phases in hydrophilic interaction chromatography. *J. Chromatogr. A* 2005; **1074**: 71–80.

71. Olsen BA. Hydrophilic interaction chromatography using amino and silica columns for the determination of polar pharmaceuticals and impurities. *J. Chromatogr. A* 2001; **913**: 113–122.

72. Jandera P. Stationary and mobile phases in hydrophilic interaction chromatography: A review. *Anal. Chim. Acta* 2011; **692**: 1–25.

73. Chirita RL, West C, Zubrzycki S, Finaru AL, Elfakir C. Investigations on the chromatographic behaviour of zwitterionic stationary phases used in hydrophilic interaction chromatography. *J. Chromatogr. A* 2011; **1218**: 5939–5963.

74. McCalley DV. Is hydrophilic interaction chromatography with silica columns a viable alternative to reversed-phase liquid chromatography for the analysis of ionisable compounds? *J. Chromatogr. A* 2007; **1171**: 46–55.

75. Edwards JL, Edwards RL, Reid KR, Kennedy RT. Effect of decreasing column inner diameter and use of off-line two-dimensional chromatography on metabolite detection in complex mixtures. *J. Chromatogr. A* 2007; **1172**: 127–134.

76. Faircjild JN, Horvath K, Gooding JR, Campagna SR, Guiochon G. Two-dimensional liquid chromatography/mass spectrometry/mass spectrometry separation of water-soluble metabolites. *J. Chromatogr. A* 2010; **1217**: 8161–8166.

77. Espinosa S, Bosch E, Roses M. Retention of ionizable compounds on HPLC. 12. The properties of liquid chromatography buffers in acetonitrile–water mobile phases that influence HPLC retention. *Anal. Chem.* 2002; **74**: 3809–3818.

78. Gagliardi LG, Castells CB, Rafols C, Roses M, Bosch E. δ conversion parameter between ph scales ($_w^s$pH and $_s^s$pH) in acetonitrile/water mixtures at various compositions and temperatures. *Anal. Chem.* 2007; **79**: 3180–3187.

79. Little JL, Wempe MF, Buchanan CM. Liquid chromatography–mass spectrometry/mass spectrometry method development for drug metabolism studies: Examining lipid matrix ionization effects in plasma. *J. Chromatogr. B* 2006; **833**: 219–230.

80. Chambers E, Wagrowski-Diehl DM, Lu Z, Mazzeo JR. Systematic and comprehensive strategy for reducing matrix effects in LC/MS/MS analyses. *J. Chromatogr. B* 2007; **852**: 22–34.

81. Park EJ, Lee HW, Ji HY, Kim HY, Lee MH, Park ES, Lee KC, Lee HS. Hydrophilic interaction chromatography-tandem mass spectrometry of donepezil in human plasma: Application to a pharmacokinetic study of donepezil in volunteers. *Arch. Pharm. Res.* 2008; **31**: 1205–1211.

82. Jian W, Edom RW, Xu Y, Weng N. Recent advances in application of hydrophilic interaction chromatography for quantitative bioanalysis. *J. Sep. Sci.* 2010; **33**: 681–697.

83. Tippens HD, Nguyen HP, Schug KA. Evaluation of matrix effects from urine for estrogen metabolites using HILIC coupled with ESI-MS. Proc 59th ASMS Conference on Mass Spectrometry and Allied Topics. June, 2011.

84. Song Q, Naidong W. Analysis of omeprazole and 5-OH omeprazole in human plasma using hydrophilic interaction chromatography with tandem mass spectrometry (HILIC-MS/MS)–eliminating evaporation and reconstitution steps in 96-well liquid/liquid extraction. *J. Chromatogr. B* 2006; **830**: 135–142.

85. Swaim LL, Johnson RC, Zhou Y, Sandlin C, Barr JR. Quantification of organophosphorus nerve agent metabolites using a reduced-volume, high-throughput sample processing format and liquid chromatography-tandem mass spectrometry. *J. Anal. Toxicol.* 2008; **32**: 774–777.

86. Qin F, Zhao YY, Sawyer MB, Li XF. Hydrophilic interaction liquid chromatography–tandem mass spectrometry determination of estrogen conjugates in human urine. *Anal. Chem.* 2008; **80**: 3404–3411.

87. Johnson RC, Zhou Y, Statler K, Thomas J, Cox F, Hall S, Barr JR. Quantification of saxitoxin and neosaxitoxin in human urine utilizing isotope dilution tandem mass spectrometry. *J. Anal. Toxicol.* 2009; **33**: 8–14.

88. Mawhinney DB, Hamelin EI, Fraser R, Silva SS, Pavlopoulos AJ, Kobelski RJ. The determination of organophosphonate nerve agent metabolites in human urine by hydrophilic interaction liquid chromatography tandem mass spectrometry. *J. Chromatogr. B* 2007; **852**: 235–243.

89. Onarato JM, Langish RA, Shipkova PA, Sanders M, Wang J, Kwagh J, Dutta S. A novel method for the determination of 1,5-anhydroglucitol, a glycemic marker, in human urine utilizing hydrophilic interaction liquid chromatography/MS^3. *J. Chromatogr. B* 2008; **873**: 144–150.

90. Kopp EK, Sieber M, Kellert M, Dekant W. Rapid and sensitive HILIC-ESI-MS/MS quantitation of polar metabolites of acrylamide in human urine using column switching with an online trap column. *J. Agric. Food Chem.* 2008; **56**: 9828–9834.

91. Matuszewski BK, Constanzer ML, Chavez-Eng CM. Strategies for the assessment of matrix effect in quantitative bioanalytical methods based on HPLC–MS/MS. *Anal. Chem.* 2003; **75**: 3019–3030.

92. Grant RP. High throughput automated LC-MS/MS analysis of endogenous small molecule biomarkers. *Clin. Lab. Med.* 2011; **31**: 429–441.

93. Mashego MR, Wu L, Van Dam JC, Ras C, Vinke JL, Van Winden WA, Van Gulik WM, Heijnen JJ. MIRACLE: Mass isotopomer ratio analysis of U-13C-labeled extracts. A new method for accurate quantification of changes in concentrations of intracellular metabolites. *Biotechnol. Bioeng.* 2004; **85**: 620–628.

94. Wu L, Mashego MR, Van Dam JC, Proell AM, Vinke JL, Ras C, Van Winden WA, Van Gulik WM, Heijnen JJ. Quantitative analysis of the microbial metabolome by isotope dilution mass spectrometry using uniformly ^{13}C-labeled cell extracts as internal standards. *Anal. Biochem.* 2005; **336**: 164–171.

CHAPTER

8

HILIC FOR FOOD, ENVIRONMENTAL, AND OTHER APPLICATIONS

MICHAEL A. KOUPPARIS, NIKOLAOS C. MEGOULAS,
and AIKATERINI M. GREMILOGIANNI

*Laboratory of Analytical Chemistry, Department of Chemistry,
University of Athens, Athens, Greece*

8.1 INTRODUCTION

Among all separation modes in liquid chromatography, reversed-phase liquid chromatography (RPLC) is by far the most frequently used in food analysis. However, several drawbacks have limited the use of RPLC, including the necessity for the determination of low molecular weight and/or very polar compounds. These kinds of compounds normally have insufficient retention in typical reversed-phase (RP) columns, making their separation from polar matrix interferences impossible. Several methods using ion-pairing reagents in the mobile phase have been used in order to increase the retention time of these compounds. There are two main disadvantages of the use of these additives: first, they are not easily removed from the packing material of the RP columns and second, most of them are not compatible with mass spectrometric detection because they are nonvolatile. Even if volatile ion-pair reagents are used in combination with mass spectrometric detection, it has been proven

Hydrophilic Interaction Chromatography: A Guide for Practitioners, First Edition.
Edited by Bernard A. Olsen and Brian W. Pack.
© 2013 John Wiley & Sons, Inc. Published 2013 by John Wiley & Sons, Inc.

that ion suppression is caused, in particular in electrospray ionization (ESI) mode. The introduction of HILIC in food and environmental analysis has helped to overcome these drawbacks, as has been pointed out in recent reviews.

Bernal et al. [1] focused on food analysis applications and their categorization according to food matrix, while Van Nuijs et al. [2] focused on polar contaminants in food and environmental matrices.

8.2   FOOD APPLICATIONS FOR HILIC

8.2.1   Review of HILIC Analytical Methods for Food Analysis

An analytical method for food analysis commonly comprises two stages. The first includes extraction and purification of the analytes from the matrix and the second concerns the chromatographic separation and detection of the target analytes. Nevertheless, the optimization of these two stages cannot be independently achieved, and their interaction always must be taken into consideration.

In the following two sections, the two stages of HILIC analytical methods in food analysis are reviewed.

8.2.1.1   Sample Preparation in HILIC Methods Applied to Food Matrices

Due to the complexity of the food matrices with high lipid or protein content, more sophisticated extraction protocols and further purification steps are required. Liquid–liquid extraction (LLE) is a common technique [3–15], which has several advantages, such as simplicity, speed, and low cost. In the cases of multicomponent determinations [16–19] or when liquid chromatography is combined with mass spectrometric detection [20–25], a solid-phase extraction (SPE) for separation and/or purification is used [21,23].

The determination of phospholipids in eggs yolks [6] is a characteristic example of LLE, using a mixture of chloroform/methanol/water (1:2:0.8, v/v) as the extraction solvent. The addition of the extraction mixture was followed by homogenization and centrifugation. The extract was dried under a nitrogen stream and the residue was redissolved in methanol. In the case of the analysis of glycosinolates from broccoli seeds by HILIC [7], the extraction medium was a mixture of equal volumes of dimethylsulphoxide, dimethylformamide, and acetonitrile (ACN). After evaporation of the organic extraction, the residue was redissolved in the mobile phase buffer.

In some methods [4,11,13–15], acidic solutions were added in order to denature and precipitate the matrix proteins and protonate the basic compounds to be transferred into the aqueous phase. However, when an acidic solution is used, a neutralization (pH 6.5–7) step is required before the extraction. This step is usually performed with the addition of solid potassium carbonate [11,15] or potassium hydroxide aqueous solution [14]. It is worth mentioning also that after the extraction and prior to injection further dilution with ACN [4,11,15] or ACN with 10% formic acid and water (1:9, v/v) was necessary.

LLE has also been used in the case of screening, identification, and confirmation of veterinary drugs belonging in different classes in food matrices [12,13]. In the later method [13], for the multiclass analysis of veterinary drugs in chicken muscle, a mixture of 2% trichloroacetic acid (TCA) aqueous solution and ACN was used for the precipitation of the proteins. The effect of the concentration of TCA solution in the mixture with ACN on the recovery of the drugs was thoroughly investigated. It was concluded that low concentrations of TCA solution (1%) led to low recoveries of streptomycin, dihydrostreptomycin, and pencillamine, whereas increased concentrations resulted in low recoveries and/or signal suppression, particularly in the case of erythromycin (recovery <20%). Finally, 2% TCA in water–ACN (1:1, v/v) was selected as a compromise for satisfactory recoveries for all analytes by minimizing the signal suppression and degradation of analytes.

As mentioned earlier, SPE is another extraction clean-up procedure adopted in food analysis based on HILIC separation mode. All kinds of SPE sorbents have been used, including silica-based C_{18}-bonded phases [16,25,26] and polymeric sorbents [21,22]. Due to the polar characteristic of most analytes, ion-exchange sorbents (strong or weak cation or anion exchange) [18,24] or mixed-mode sorbents [17,23] have also been used.

In the determination of seven biogenic amines (BAs) in cheese [16], after the precipitation of proteins and the removal of fats with hydrochloric acid solution (0.1 M), a clean-up step with the use of SPE cartridges followed. The extract containing the BAs was alkalized to pH 11 in order to deprotonate the amines and facilitate their adsorption on the C_{18} sorbent (Phenomenex). Their elution was achieved by the use of a methanol/ACN mixture (80:20 v/v) adjusted to pH 3 with the addition of formic acid. It is worthwhile to mention that the removal step of the elution solvent by evaporation under nitrogen stream is avoided, due to the compatibility of the elution solvent with the mobile phase of this HILIC application. The compatability of HILIC with extraction solvents rich in organic solvent is a significant advantage of HILIC over RPLC in which distorted or split peaks can result from mismatches between the injection solvent and the primarily aqueous mobile phase.

In the determination of chlormequat and mepiquat in food samples [17], the elution solvent required further dilution with ACN (mobile phase component) prior to injection, without evaporation to dryness. In this particular method, a mixture of methanol/water (20:80 v/v) was used for the elution of the two quaternary ammonium compounds from Supelclean™ ENVI™-18 SPE cartridges. This particular elution mixture was found to have high elutropic strength in the HILIC separation, so further dilution with ACN (1:1, v/v) was needed to improve the chromatographic peak shape.

In methods developed for the determination of perchlorate [23] and acrylamide [21], an SPE step was added in the sample preparation for "chemical filtration" of the interfering components of the matrix. In the determination of perchlorate in milk and milk powder [23], a Strata X SPE cartridge was used to retain a wide range of matrix interferences (basic, neutral, and acidic compounds). Perchlorate ions simply pass though the cartridge and their recovery was assisted by a washing step of the cartridge with water. In the case of the determination of acrylamide [21], which can be formed during heat treatment of food, two SPE steps were included for extraction and purification. The first step was performed on an Isolute Multimode cartridge for the removal of matrix components from the water extracts. The second SPE step was applied for enrichment and further purification using Isolute ENV+ (hydroxylated polystyrene-divinylbenzene copolymer) column, selected after investigation of a large variety of cartridges [21].

8.2.1.2   HILIC Methods Applied to Food Matrices: Chromatographic Parameters and Detection

The term "chromatographic parameters" usually refers to the type and the temperature of the stationary phase, and the composition and flow rate of the mobile phase. In general, the stationary phases that are used in the HILIC mode can be divided in two categories on the basis of the column support: silica-based and the polymer-based. Different types of moieties have been bonded to these supports, giving a wide variety of commercially available columns with different functionalities for HILIC applications [27–29, also see Chapter 2].

Different types of mobile phases have been applied for HILIC applications, consisting of a high content of organic solvent, usually ACN, and a lower content of an aqueous solution. The aqueous solution could be pure water or a buffer solution with a pH-affecting additive. Several chromatographic factors are optimized during method development for HILIC assays [30, also see Chapter 3], but very few optimization experimental data are shown in the published methods. Some selected examples of optimization experiments and the conclusions derived from them are briefly discussed below.

Esparza et al. [17] investigated the chromatographic behavior of two quaternary ammonium compounds using aqueous buffer solutions with varying composition, pH value, and ionic strength, in combination with ACN as the organic solvent. The separation was performed on a bare silica column (Atlantis HILIC, Waters), and it was found that in general, at pH values of the aqueous buffer higher than 4.25, the increased interaction of the cationic analytes with the deprotonated silanol groups resulted in higher retention times and wider peaks. A similar effect was observed with a pH of 3.75 using a buffer with lower ionic strength. From the detailed investigation of the effect of pH and ionic strength, a pH of 3.5 using ammonium formate buffer was found as optimum.

Similar parameters were investigated in the study of Mora et al. [15] for the determination of adenosine and its metabolites in pork muscles, where ACN was the chosen organic solvent in combination with aqueous ammonium acetate. The separation was performed on a zwitterionic stationary phase (ZIC-pHILIC). The study highlighted that ammonium acetate was found as optimum regarding its solubility in the organic solvent of the mobile phase and its interaction with the analytes, resulting in best results in selectivity and reproducibility. In addition, in this study a comparison between the HILIC and a RPLC method for the determination of the analytes under the same extraction procedure was performed. In the RPLC method, the mobile phase contained an ion-pairing reagent (tetrabutyl ammonium hydrogen sulfate) and the separation was performed on a Zorbax Eclipse XDB-C18 (Agilent Technologies) column. From the results, the two methods were concluded to be equivalent, rendering the HILIC method as a reliable alternative to the RPLC methods commonly used for these analytes [31,32]. It is worthwhile to mention that the HILIC method has the further advantage that can be combined with mass spectrometric detection.

In the case of the determination of two aminoglycoside antibiotics (streptomycin and dihydrostreptomycin) in milk [18], the comparison between a HILIC and an RPLC method, both with mass spectrometric detection, showed that the HILIC method was 80 and 210 times more sensitive for streptomycin and dihydrostreptomycin, respectively. Ion suppression occurred in ESI mode, when ion-pairing reagents are used, proving once again that HILIC can have an extra advantage when combined with mass detection [33, also see Chapter 7] on mass spectrometric detection.

8.2.2 Selected Detailed Examples of HILIC Applications in Food Analysis

In the section that follows, selected examples from the field of food analysis are presented. The categorization of the presented studies was based on the target molecules. Moreover, these molecules were selected according to their significance in terms of food safety or human health. As in the previous section, the extraction procedures used and the chromatographic parameters that were investigated are presented.

8.2.2.1 Melamine (MEL) and Cyanuric Acid (CYA)

MEL (Fig. 8.1) is a triazine ring with three amino groups and a widely used industrial compound in the manufacture of plastics, flame retardants, general heat tolerant products, fertilizers, and glues. However, the adulteration of the food commodities with these two nitrogen-rich chemicals fraudulently increases the apparent protein content. In the spring of 2007, MEL and another triazine compound, CYA (Fig. 8.1), were notoriously inserted in pet food and resulted in the death of many pets. Also, in September 2008, an enormous scandal concerned with melamine-tainted, protein-based food commodities such as liquid milk and milk powder for infants, biscuits, desserts, and dairy products was revealed in a number of countries. As a consequence, thousands

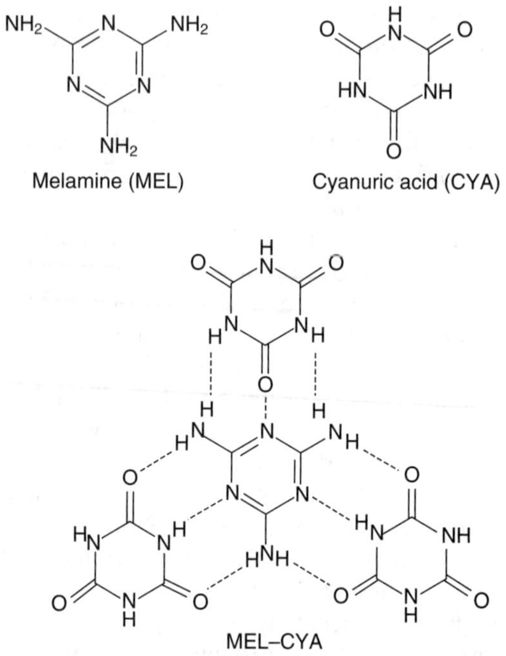

Figure 8.1. The chemical structures of melamine, cyanuric acid, and MEL-CYA. Reprinted from Reference 34 with permission from Elsevier.

of young Chinese children required medical treatment; moreover, six deaths of infants were reported by Ministry of Health of People's Republic of China.

The combination of MEL and CYA results in the formation of stable and in-soluble crystals that precipitate in the kidneys and cause renal failure. These serious impacts of the presence of the specific thiazine compounds in food stuffs led the U.S. Food and Drug Administration (FDA) and the European Food Safety Authority (EFSA) to issue tolerable daily intake (TDI) for MEL and its structural analogs, of 0.63 mg/kg (FDA) and 0.20 mg/kg (EFSA) body weight for adults, respectively. Concern about melamine adulteration of drug product ingredients has also been raised as described in Chapter 4.

Due to the dangerous consequences of melamine's presence in foods and the subsequent necessity for its accurate determination, several studies based on HILIC methods have been reported. MEL has been generally analyzed in different matrices including animal feeds [35], and in food stuffs, such as fish [36], meat [35,37], milk and milk products [38–41], flour and bakery goods, and honey [37,42]. Being a small and polar molecule, MEL is easily and accurately determined by HILIC. The study of Andersen et al., for the determination of MEL in multiple species of fish and shrimp, is a representative example [36]. In order to validate the method, they used extracts from fish muscles; this fish had been intensively fed with MEL, CYA, or a combination of MEL and CYA.

The elution of the target compounds was performed on a silica-based column (Atlantis HILIC, Waters) with gradient elution. The extraction procedure included precipitation of the proteins with 24 mL of ACN/water (50/50 v/v) and 1 mL of a solution 1 M hydrochloric acid.

One of the most significant problems in mass spectrometric detection is the matrix effect caused by co-eluted compounds from the matrix that causes suppression or enhancement of the ionization. In this case, the use of isotopic internal standards that have the same elution characteristics as the analytes is highly desirable. In the study of Varelis and Jeskelis [35], isotopic noncommercial internal standards were prepared and used for the accurate quantitative determination of both MEL and CYA in chicken, pork, catfish, and pet food. The extraction procedure involved two SPE steps, one with a strong cation exchange cartridge for the extraction of the MEL and another with a strong anion exchange cartridge for the extraction of the CYA.

Matrix effect was also studied in the work of MacMahon et al. [41], which focused on the determination of six compounds—among them MEL—that can be used in economic adulteration of protein-containing foods. From the comparison of 5-point calibration curves constructed using standard solutions in pure solvents and fortified matrix extracts, the matrix effect was studied and the results are presented in Table 8.1. As a consequence, the use of matrix-matched calibration standards is considered necessary for valid analytical results.

Simpler extraction procedures involving LLE were presented for the determination of MEL in dairy products [39] and bakery goods [38]. Acetonitrile or a mixture of ACN and water were used both for protein precipitation and as mobile phase component. In both cases, the extraction was ultrasonically assisted. In the case of the bakery goods, a purification step for the removal of the fat with the addition of dichloromethane followed the ACN extraction step. The chromatographic conditions and the detectors used are presented for most of the above-mentioned methods in Tables 8.2 and 8.3.

Table 8.1. Matrix Effect Results for Six Compounds: Cyromazine (CY), Triuret (TU), Biuret (BU), Dicyandiamine (DC), Melamine (MEL), Amidinourea (AU), in Different Food Matrices

Matrix	CY (%)	TU (%)	BU (%)	DC (%)	MEL (%)	AU (%)
Wheat gluten	93	86	58	40	67	65
Soy protein	86	82	86	92	80	45
Skim milk	101	102	102	104	102	105
Skim milk powder	102	106	104	107	106	108
Wheat flour	102	98	96	95%	92	69
Corn gluten meal	92	92	78	86	58	78

Source: Reprinted from Reference 40 with permission from Elsevier.

Table 8.2. Selected Examples of HILIC Applications in Food of Animal Origin.
Chromatographic Parameters and Sample Preparation Technique

Target Compound(s)	Analytes	Sample Matrix	Sample Pretreatment	Stationary Phase/Mobile Phase	Detection	Reference
Biogenic amines	Cadaverine, histamine, putrescine, spermidine, spermine, tryptamine, and tyramine	Cheese	SPE	Atlantis HILIC, Waters (150 × 2.1 mm ID, 3 μm). Gradient elution: acetonitrile (ACN) and aqueous ammonium formate 50.0 mM, pH 4 (with formic acid)	APCI-MS/MS TIS-MS/MS	[16]
		Fish tissues	LLE	TSKgel Amide-80 (150 × 2.0 mm ID, 3 μm). Gradient elution: ACN and aqueous ammonium formate 30.0 mM, pH 4 (with formic acid)	ESI-MS/MS	[47]
Small polar compounds	Free amino acids and small polar peptides	Parmesan cheese	Soxhlet extraction	TSKgel Amide-80 (250 × 1.5 mm ID, 5 μm) Gradient elution: Solvent A, ammonium acetate buffer (0.65 mM, pH 5.5) in ACN (90:10 v/v) Solvent B, ACN and ammonium acetate buffer (2.6 mM, pH 5.5) (60:40 v/v)	ESI-MS/MS	[84]
Water-soluble vitamins	Vitamin B1	Dry-curved sausages	LLE followed by SPE	Luna HILIC Phenomenex (150 × 3.0 mm ID, 3 μm). Gradient elution: Solvent A, ACN and aqueous ammonium acetate 50.0 mM pH 5.8 (90:10 v/v) Solvent B, ACN and aqueous ammonium acetate 10.0 mM pH 5.8 (50:50 v/v)	DAD	[45]
	Vitamin C and related compounds	Fish tissues	LLE	APS-2 Hypersil min (150 × 3.0 mm ID, 3 μm). Isocratic elution: ACN and aqueous ammonium acetate 100.0 mM (90:10 v/v)	UV-vis	[43]
		Ham	LLE	TSKgel Amide-80 (100 × 4.6 mm ID, 5 μm). Gradient elution: ACN and aqueous 0.1% trifluoroacetic acid	PDA	[44]

246

Food contaminants	Melamine	Fish tissues	LLE followed by SPE	Atlantis HILIC, Waters (50 × 3.0mm ID, 3μm). Gradient elution: ACN and aqueous ammonium formate 20.0mM	ESI-MS/MS	[35]
		Royal jelly	SPE	Zorbax Rx-Sil, Agilent (50 × 3.0mm ID, 3μm). Isocratic elution: ACN and aqueous ammonium formate 5.0mM (88:12 v/v)	ESI-MS/MS	[41]
		Milk, milk products	Ultrasound treatment and LLE	AQUITY™ BEH (100 × 2.1 mm ID, 1.7μm). Isocratic elution: ACN and water (90:10 v/v)	ESI-MS/MS	[37]
		Dairy products	Ultrasound treatment and LLE	Prevail Amino column, Alltech (150 × 4.6 mm ID, 5μm). Isocratic elution: ACN–H_3PO_4 (15mM adjusted pH6.5, 88:12 v/v)	UV-vis	[38]
	Melamine and cyanuric acid	Chicken, pork, and catfish	LLE followed by SPE	BioBasic AX, Thermo (150 × 2.0mm ID, 5μm). Gradient elution: ACN, isopropanol and aqueous ammonium formate 20.0mM	ESI-MS/MS	[34]
		Egg, pork, liver kidney, shrimp, sausage casing, honey, soybean milk, soybean powder and dairy products	Mixed-mode SPE	Atlantis HILIC, Waters (50 × 3.0 mm ID, 3μm). Gradient elution: ACN and aqueous ammonium formate 20.0 mM		[36]

(Continued)

247

Table 8.2. (Continued)

Target Compound(s)	Analytes	Sample Matrix	Sample Pretreatment	Stationary Phase/Mobile Phase	Detection	Reference
Marine biotoxins	PSP toxins	Mussels	LLE	TSKgel Amide-80, TosoHaas (250 × 2.0 mm ID, 5 μm). Isocratic elution: (A) 35% and (B) 65%. Where (A) is water and (B) ACN–water (95:5), both containing 2.0 mM ammonium formate and 3.6 mM formic acid (pH3.5)	ESI-MS/MS	[52,57]
	Domoic acid and PSP toxins	Mussels	LLE	Acquity BEH Amide, Waters (100 × 2.1 mm ID, 1.7 μm). Gradient elution: (A) aqueous ammonium formate 2 mM. pH3.5 (B) 0.1% formic acid in ACN	MS/MS	[55]
	Tetrodoxin and its analogs	Puffer fish	LLE	ZIC-HILIC, SeQuant (20 × 2.1 mm ID, 5 μm). Gradient elution of (A) 10 mM ammonium formate and 10 mM formic acid in water and (B) 80% ACN and 20% water with a final concentration of 5 mM ammonium formate and 2 mM formic acid	MS/MS	[58]

248

Table 8.3. Selected Examples of HILIC Applications on Beverages and Food of Plant Origin. Chromatographic Parameters and Sample Preparation Technique

Target Compound(s)	Analytes	Sample Matrix	Sample Pre-Treatment	Stationary Phase/Mobile Phase	Detection	Reference(s)
Water-soluble vitamins	Vitamin C and related compounds	Tea drinks, dried fruits	LLE	Intersil Diol (150 × 2.0 mm ID, 3 μm). Isocratic elution: ACN–water—with 66.7 mM ammonium acetate (85:15, v/v).	Vitamin C and related compounds	[42]
	Vitamins B1, B2, B3, B6, C	Fruits juices, chestnuts energy drinks	LLE	TSKgel Amide-80 (100 × 4.6 mm ID, 5 μm). Gradient elution: ACN and aqueous 0.1% trifluoroacetic acid	Vitamins B1, B2, B3, B6, C	[44]
Food contaminants	Melamine	Bakery goods and flour	Ultrasound treatment and LLE	AQUITY™ BEH (100 × 2.1 mm ID, 1.7 μm). Isocratic elution: ACN and water (90:10 v/v).	MS/MS	[37]

249

8.2.2.2   Water-Soluble Vitamins

Vitamins are organic compounds essential for the normal growth and development of the human body. They cannot be produced by the human body and,
as a result, can only be obtained from the diet. Vitamins are generally separated
into two categories: water-soluble and lipophilic. The first category includes
ascorbic acid (AA) and related compounds, biotin, folic acid, thiamine, riboflavin, pyridoxine, and related compounds. Lipophilic vitamins, for example,
tocopherols, cholecalciferol (vitamin D), and retinol (vitamin A) because of
their chemical structure and properties, are not appropriate for HILIC analysis. On the contrary, due to the diverse polarities of the water-soluble vitamins
(including their related compounds), HILIC is useful for the determination of
these analytes in food and beverage matrices, dietary supplements, and fortified foods [43–48].

The first and characteristic example for the adoption of the HILIC technique in vitamin analysis is the work of Tai and Gohda [43]. In this work,
a simultaneous determination of AA, D-isoascorbic acid (EA), and two
derivatives 2-*O*-α-D-glycopyranosyl-L-ascorbic acid (AA-2G) and 2-*O*-β-D-
glycopyranosyl-L-ascorbic acid (AA-2βG) in tea beverages and dried fruits,
was developed. All these compounds act as antioxidants and often are simultaneously present in foods. The selection of HILIC against RP-HPLC was
based mainly on the "precise quantitation for the AA and related compounds
owing to the rapid elution of hydrophobic sample components and the late
elution of polar components." The separation was performed on an Inertsil
Diol column (Table 8.3) and the suitability of ACN against methanol as the
organic solvent in the mobile phase was documented. Acetonitrile was optimal
due to the poor retention and peak shapes obtained with methanol. The
dependence of the retention times of the analytes on the ACN content in the
mobile phase was also examined and optimized (Fig. 8.2).

One of the main issues that make the determination of AA and its related
compounds challenging is the lack of stability in solution. Parameters affecting
the stability of these compounds are exposure to light, increases of temperature and pH value, and the presence of oxygen and metal ions [44]. Subsequently,
the use of different stabilizing agents, such as oxalic acid, *o*- and *m*-phosphoric
acid and ethylenediaminetetraacetic acid (EDTA), acting either as metal
masking agents or as reducing agents, is necessary. In all the studies concerned
with AA and its related compounds, the stability of their solutions and the
effect of the various experimental parameters of the analytical procedure have
to be examined. The use of such stabilizers in HILIC methodology is limited
to those that can be easily dissolved in high concentrations of organic solvents,
commonly in ACN. The similarity of the dilution solvent of the analytes in the
injected solutions with the mobile phase is desirable in liquid chromatography,
resulting in better peak shapes and reproducible retention times.

The sum of AA and dehydroascorbic acid (DHA) is defined as "total
vitamin C" in food commodities and has to be differentiated from the EA
isomer, which has almost the same antioxidant activity but low "vitamin C"

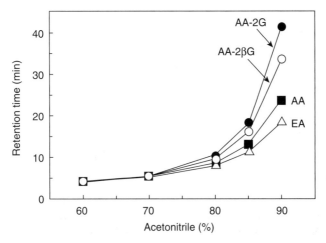

Figure 8.2. The effect of ACN content on the retention of AA, EA, AA-2G, and AA-2Bg on Inertsil Diol column. Mobile phase, ACN–water (v/v) containing 10 mM ammonium acetate; flow rate, 0.7 mL/min; detection UV, 260 nm. Reprinted from Reference 43 with permission from Elsevier.

activity. In the work of Baros et al. [45], a fast and simple method was developed for the quantitative determination of AA and DHA and their differentiation from EA. An aqueous solution containing 5% *m*-phosphoric acid, 2 mM tris(2-carboxyethyl)-phosphine hydrochloride, and 2 mM EDTA was used both as an extraction and as a reduction medium. The TSKgel Amide-80 column proved to be compatible with this extraction/dilution solvent, even thought it does not contain organic solvent.

Thiamine or vitamin B1 is also a water-soluble vitamin that has been determined by HILIC using diode array detection [46]. The development and validation of the method were carried out on dry-cured sausages because the total content of thiamine in meat products depends on their treatment. A more demanding extraction procedure was applied in order to resolve the vitamin B1 from polar interferences. This procedure consisted of an extraction and a clean-up step. The extraction of vitamin B1 involved an acid hydrolysis with boiling diluted acid (0.1 M HCl and incubation at 100°C for 30 min in a water bath) to denature the proteins and to release vitamin B1 from the matrix, followed by the use of an enzymatic solution (10% w/v aqueous takadiastase solution) in order to hydrolyze phosphate esters. An efficient purification step for the crude extracts was necessary and was achieved by an SPE step by using weak cation exchange cartridges.

HILIC has also been used as an analytical tool in the determination of multicomponent water-soluble vitamins [45,47]. In the work of Karatapanis et al. [47], three experimental parameters were investigated separately to evaluate their influence on resolution and peak shape of selected water-soluble vitamins, namely ACN content in the mobile phase, pH value of the aqueous part of the mobile phase, and sample dilution medium (Fig. 8.3). From this

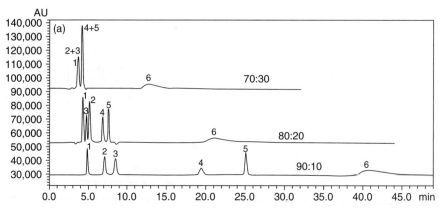

Figure 8.3a. Effect of ACN content on separation of selected water-soluble vitamins WSVs). Conditions: Inertsil Diol (150 × 4.6 mm, 5 μm particles); column temperature, 25°C; flow rate, 0.6 mL/min; detection wavelength, 272 nm; mobile phase, ACN–H$_2$O at various compositions v/v containing ammonium acetate 10 mM, pH 5. Peak assignment: (1) nicotinamide, (2) pyridoxine, (3) riboflavin, (4) nicotinic acid, (5) L-ascorbic acid, and (6) thiamine.

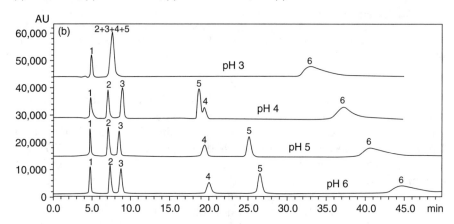

Figure 8.3b. Effect of pH on separation of selected WSVs. Conditions: Mobile phase, ACN–H$_2$O 90:10 v/v containing ammonium acetate 10 mM, with the aqueous buffer adjusted at various pH values. Other conditions and peak assignment as in Figure 8.3a.

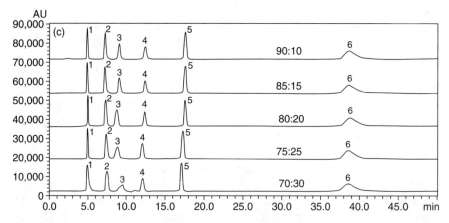

Figure 8.3c. Effect of sample diluent on peak shape of selected WSVs. Mobile phase, ACN–H$_2$O 90:10 v/v; ammonium acetate, 10 mM containing 20 mM of triethylamine, pH 5.0. Other conditions and peak assignment as in Figure 8.3a. Reprinted from Reference 47 with permission from Wiley.

study it was concluded that mobile phases based on alcohols as the organic part led to less retention of the analytes and poor or even absence of separation. On the other hand, ACN as organic solvent provided good separation for the six compounds, even using isocratic elution mode (Fig. 8.3a). In Figure 8.3b it is illustrated that variations in the mobile phase pH is a tool of great importance for achieving separation of ionized compounds.

8.2.2.3   Seafood and Other Toxins

Marine toxins are poisonous complex organic compounds produced from *phytoplankton* (microscopic algae) by transforming simple inorganic compounds. These toxins are naturally present in nonharmful quantities. However, they can be harmful when the population of these microorganisms is unexpectedly increased in locations near shellfish production areas. This phenomenon is known as "harmful algae blooms" or "red tides" because of the red coloration of sea water from the high density of the algae. Climate changes, such as the increase of the global temperature, eutrophication, and the steadily emerging density of commercial shipping contribute to this phenomenon.

These kinds of toxins accumulate into the edible tissues of bivalve-filter feeding shellfishes and in this way become harmful for humans or other consumers. They are categorized, according to the main symptoms that appear after the consumption of poisonous seafood, into paralytic, diarrhetic, and amnesic shellfish poisons (PSP, DSP, and ASP). As the toxins are characterized by diverse molecular masses and polarities (from highly polar to lipophilic), their identification and quantification are analytical challenges [49]. The official testing of marine biotoxins worldwide involves animal bioassays [50]. However, these bioassays raise a number of ethical and technical concerns [51] and because of this instrumental analytical techniques have been sought.

PSP toxins are a group of compounds based on saxitoxin (STX) and structured by a tetrahydropurine moiety with a five-membered ring fused at an angular position and a ketone hydrate stabilized by two neighboring electron-withdrawing guanidium groups (Fig. 8.4). As they are highly hydrophilic, nonvolatile, and thermally labile, their determination by gas chromatography or RPLC without the use of ion-pairing additives in the mobile phase is impossible. A significant number of studies have been published for the determination of STX and its derivatives by HILIC, with mass spectrometric detection in most of the cases [52–58]. In the work of Diener et al. [54], two different combinations of HILIC separation mode with mass spectrometric detection (MS/MS) and fluorescence detection (FD) after postcolumn derivatization were compared. The HILIC-FD method appeared to have higher sensitivity for most of the PSP toxins. However, the use of MS/MS detection is more desirable due to the possibility of using selected reaction monitoring (SRM) for identification of STX derivatives. Another interesting method was developed by Marchesini et al. [55], in which, for the first time, the combination of biosensor-based bioanalysis with nano-HILIC-time-of-

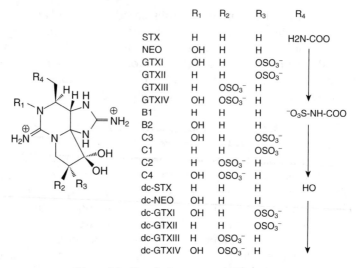

	R₁	R₂	R₃	R₄
STX	H	H	H	H2N-COO
NEO	OH	H	H	
GTXI	OH	H	OSO₃⁻	
GTXII	H	H	OSO₃⁻	
GTXIII	H	OSO₃⁻	H	
GTXIV	OH	OSO₃⁻	H	
B1	H	H	H	⁻O₃S-NH-COO
B2	OH	H	H	
C3	OH	H	OSO₃⁻	
C1	H	H	OSO₃⁻	
C2	H	OSO₃⁻	H	
C4	OH	OSO₃⁻	H	
dc-STX	H	H	H	HO
dc-NEO	OH	H	H	
dc-GTXI	OH	H	OSO₃⁻	
dc-GTXII	H	H	OSO₃⁻	
dc-GTXIII	H	OSO₃⁻	H	
dc-GTXIV	OH	OSO₃⁻	H	

Figure 8.4. Chemical structure of PSP toxins.

flight (TOF)-MS and a carbamoyl-funtionalized silica nanocolumn were employed for the determination of PSP toxins in mussels and cockle flesh.

Usually, the extraction of PSP toxins from shellfish, particularly mussels, was achieved by either aqueous hydrochloric acid solutions [54,56] or a mixture of ACN and aqueous formic acid solution 0.1% (80:20, v/v) [53]. In the last work, further purification with SPE cartridges before the injection to the chromatographic system was applied.

Apart from toxins that are identified in shellfish, other compounds with high toxicity are presented in marine life. Tetrodotoxin (TTX) and its analogs (Fig. 8.5) are powerful and specific sodium channel blockers, just like PSP compounds, and they are the toxic agents of puffer fish, *Takifugu oblongus* [59,60]. In the work of Diener et al. [54], a ZIC-HILIC column was applied and good separation of the principal TTX analogs with reproducible retention times of toxins in different sample materials was achieved.

Selected examples of HILIC applications, including chromatographic parameters, sample preparation techniques, and detection are presented in Table 8.2 for food of animal origin and Table 8.3 for beverages and food of plant origin.

8.3   ENVIRONMENTAL AND OTHER APPLICATIONS OF HILIC

8.3.1   Review of Environmental Applications Based on the Stages of Method Development

Environmental analysis is an inseparable part of analytical chemistry and new trends in analytical techniques are quickly adopted for environmental

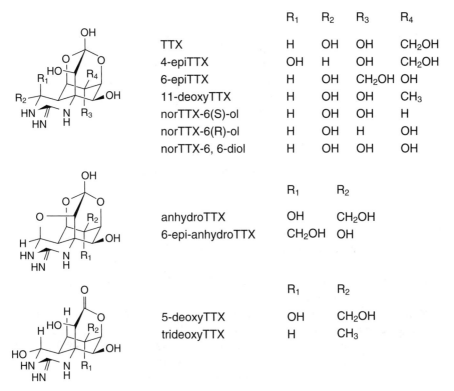

	R₁	R₂	R₃	R₄
TTX	H	OH	OH	CH₂OH
4-epiTTX	OH	H	OH	CH₂OH
6-epiTTX	H	OH	CH₂OH	OH
11-deoxyTTX	H	OH	OH	CH₃
norTTX-6(S)-ol	H	OH	OH	H
norTTX-6(R)-ol	H	OH	H	OH
norTTX-6, 6-diol	H	OH	OH	OH

	R₁	R₂
anhydroTTX	OH	CH₂OH
6-epi-anhydroTTX	CH₂OH	OH

	R₁	R₂
5-deoxyTTX	OH	CH₂OH
trideoxyTTX	H	CH₃

Figure 8.5. Chemical structure of tetrodotoxin (TTX) and its analogs.

applications. In this sense, HILIC has gained merit in this field [2]. Analysis of environmental samples is challenging, not only due to matrix complexity but also due to the trace level concentration of the target compounds that must be detected.

Samples from aqueous systems contaminated with polar pollutants are the most common samples in environmental HILIC applications. A characteristic example of this category of application is the determination of endocrine-disrupting compounds, such as estrogens, in aqueous systems [61,62]. In the work of Qin et al. [62], the determination of free estrogens and their conjugates in river water was achieved by the combination of RPLC and HILIC. The free estrogens were separated after derivatization with dansyl chloride on the RP column, whereas the conjugates were separated on the HILIC column after switching. Other applications of HILIC in aqueous systems, such as river and lake water, include the determination of toxins produced by cyanobacteria [63,64], some of which are in the same category as those determined in food matrices.

Aromatic amines, such as anilines, 1-naphthylamine, *N,N*-dimethylaniline, *N,N*-diethylaniline, and benzidine were also determined in environmental water [65]. The analysis was conducted on a bare silica column and different chromatographic parameters affecting the efficiency of the separation were examined. From the four organic solvents tried as the organic component of the mobile phase (ACN, methanol, isopropanol, and tetrahydrofuran), ACN provided the best resolution of the amines.

8.3.2   Selected Detailed Examples of Environmental and Other HILIC Applications

8.3.2.1   Metals and Their Related Organic Compounds

Metals and metalloids and their compounds in environmental and other samples are of interest because of their high toxicity. Examples include organic arsenic compounds in marine organisms [66,67], and selenium compounds [68,69], which are essential trace elements, the deficiency of which can cause several health disorders.

Additionally, gadoliniumn compounds have been used since 1988 as contrast agents in magnetic resonance imaging (MRI). Anthropogenic gadolinium (Gd) and its complexes are determined in hospital waste waters [70] and in drinking water of urban areas [71]. In the case of the determination of Gd chelates in hospital waste water, the samples were collected in silylated glass bottles and filtered. An internal standard solution was added (rhodium diethylenetriaminepentaacetate, Rh-DTPA) and the sample was analyzed by HILIC-ICP (inductively coupled plasma)-MS. It is worthwhile to mention that the combination of HILIC with ICP-MS for the determination of elements is a new trend in environmental analysis [72].

8.3.2.2   Pharmaceutical Compounds in Aqueous Environmental Samples

The determination of pharmaceuticals in environmental samples is of high interest. Their determination requires a more demanding sample preparation, due to the very low levels of concern and their potential toxicity [73]. Several drugs have been determined in environmental samples such as illicit drugs [74], cocaine, benzoylecgonine and other metabolites [75–77], cytostatics and metabolites [78], drugs of abuse [77], and top prescribed pharmaceuticals [79].

In the study of Kovalova et al. [78], the retention behavior of cytostatics and their metabolites on a ZIC-pHILIC column was investigated. It was proved that the effect of the aqueous part in the mobile phase was stronger than the influence of the other parameters (pH, salt concentration, organic solvent [ACN and methanol]). The use of HILIC column achieved good separation performance for nine drugs of abuse [77], particularly for the hydrophilic compounds, which elute early (amphetamine-like stimulants) or show no retention (ecgonine methyl ester) when RP column are used.

8.3.2.3   Other Applications

Besides food and environmental analyses, there are several other applications of the HILIC, such as the determination of hexapeptides in anti-wrinkle cosmetics [80], low molecular weight compounds in fish meal [81], glyphosate and glufosinate residues [82], and saponins in plants [83].

Selected examples of environmental HILIC applications, including chromatographic parameters and sample preparation techniques, are presented in Table 8.4 on p. 258.

8.4   CONCLUSIONS

The utility of HILIC in food analysis, environmental analysis, and other applications (cosmetics, plants) provides the separation and determination of several polar low molecular weight components of great importance in the aforementioned areas. Some very interesting examples of analytes, for which HILIC is the method of choice, include BAs, amino acids and small polar peptides, water-soluble vitamins, melamine, marine biotoxins, endocrine, disrupting compounds, antibiotics (including aminoglycosides), pharmaceuticals, acrylamid, perchlorate, quaternary ammonium compounds, aromatic amines, and metal compounds.

In most of the HILIC applications in food and environmental analysis, a bare silica column has been found suitable; however, a recent trend is the use of polymeric zwitterionic columns, such as the ZIC-pHILIC, which is the ideal choice for polar and hydrophilic compounds. Concerning the mobile phase, in most cases mixtures of ACN with aqueous ammonium acetate or formate buffers are used. The pH and the buffer concentration should be optimized for the target analytes and the selected column.

The selection of sample preparation methodology should be based on the characteristics of the sample matrix, the analytes, and the detector used (in particular, MS detector). Both LLE and SPE have been found adequate. The combination of LLE and SPE provides very clean sample working solutions.

The hyphenation of HILIC with MS detection achieves increased sensitivity required in trace analysis due to the absence of the ion-pairing reagents that are sometimes necessary in classical RP chromatography, which causes ion suppression with the ESI mode. The high content of organic solvent in the mobile phase also enhances the sensitivity due to increased mobile phase volatility relative to typical RPLC mobile phases. The selectivity obtained by HILIC is also increased due to the efficient separation of polar hydrophilic analytes from the lipophilic ones, which is added to the high selectivity of MS detection, achieving very selective analytical methods for polar hydrophilic analytes.

Table 8.4. Selected Examples of Environmental HILIC Applications, Chromatographic Parameters and Sample Preparation Technique

Target Compound(s)	Analytes	Sample Matrix	Sample Pretreatment	Stationary Phase/Mobile Phase	Detection	Reference(s)
Endocrine-disrupting compounds	Free and conjugated estrogens	River water	SPE	ZIC-pHILIC, SeQuant (100 × 2.1 mm ID, 5 μm). Gradient elution of (A) 95% acetonitrile (ACN) and 5% ammonium formate buffer (5 mM, pH 6.80) and (B) 75% ACN and 25% ammonium formate buffer (5 mM, pH 6.80)	MS/MS	[60]
Antibiotics	Spectinomycin and lincomycin	Hog manure supernatant and run-off from cropland	SPE	Altima HP, Altech Associates (150 × 2.1 mm ID, 3 μm). Isocratic elution: (A) 65% and (B) 35%. Where (A) is ACN–water (10:90) and (B) ACN–water (90:10), both containing 0.1% formic acid	MS/MS	[85]
Pharmaceutical compounds	Cytostatics and their metabolites	Hospital wastewater	SPE	ZIC-pHILIC, SeQuant (150 × 2.1 mm ID, 3.5 μm). Gradient elution of (A) ACN and 30 mM ammonium acetate buffer (3/2, v/v) and (B) ACN	MS/MS	[77]
	13 top-prescribed pharmaceuticals	Influent wastewater	SPE	Luna HILIC, Phenomenex (150 × 2.1 mm ID, 3.5 μm). Gradient elution of (A) 5 mM ammonium acetate buffer (3/2, v/v) and (B) ACN/methanol (87.5/12.5, v/v)	MS/MS	[78]
	Cocaine and metabolites	Waste and surface water	SPE	Zorbax Rx-SIL (150 × 2.1 mm ID, 5 μm). Gradient elution of (A) 2 mM ammonium acetate/acetic acid buffer (pH 4.5) and (B) ACN	MS/MS	[86]

REFERENCES

1. Bernal J, Ares AM, Jaroslav P, Wiedmer SK. Hydrophilic interaction liquid chromatography in food analysis. *J. Chromatogr. A* 2011; **1218**: 7438–7452.

2. Van Nuijs ALN, Tarcomnicu I, Covaci A. Application of hydrophilic interaction chromatography for the analysis of polar contaminants in food and environmental samples. *J. Chromatogr. A* 2011; **1218**: 5964–5974.

3. Bruce SJ, Guy PA, Rezzi S, Ross AB. Quantitative measurement of betaine and free choline in plasma, cereals and cereal products by isotope dilution LC-MS/MS. *J. Agric. Food Chem.* 2010; **58**: 2055–2061.

4. Mora L, Sentandreu MA, Toldra F. Effect of cooking conditions on creatinine formation in cooked ham. *J. Agric. Food Chem.* 2008; **56**: 11279–11284.

5. Kalili KM, de Villiers A. Off-line comprehensive 2-dimensional hydrophilic interaction × reversed phase liquid chromatography analysis of procyanidins. *J. Chromatogr. A* 2009; **1216**: 6274–6284.

6. Zhao Y-Y, Xiong Y, Curtis JM. Measurement of phospholipids by hydrophilic interaction liquid chromatography coupled to tandem mass spectrometry: The determination of choline containing compounds in foods. *J. Chromatogr. A* 2011; **1218**: 5470–5479.

7. Troyer JK, Stephenson KK, Fahey JW. Analysis of glycosinolates from broccoli and other cruciferous vegetables by hydrophilic interaction liquid chromatography. *J. Chromatogr. A* 2001; **919**: 299–304.

8. De Person M, Hazotte A, Elfakir C, Lafosse M. Development and validation of a hydrophilic interaction chromatography-mass spectrometry assay for taurine and methionine in matrices rich in carbohydrates. *J. Chromatogr. A* 2005; **1081**: 174–181.

9. Arias F, Li L, Huggins ThG, Keller PR, Suchanek PM, Wehmeyer KR. Trace analysis of bromate in potato snacks using high-performance liquid chromatography–tandem mass spectrometry. *J. Agric. Food Chem.* 2010; **58**: 8134–8138.

10. Crnogorac G, Schmauder S, Schwack W. Trace analysis of dithiocarbamate fungicide residues on fruits and vegetables by hydrophilic interaction liquid chromatography/tandem mass spectrometry. *Rapid Commun. Mass Spectrom.* 2008; **22**: 2539–2546.

11. Mora L, Hernandez-Cazares AS, Sentandreu MA, Toldra F. Creatine and creatinine evolution during the processing of the dry-cured ham. *Meat Sci.* 2010; **84**: 384–389.

12. Martos PA, Jayasundara F, Dolbeer J, Jin W, Spilsbury L, Mitchell M, Varilla C, Shurmer B. Multiclass, multiresidue drug analysis, including aminoglycosides, in animal tissue using liquid chromatography coupled to tandem mass spectrometry. *J. Agric. Food Chem.* 2010; **58**: 5932–5944.

13. Chiaochan C, Koesukwiwat U, Yudthavorasit S, Leepipatpiboon N. Efficient hydrophilic interaction liquid chromatography-tandem mass spectrometry for the multiclass analysis of veterinary drugs in chicken muscle. *Anal. Chim. Acta* 2010; **682**: 117–129.

14. Broncano JM, Otte J, Petrón MJ, Parra V, Timón ML. Isolation and identification of low molecular weight antioxidant compounds from fermented "chorizo" sausages. *Meat Sci.* 2012; **90**(2): 499–501.

15. Mora L, Hernadez-Cazares AS, Aristoy M, Toldra F. Hydrophilic interaction chromatographic determination of adenosine triphosphate and its metabolites. *Food Chem.* 2010; **123**: 1282–1288.

16. Giannotti V, Chiuminato U, Mazzucco E, Gosetti F, Bottaro M, Frascarolo P, Gennaro MC. A new hydrophilic interaction liquid chromatography tandem mass spectrometry method for the simultaneous determination of seven biogenic amines in cheese. *J. Chromatogr. A* 2008; **118**: 296–300.

17. Esparza X, Moyano E, Galceran MT. Analysis of chlormequat and mepiquat by hydrophilic interaction chromatography coupled to tandem mass spectrometry in food samples. *J. Chromatogr. A* 2009; **1216**: 4402–4406.

18. Gremilogianni AM, Megoulas NC, Koupparis MA. Hydrophilic interaction vs ion pair liquid chromatography for the determination of *streptomycin* and *dihydrostreptomycin* residues in milk based on mass spectrometric detection. *J. Chromatogr. A* 2010; **1217**: 6646–6651.

19. Zheng M-M, Zhang M-Y, Peng G-Y, Feng Y-Q. Monitoring of sulphonamide antibacterial residues in milk and egg by polymer monolith microextraction coupled hydrophilic interaction chromatography/mass spectrometry. *Anal. Chim. Acta* 2008; **625**: 160–172.

20. Calvano CD, Aresta A, Zambonin CG. Detection of hazelnut oil in extra-virgin olive oil by analysis of polar components by micro-solid phase extraction based on hydrophilic liquid chromatography and MALDI-ToF mass spectrometry. *J. Mass Spectrom.* 2010; **45**: 981–988.

21. Rosén J, Nyman A, Hellenäs K-E. Retention studies of acrylamide for the design of a robust liquid chromatography-tandem mass spectrometry method for food analysis. *J. Chromatogr. A* 2007; **1172**: 19–24.

22. Bohm DA, Stachel CS, Gowik P. Confirmatory method for the determination of streptomycin in apples by LC-MS/MS. *Anal Chim Acta* 2010; **672**: 103–106.

23. Cheng L, Chen H, Shen M, Zhou Z, Ma A. Analysis of perchlorate in milk powder and milk by hydrophilic interaction chromatography combined with tandem mass spectrometry. *J. Agric. Food Chem.* 2010; **58**: 3736–3740.

24. Sørensen JL, Nielsen KF, Thrane U. Analysis of moniliformin in maize plants using hydrophilic interaction chromatography. *J. Agric. Food Chem.* 2007; **55**: 9764–9768.

25. Fenaille F, Parison V, Vuichoud J, Tabet J-C, Guy PA. Quantitative determination of dityrosine in milk powders by liquid chromatography coupled to tandem mass spectrometry using isotope dilution. *J. Chromatogr. A* 2004; **1052**: 77–84.

26. Wölwer-Rieck U, Tomberg W, Wawrzun A. Investigations on the stability of stevioside and rebaudioside A in soft drinks. *J. Agric. Food Chem.* 2010; **58**: 12216–12220.

27. Jandera P. Stationary and mobile phases in hydrophilic interaction chromatography: A review. *Anal. Chim. Acta* 2011; **692**: 1–25.

28. Chirita R-I, West C, Zubrzycki S, Finaru A-L, Elfakir C. Investigations on the chromatographic behaviour of zwitterionic stationary phases used in hydrophilic interaction chromatography. *J. Chromatogr. A* 2011; **1218**: 5939–5963.

29. Buszewski B, Noga S. Hydrophilic interaction liquid chromatography (HILIC)-a powerful separation technique. *Anal. Bioanal. Chem.* 2012; **402**(1): 231–247.

30. Dejaegher B, Mangelings D, Heyden YV. Method development for HILIC assays. *J. Sep. Sci.* 2008; **31**: 1438–1448.

31. Meynial I, Paquet V, Combes D. Simultaneous separation of nucleotides and nucleotide sugars using an ion-pair reversed-phase HPLC—Application for assaying glycosyltransferase activity. *Anal. Chem.* 1995; **67**: 1627–1631.

32. Veciana-Nogues MT, Izquierdo-Pulido M, Vidal-Carou MC. Determination of ATP related compounds in fresh and canned tuna fish by HPLC. *Food Chem.* 1997; **59**: 467–472.

33. Nguyen H, Schug KA. The advantages of ESI-MS detection in conjunction with HILIC mode separations: Fundamentals and applications. *J. Sep. Sci.* 2008; **31**: 1465–1480.

34. Sun F, Ma W, Xu L, Zhu Y, Liu L, Peng C, Wang L, Kuang H, Xu C. Analytical methods and recent developments in the detection of melamine. *Trends Anal. Chem.* 2010; **29**: 1239–1249.

35. Varelis P, Jeskelis R. Preparation of [^{13}C$_3$]-melamine and [^{13}C$_3$]-cyanuric acid in meat and pet food using liquid chromatography-tandem mass spectrometry. *Food Addit. Contam.* 2008; **25**: 1208–1215.

36. Andersen WC, Turnipseed SB, Karbiwnyk CM, Clark SB, Madson MR, Gieseker CM, Miller RA, Rummel NG, Reimschuessel R. Determination and confirmation of melamine residues in catfish, trout, tilapia, salmon and shrimp by liquid chromatography with tandem mass spectrometry. *J. Agric. Food Chem.* 2008; **56**: 4340–4347.

37. Deng X, Guo D, Zhao S, Han L, Sheng Y, Yi X, Zhou Y, Peng T. A novel mixed-mode solid phase extraction for simultaneous determination of melamine and cyanuric acid in food by hydrophilic interaction chromatography coupled to tandem mass chromatography. *J. Chromatogr. B* 2010; **878**: 2839–2844.

38. Goscinny S, Hanot V, Halbardier J-F, Michelet J-Y, van Loco J. Rapid analysis of melamine residue in milk, milk products, bakery goods and flour by ultra-performance liquid chromatography/tandem mass spectrometry: From crisis to acreditation. *Food Control* 2011; **22**: 226–230.

39. Zheng X, Yu B, Li K, Dai Y. Determination of melamine in dairy products by HILIC-UV with NH$_2$ column. *Food Control* 2012; **23**: 245–250.

40. Ihunegbo FN, Tesfalidet S, Jiang W. Determination of melamine in milk powder using zwitterionic HILIC stationary phase with UV detection. *J. Sep. Sci.* 2010; **33**: 988–995.

41. MacMahon S, Begley TH, Diachenko GW, Stromgren SA. A liquid chromatography-tandem mass spectrometry method for the detection of economically motivated adulteration in protein-containing foods. *J. Chromatogr. A* 2012; **1220**: 101–107.

42. Zhou J, Zhao J, Xue X, Zhang J, Chen F, Li Y, Wu L, Li C, Mi J. Hydrophilic interaction chromatography/tandem mass spectrometry for the determination of melamine in royal jelly and royal jelly lyophilized powder. *J. Chromatogr. B* 2009; **877**: 4164–4170.

43. Tai A, Gohda E. Determination of ascorbic acid and related compounds in food and beverages by hydrophilic interaction liquid chromatography. *J. Chromatogr. B* 2007; **853**: 214–220.

44. Drivelos S, Dasenaki ME, Thomaidis NS. Determination of isoascorbic acid in fish tissue by hydrophilic interaction chromatography-ultraviolet detection. *Anal. Bioanal. Chem.* 2010; **397**: 2199–2210.

45. Baros AIRNA, Silva AP, Goncalves B, Nunes FM. A fast, simple, and reliable hydrophilic interaction liquid chromatography method for the determination of ascorbic and isoascorbic acids. *Anal. Bioanal. Chem.* 2010; **396**: 1863–1875.

46. Gratacos-Cubarsi M, Sarraga C, Clariana M, Garcia Regueiro JA, Castellari M. Analysis of vitamin B1 in dry-cured sausages by hydrophilic interaction chromatography (HILIC) and diode array detection. *Meat Sci.* 2011; **87**: 234–238.

47. Karatapanis AE, Fiamegos YC, Stalikas CD. HILIC separation and quantitation of water-soluble vitamins using diol column. *J. Sep. Sci.* 2009; **32**: 909–917.

48. Ito S, Nakata F. *Direct Analysis of Biogenic Amines in Food by HILIC-MS, Customer Support Center.* Kanagawa, Japan: Tosoh Corporation; 2011.

49. Quilliam MA. The role of chromatography in the hunt for red tide toxins. *J. Chromatogr. A* 2003; **1000**: 527–548.

50. European Commission. 2004), Regulation (EC) No 853/2004 of the European Parliament and of the Council of 29 April 2004 laying down specific hygiene rules for food of animal origin. Off. Eur. Union, Vol. **L 139** pp. 55–205.

51. Campbell K, Vilariňo N, Botana LM, Elliott CTA. European perspective on progress in moving away from the mouse bioassay for marine-toxin analysis. *Trends Anal. Chem.* 2011; **30**(2): 239–253.

52. Dell'Aversano C, Walter JA, Burton IW, Stirling DJ, Fattorusso E, Quilliam MA. Isolution and structure elucidation of new and unusual saxitoxin analogues from mussels. *J. Nat. Prod.* 2008; **71**: 1518–1523.

53. Dell'Aversano C, Hess P, Quilliam MA. Hydrophilic interaction liquid chromatography–mass spectrometry for the analysis of paralytic shellfish poisoning (PSP) toxins. *J. Chromatogr. A* 2005; **1081**: 190–201.

54. Diener M, Erler K, Christian B, Luckas B. Application of a new zwitterionic hydrophilic interaction chromatography column for determination of paralytic shellfish poisoning toxins. *J. Sep. Sci.* 2007; **30**: 1821–1826.

55. Marchesini GR, Hooijerink H, Haasnoot W, Buijs J, Campbell K, Elliott CT, Nielen MWF. Towards surface plasmon resonance biosensing combined with bioaffinity-assisted nano HILIC liquid chromatography / time-of-flight mass spectrometry identification of paralytic shellfish poisons. *Trends Anal. Chem.* 2009; **28**(6): 792–803.

56. Blay P, Hui JPM, Chang J, Melanson JE. Screening for multiple classes of marine biotoxins by liquid chromatography-high-resolution mass spectrometry. *Anal. Bioanal. Chem.* 2011; **400**: 577–585.

57. Turrell E, Stobo L, Lacaze JP, Piletsky S, Piletska E. Optimization of hydrophilic interaction liquid chromatography/mass spectrometry and development of solid-phase extraction for the determination of paralytic shellfish poisoning toxins. *J. AOAC Int.* 2008; **91**(6): 1372–1386.

58. Ciminiello P, Dell'Aversano C, Fattorusso E, Forino M, Magno GS, Tartaglione L, Quilliam MA, Tubaro A, Poletti R. Hydrophilic interaction liquid chromatography/mass spectrometry for determination of domoic acid in Adriatic shellfish. *Rapid Commun. Mass Spectrom.* 2005; **19**: 2030–2038.

59. Diener M, Christian B, Ahmed MS, Luckas B. Determination of tetrodoxin and its analogs in the puffer fish *Takifugu oblongus* from Bangladesh by hydrophilic interaction chromatography and mass-spectrometric detection. *Anal. Bioanal. Chem.* 2007; **389**: 1997–2002.

60. Nakagawa T, Jang J, Yotsu-Yamashita M. Hydrophilic interaction liquid chromatography–electrospray ionization mass spectrometry of tetrodoxin and its analogs. *Anal. Biochem.* 2006; **352**: 142–144.

61. LaFleur AlD, Schug KA. A review of separation methods for the determination of estrogens and plastics-derived estrogen mimics from aqueous systems. *Anal. Chim. Acta* 2011; **696**: 6–26.

62. Qin F, Zhao YY, Sawyer MB, Li X-F. Column-switching reversed phase-hydrophilic interaction liquid chromatography/tandem mass spectrometry method for determination of free estrogens and their conjugates in river water. *Anal. Chim. Acta* 2008; **627**: 91–98.

63. Kubo T, Kato N, Hosoya K, Kaya K. Effective determination of a cyanobacterial neurotoxin, *N*-methylamino-L-alanine. *Toxicon* 2008; **51**: 1264–1268.

64. Lajeunesse A, Segura PA, Gelinas M, Hudon C, Thomas K, Quilliam MA, Gagnon C. Detection and confirmation of saxitoxin analogues in freshwater benthic *Lyngbya wollei* algae collected in the St. Lawrence River (Canada) by liquid chromatography-tandem mass spectrometry. *J. Chromatogr. A* 2012; **1219**: 93–103.

65. Li R, Zhang Y, Lee ChC, Lu R, Huang Y. Development and validation of a hydrophiliv interaction liquid chromatographic method for determination of aromatic amines in environmental water. *J. Chromatogr. A* 2010; **1217**: 1799–1805.

66. Xie D, Mattusch J, Wennrich R. Separation of organoarsenicals by means of zwitterionic Hydrophilic Interaction Chromatography (ZIC-HILIC) and parallel ICP-MS/ESI-MS detection. *Eng. Life Sci.* 2008; **8**(6): 582–588.

67. Xie D, Mattusch J, Wennrich R. Retention of arsenic species on zwitterionic stationary phase in hydrophilic interaction chromatography. *J. Sep. Sci.* 2010; **33**: 817–825.

68. Dernovics M, Lobinski R. Speciation analysis of selenium metabolites in yeast-based food supplements by ICPMS–assisted hydrophilic interaction HPLC–hybrid linear ion trap/Orbitrap MS^n. *Anal. Chem.* 2008; **80**: 3975–3984.

69. Far J, Preud'homme H, Lobinski R. Detection and identification of hydrophilic selenium compounds in selenium-rich yeast by size exclusion-microbore normal-phase HPLC with the on-line ICP-MS and electrospray Q-TOF-MS detection. *Anal. Chim. Acta* 2010; **657**: 175–190.

70. Künnemeyer J, Terborg L, Meermann B, Brauckmann C, Möller I, Scheffer A, Karst U. Speciation analysis of gadolinium chelates in hospital effluents and wastewater treatment plant sewage by a novel HILIC/ICP-MS method. *Environ. Sci. Technol.* 2009; **43**: 2884–2890.

71. Kulaksiz S, Bau M. Anthropogenic gadolinium as a microcontaminant in a tap water used as drinking water in urban areas and megacities. *Appl. Geochem.* 2011; **26**: 1877–1885.

72. Popp M, Hann S, Koellensperger G. Environmental application of elemental speciation analysis based on liquid or gas chromatography hyphenated to inductively coupled plasma mass spectrometry—A review. *Anal. Chim. Acta* 2010; **668**: 114–129.

73. Kostopoulou M, Nikolaou A. Analytical problems and the need for sample preparation in the determination of pharmaceuticals and their metabolites in aqueous environmental matrices. *Trends Anal. Chem.* 2008; **27**(11): 1023–1035.

74. Van Nuijs ALN, Castiglioni S, Tarcomnicu I, Postigo C, de Alda ML, Neels H, Zuccato E, Barcelo D, Covaci A. Illicit drug consumption estimations derived from wastewater analysis: A critical review. *Sci. Total Environ.* 2011; **409**: 3564–3577.

75. Van Nuijs ALN, Pecceu B, Theunis L, Dubois N, Charlier C, Jorens PhG, Bervoets L, Blust R, Neels H, Covaci A. Spatial and temporal vacations in the occurrence of cocaine and benzoylecgonine in waste and surface water from Belgium and removal during wastewater treatment. *Water Res.* 2009; **45**: 1341–1349.

76. Castiglioni S, Bagnati R, Melis M, Panawennage D, Chiarelli P, Fanelli R, Zuccato E. Identification of cocaine and its metabolites in urban wastewater and comparison with the human excretion profile in urine. *Water Res.* 2011; **45**: 5141–5150.

77. Van Nuijs ALN, Tarcomnicu I, Bervoets L, Blust R, Jorens PhG, Neels H, Covaci A. Analysis of drugs of abuse in wastewater by hydrophilic interaction liquid chromatography-tandem mass spectrometry. *Anal. Bioanal. Chem.* 2009; **395**: 819–828.

78. Kovalova L, McArdell CS, Hollender J. Challenge of the high polarity and low concentrations in the analysis of cytostatics and metabolites in wastewater by hydrophilic interaction chromatography/tandem mass spectrometry. *J. Chromatogr. A* 2009; **1216**: 1100–1108.

79. Van Nuijs ALN, Tarcomnicu I, Simons W, Bervoets L, Blust R, Jorens PhG, Neels H, Covaci A. Optimization and validation of a hydrophilic interaction liquid chromatography-tandem mass spectrometry method for the determination of 13 top-prescribed pharmaceuticals in influent wastewater. *Anal. Bioanal. Chem.* 2010; **398**: 2211–2222.

80. Zhou W, Wang PG, Krynitsky AJ, Rader JI. Rapid and simultaneous determination of hexapeptides (Ac-EEMQRR-amide and H_2N-EEMQRR-amide) in anti-wrinkle cosmetics by hydrophilic interaction liquid chromatography–solid phase extraction preparation and hydrophilic interaction liquid chromatography with tandem mass spectrometry. *J. Chromatogr. A* 2011; **1218**: 7956–7963.

81. Wu TH, Bechtel PJ. Screening for low molecular weight compounds in fish meal soluble by hydrophilic interaction liquid chromatography coupled to mass spectrometry. *Food Chem.* 2012; **130**: 739–745.

82. Li X, Xu J, Jiang Y, Chen L, Xu Y, Pan C. Hydrophilic-interaction liquid chromatography (HILIC) with DAD and mass spectroscopic detection for direct analysis of glyphosate and glufosinate residues and for product quality control. *Acta Chromatogr.* 2009; **21**(4): 559–576.

83. Wang Y, Lu X, Xu G. Development of a comprehensive two-dimensional hydrophilic interaction chromatography/quadrupole time-of-flight mass spectrometry system and its application in separation and identification of saponins from *Quillaja saponaria. J. Chromatogr. A* 2008; **1181**: 51–59.

84. Schlichtherle-Cerny H, Affolter M, Cerny C. Hydrophilic interaction liquid chromatography coupled to electrospray mass spectrometry of small polar compounds in food analysis. *Anal. Chem.* 2003; **75**: 2349–2354.

85. Peru KM, Kuchta SL, Headley JV, Cessna AJ. Development of hydrophilic interaction chromatography-mass spectrometry assay for spectinomycin and lincomycin in liquid hog manure supernatant and run-off from cropland. *J. Chromatogr. A* 2006; **1107**: 152–158.

86. Gheorghe A, van Nuijs A, Pecceu B, Bervoets L, Jorens PG, Blust R, Neels H, Covaci A. Analysis of cocaine and its principal metabolites in waste and surface water using solid-phase extraction and liquid chromatography–ion trap tandem mass spectrometry. *Anal. Bioanal. Chem.* 2008; **391**: 1309–1319.

CHAPTER

9

THEORY AND PRACTICE OF TWO-DIMENSIONAL LIQUID CHROMATOGRAPHY SEPARATIONS INVOLVING THE HILIC MODE OF SEPARATION

STEPHEN R. GROSKREUTZ and DWIGHT R. STOLL

Department of Chemistry, Gustavus Adolphus College, St. Peter, MN

9.1 FUNDAMENTALS OF MULTIDIMENSIONAL LIQUID CHROMATOGRAPHY

9.1.1 Scope

Despite very slow growth in the previous 20 years, in the most recent decade we have observed increasing interest and significant advances in the development of multidimensional liquid chromatography (MDLC), particularly two-dimensional separations (2DLC). The intent of this chapter is not to comprehensively review 2DLC, but to provide a brief introduction to the fundamentals of the methodology, followed by a more detailed discussion of

Hydrophilic Interaction Chromatography: A Guide for Practitioners, First Edition.
Edited by Bernard A. Olsen and Brian W. Pack.
© 2013 John Wiley & Sons, Inc. Published 2013 by John Wiley & Sons, Inc.

some theoretical, but mostly practical issues that are germane to the use of hydrophilic interaction liquid chromatography (HILIC) in at least one of the dimensions of a 2DLC system. For a broader view of the fundamentals and applications of MDLC, the reader is referred to a number of books [1–4] and selected excellent reviews published in this area [5–8]. In the interest of focusing on the chromatographic aspects of the technique, we have deliberately not discussed problems, opportunities, and solutions related to data analysis that are unique to 2DLC. Readers interested in this topic are directed to a recent thorough treatment of the subject in a recent book chapter [9]. To our knowledge, at the time of preparation of this chapter, there was no existing review specifically focused on 2DLC involving the HILIC mode of separation, although Jandera has discussed some of the potential advantages of using HILIC in combination with other separation modes in 2DLC [10] and reviewed the handful of applications that had been described at the time of his article [11].

9.1.2   Potential Advantages of 2D Separations
over Conventional Separations

Advances in instrumentation (elevated pressure and temperature capabilities) and column technology (e.g., superficially porous particles [12]) continue to result in improvements in the ability to resolve more peaks in a given time, or resolve a given number of peaks in a shorter time. However, the absolute number of peaks that can be resolved in practical analysis times pale in comparison to the demands made of chromatographic separations in a number of application areas. As pointed out recently by Guiochon [13], after a century of development of column LC, we are quickly approaching some fundamental limits of the resolving power of one-dimensional (1D) LC. Peak capacity is the most commonly used metric for quantitative comparisons of the resolving power of 1D and 2D methods. Given the clear limitations of 1D methods [14,15], 2D methods are most commonly considered as alternatives to conventional 1D work because of the potential for much higher peak capacity in the 2D case, which is particularly valuable in the analysis of complicated samples such as those encountered in food and beverage analysis, environmental analysis, and biological applications such as metabolomics and proteomics. It is common in these cases to encounter mixtures of hundreds and even thousands of compounds. The methods of estimation and the limits of peak capacity in 1DLC and 2DLC have been reviewed extensively elsewhere [9,15], and will not be discussed in detail here. Rather, we point to important aspects of this area that are of major practical significance.

Despite the fact that the concept of peak capacity was introduced by Giddings over 40 years ago [16], there has been considerable debate in the last decade or so over methods of estimating the peak capacity of both 1D and 2D separations. This certainly is important to the development of each technique in its own right, but it is critically important to practitioners faced with decisions about which mode of separation to use (i.e., 1D or 2D?), because

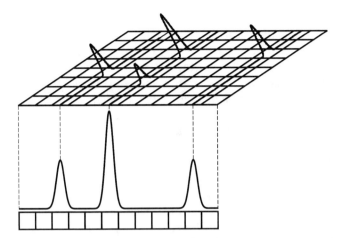

Figure 9.1. Diagram of the simplistic "bin model" of peak capacity in one- and two-dimensional liquid chromatography. The lower panel shows how each unit of peak capacity corresponds to one "bin" into which a peak can be placed in 1DLC. The upper panel illustrates the impact of the "product rule" (see Eq. 9.3) on peak capacity in LC × LC. The overall peak capacity becomes the product of the individual first and second dimension separation peak capacities.

unfair comparisons of the two techniques obviously lead to poor decisions concerning the best method for the problem at hand. The lower panel of Figure 9.1 shows a simplistic view of the concept of peak capacity in the 1D case, that a unit of peak capacity constitutes a "bin" into which a peak (typically the 4σ width is used) can be placed, and that the peak capacity of the separation is simply an estimate of the number of bins contained by the entire separation space. Much of the recent variability in the means of estimating 1D peak capacities has resulted from differing views on what exactly constitutes the separation space [7,17]. For example, when gradient elution is used in 1DLC, the column dead time and the gradient time are convenient time markers that can be used as bounds for the separation space. This concept yields Eq. 9.1, where n_c is the peak capacity, t_g is the gradient time, and w is the 4σ peak width.

$$n_c \cong \frac{t_g}{w} \qquad (9.1)$$

However, if under a particular set of separation conditions no peaks appear in the last half of the gradient time, then one can reasonably argue that the latter half of the gradient time is not used and therefore should not be considered as part of the separation space when estimating the peak capacity. This yields a more conservative estimate of the peak capacity, represented by Eq. 9.2, where $t_{R,first}$ and $t_{R,last}$ are the retention times of the first and last peaks of interest in the chromatogram.

$$n_c' = \frac{t_{R,last} - t_{R,first}}{w} \qquad (9.2)$$

There seems to be more agreement concerning the other element of the peak capacity calculation in the 1D case, namely the estimation of peak widths. This issue has been discussed extensively recently, and details can be found in excellent reviews and articles [13,15]. Once one arrives at estimates for the peak width and the size of the separation space, the limits of 1D peak capacity can be estimated as a function of analysis time, which is particularly important to the discussion presented in this chapter. Wang and coworkers [17,18] have made such estimates for high-performance 1D separations of peptides (<3000 Da molecular weight) on the time scale of about an hour. Supported by experimental results obtained using instrumentation with a 400 bar pressure limit, the authors found an upper limit to the peak capacity of about 500 in 2 h. We can expect that increasing both the available pressure and the analysis time will increase the achievable peak capacity, but these dependencies are relatively weak, with just fourth root dependencies on both the pressure and analysis time [13,19]. The prospects for smaller molecules (i.e., <500 Da compounds) are even less encouraging, with peak capacity limits on the order of a few hundred in 2 h using existing technology [15] and reasonable conditions. Thus, if we are conservative we can generalize the limitations of current gradient elution 1DLC by stating that peak capacities exceeding 1000 are not achievable within analysis times of a few hours [7].

Bearing in mind the limitations discussed above for 1D separations, the upper panel of Figure 9.1 shows the principal advantage of a so-called comprehensive 2D separation over a 1D one, that the separation space is extended into a second dimension, allowing further separation of compounds that are not resolved by the first dimension separation (e.g., see the middle peak in the 1D separation). The differences between comprehensive 2D separations, denoted hereafter as LC × LC, and other modes of 2D separation will be discussed in more detail below. The great advantage of LC × LC over 1D separations is expressed in terms of peak capacity by Eq. 9.3, where $n_{c,2D}$, ^1n_c, and ^2n_c are the peak capacities of the 2D separation and the constituent 1D first and second dimensions, respectively [20–23].

$$n_{c,2D} = {}^1n_c \times {}^2n_c \qquad (9.3)$$

The LC × LC peak capacity is derived from the product of the constituent 1D peak capacities rather than the sum, which is what gives the LC × LC method so much potential resolving power. It is important at this stage to realize that the LC × LC peak capacity expressed by Eq. 9.3 is a theoretical maximum and that practically achievable LC × LC peak capacities are frequently smaller due to a number of factors, as described in Sections 9.1.4 and 9.1.5.

The comparison of the resolving power of 1DLC and LC × LC is of paramount practical importance; such comparisons require departures from Eq. 9.3 to be taken into account. It is useful here, though, to discuss our current understanding of the relative performance of 1DLC and LC × LC. We are only aware of two head-to-head experimental comparisons of this kind, one of which compares 1D gas chromatography (GC) and GC × GC [24,25]. Our view is that at very short analysis times (i.e., less than a few minutes), 1DLC will always be superior to LC × LC, largely because second dimension separations are too slow to adequately sample very narrow first dimension peaks (see Section 9.1.4). Conversely, at very long analysis times (i.e., greater than an hour), LC × LC will always be superior to 1DLC because the increase in the resolving power of 1DLC stagnates quickly in the region of tens of minutes, and the added separation of the second dimension quickly multiplies the resolving power of an existing 1DLC separation (see Eq. 9.3). Based on our own experimental results [25,26] we find that the range of analysis times where LC × LC first becomes superior to 1DLC is in the range of about 5 to 15 min for separations of low molecular weight metabolites (e.g., less than about 1000 Da). This transition time certainly is influenced by the nature of the analytes being separated and a number of experimental factors that govern performance (e.g., pressure, temperature, particle size), but this variation will largely be captured by the time range stated here. Of course, the immediate practical consequences of this realization are clear; that is, it is desirable to use conventional 1DLC methods for very rapid analyses, and LC × LC for very long analyses where high resolving power is needed. The more desirable method in the middle ground (ca. 10 min to 1 h) will be determined on a case by case basis and will be influenced by factors including robustness and ease of data analysis.

At this point it is particularly useful to understand how data are collected and formatted in 2DLC experiments, both as a basis for appreciation of the different modes of 2DLC and the different ways in which the data are represented. Figure 9.2 from one of the early and highly instructive reviews of GC × GC by Beens and coworkers [27] is reproduced in this chapter, showing the evolution of 2D data from the point of collection through reformatting and different modes of visual representation. Most of the figure is self-explanatory, but we emphasize the point that in the best case in comprehensive 2D separations, each peak appearing in the first dimension separation is sampled and transferred multiple (typically two or three) times across its width so that each chemical component shows up in multiple consecutive second dimension chromatograms. The intensities of these multiple appearances form the basis for the peak profiles in both dimensions of the 2D chromatograms, whether they are represented as contour plots or three-dimensional (3D) surface plots.

9.1.3   Modes of 2D Separation

All implementations of 2DLC involve a transfer of a portion of first dimension (^{1}D) column effluent to the second dimension (^{2}D) column for further separation. The mechanics and hardware involved in this process will be discussed in detail in Section 9.3. However, it is useful as an introduction here in order

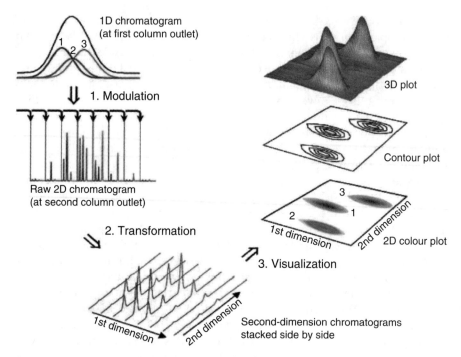

Figure 9.2. Depiction of the flow of information as it is collected and analyzed in a comprehensive 2DLC experiment. First dimension sampling or modulation is shown in step 1, where fractions of ^{1}D effluent are collected, transferred, and fractions are separated sequentially by a ^{2}D column. Step 2 shows how the sequential 2D chromatograms are reformatted into a 2D chromatogram. Visualization in step 3 depicts how the 2D data set can be represented as either a 2D contour plot or a 3D surface plot. Reprinted from Reference 27 with permission from Elsevier.

to discuss in conceptual terms the ways in which this can be carried out, both physically and in terms of the frequency of the transfer process.

9.1.3.1   Offline Fraction Transfer

The simplest approach to transferring fractions of ^{1}D effluent to a ^{2}D column is to collect the effluent in vials or other containers (e.g., well plates) as it comes out of the ^{1}D column, and then reinject those fractions into the ^{2}D column after some time has passed and/or after some processing steps (e.g., solvent exchange). The two most significant advantages of the offline approach are the simplicity of operation (i.e., sophisticated coupling of the two dimensions is not required) and the high absolute peak capacities that can be achieved [28]. However, these advantages come at the cost of increased total analysis time compared with online methods [7], and a lack of robustness due to the potential for sample contamination during the fraction transfer process [29].

9.1.3.2   Online Fraction Transfer

The online approach to fraction transfer involves the use of an interface between the two dimensions of a 2D system that provides a means of transferring fractions of ^{1}D effluent to the ^{2}D column either in real time (i.e., the two columns are actually serially coupled for a short time), or with a slight delay time (e.g., a few seconds to a few minutes). In the case where the transfer is time delayed, the delay is much shorter relative to the overall analysis time than it is in the offline case. Online transfers are done in sealed fluidic systems such that the ^{1}D effluent fraction is not exposed to the atmosphere and only experiences minimal exposure to connecting tubing and/or trapping columns.

9.1.3.3   Conceptual Comparison of Different 2D Separation Modes

In addition to characterizing 2D methods according to the method of fraction transfer (offline/online), it is also convenient to characterize them by the frequency of fraction transfer. In practice, the number of transfers has varied from a single transfer up to several hundred transfers in a single 2D separation [30]. 2D separations involving just one or a small number of transfers are typically referred to as "heartcutting" (LC-LC) methods, whereas methods involving tens or hundreds of transfers executed at regular intervals, usually across the entire ^{1}D separation, are usually referred to as "comprehensive" (LC × LC) 2D separations. Murphy and Schure [31] have suggested that separations involving less than 10 transfers should be referred to as heartcutting, and more than ten transfers as comprehensive. These modes of separation are represented graphically in Figure 9.3A and Figure 9.3C, respectively. In between these extremes lie two other methods, namely "stop and go" [7] and "selective comprehensive" (sLC × LC), a method introduced by us recently [32,33] and represented graphically in Figure 9.3B.

The heartcutting mode of separation is the simplest to execute in practice (see Section 9.3 for details), has the longest history of use dating to the 1970s in LC [34] and the late 1950s in GC [35], and has been extended to three dimensions of separation [36,37]. Figure 9.3A shows that this approach is particularly effective for highly targeted analysis, where the additional resolving power provided by the ^{2}D separation can be leveraged to separate one or a few target compounds from other matrix constituents that are not resolved by the ^{1}D separation alone. In these cases, the rest of the ^{1}D separation is ignored or provides the only source of chromatographic information about compounds eluting outside of the heartcut window.

The comprehensive mode of separation provides the largest potential for chromatographic resolution of hundreds of compounds. Figure 9.3C shows that the resolution of two adjacent peaks in the 2D separation plane can often be attributed to some separation in both the first and the second dimensions of the system. As discussed in Section 9.1.4, the relationship between the

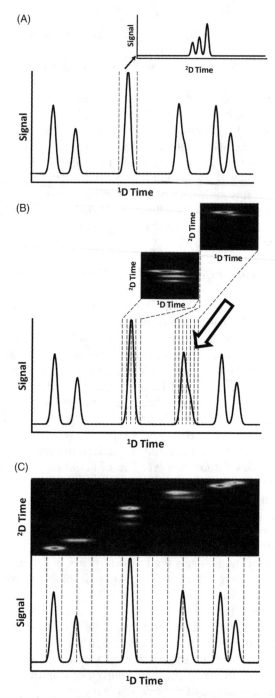

Figure 9.3. Diagrams for the three different 2D separation modes. Panel A shows the "heartcutting" mode (LC-LC), where targeted regions of the ^{1}D chromatogram are collected in a single fraction and sent to the ^{2}D separation, where a single second 1D chromatogram results. Panel B describes the "selective comprehensive" (sLC × LC) method where again targeted regions of the ^{1}D chromatogram are sampled, but in a "comprehensive" fashion resulting in a 2D chromatogram for the selected region. Panel C shows the "fully comprehensive" (LC × LC) method where every portion of the ^{1}D chromatogram is sent to the ^{2}D column for separation. Note the subtle differences between the contour plots of Panels B and C. The high sampling rate of sLC × LC methodology maintains more of the resolution achieved in the ^{1}D column than does the LC × LC method.

duration of each ^{2}D separation, the width of ^{1}D peaks prior to sampling, and the width (in time) of the sampling window influences the effective contribution of the ^{1}D separation to the overall 2D resolution. This topic has been studied in detail over the last decade, and we now have a good understanding of the influence of these variables on the quality of LC × LC separations (see Section 1.4 for details).

The ultimate value of the selective comprehensive mode of separation is not yet clear because it was introduced relatively recently, and there are a large number of variables that need to be explored before we can appreciate the ultimate role of sLC × LC relative to conventional 1D and LC × LC separations. Some of the potential advantages over 1D and LC × LC methods are communicated by way of subtle differences in Figure 9.3A–C. One of the fundamental problems associated with the heartcutting approach is that the action of collecting a single fraction of ^{1}D effluent inevitably recombines the target analyte(s) with other constituents of the sample matrix that were separated by the ^{1}D separation prior to the sampling step. This problem is mitigated somewhat in the LC × LC approach, where first dimension peaks are typically sampled multiple times across their width, which results in the appearance of an individual sample constituent in multiple consecutive second dimension chromatograms. From a theoretical perspective (see Section 1.4) sampling ^{1}D peaks multiple times is at least potentially beneficial and under most conditions improves the effectiveness of the overall 2D separation as measured by peak capacity. However, in practice the slow speed of ^{2}D separations invariably forces us to make a compromise between sampling frequency and the peak capacity of each ^{2}D separation. We can increase the peak capacity of each ^{2}D separation by increasing the analysis time, but this in turn corresponds to lower ^{1}D sampling frequencies. This compromise is represented visually in Figure 9.3C, where each dashed vertical line in the lower pane indicates the transfer of a fraction of ^{1}D effluent to the second dimension. Recognizing these limitations of the heartcutting and LC × LC approaches sets the stage for the sLC × LC approach. Figure 9.3B shows that in this approach, the sampling frequency can be increased substantially compared with an LC × LC separation of the same sample because the separation of a set of collected fractions of ^{1}D effluent is deferred for a short time and executed in the intervening time window between sampling events. This approach allows *both* high-frequency sampling of the ^{1}D separation *and* high-resolution ^{2}D separation of each ^{1}D fraction. Comparison of Figure 9.3B and Figure 9.3C for the separation of the multiplet indicated by the arrow in Figure 9.3B shows the value of a high sampling frequency, which preserves most of the separation achieved in the first dimension prior to sampling, whereas in the LC × LC case there is some remixing of those constituents and thus a loss of resolution. Although the limits of sLC × LC are not yet established, there certainly will be a threshold in terms of the number of target analytes beyond which it will be more efficient and experimentally far simpler to use the LC × LC approach.

9.1.4   Undersampling

The process we refer to as undersampling is particularly problematic in LC × LC and is a practical factor that effectively leads to not obtaining the theoretical peak capacities predicted by Eq. 9.3. Although undersampling is certainly relevant in modes of 2D separation other than LC × LC, its impact is most easily understood in the context of LC × LC and will be discussed here as such. Appreciating the significance of undersampling requires a clear understanding of the discrete steps involved in executing an LC × LC separation (described here in terms of an online LC × LC separation):

1. The first dimension separation is executed in a manner very similar to that of a conventional 1D separation.
2. As a peak elutes from the ^{1}D column, a portion of the ^{1}D effluent is collected in some kind of sampling interface where it is stored prior to reinjection into the ^{2}D column.
3. The stored portion of ^{1}D effluent is reinjected into the ^{2}D column for further separation of sample constituents that were nominally not separated by the ^{1}D separation.

The concept of undersampling is concerned with the fact that in step 2 of this process, sample constituents that were previously separated by the ^{1}D column are *recombined* during the transfer of the ^{1}D effluent fraction to the ^{2}D column. From the perspective of resolving power this is clearly a step backwards, and the impact on the quality of the LC × LC separation is negative. In all real LC × LC separations this negative effect is offset by the positive impact of further separation of sample constituents by the ^{2}D column that were previously unresolved by the ^{1}D column. The quantification of the magnitudes of these negative and positive impacts on the resolving power of 2D separations has been studied in detail, and the salient outcomes are summarized below.

Before discussing the quantitative impact of undersampling on 2D separation performance, it is instructive to visually examine the effect of undersampling by looking at a reconstructed representation of the ^{1}D chromatogram that simulates the effect of undersampling on the performance of the ^{1}D separation. Figure 9.4 shows a comparison of two scenarios:

1. Hypothetical LC × LC separation in which the transfer of effluent fractions from the first to the second dimension is very fast (<1 s), and a large number of these fractions is transferred over the course of a full LC × LC separation.
2. More practically realistic case where the sampling time (t_s) is four times as large as the standard deviation of a ^{1}D peak ($^1\sigma$) [38].

The contour plots show the two corresponding LC × LC separations of the same four-component mixture, and the lower profiles show the effective ^{1}D chromatograms, where the lower right profile is a reconstructed one that accounts for the negative impact of constituent remixing during the effluent fraction transfer process.

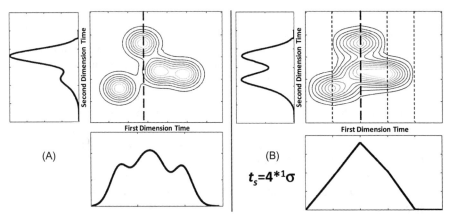

Figure 9.4. Comparison of ideally sampled (A) and undersampled LC × LC separations (B). In (A) the lower panel shows a simulated ^{1}D separation of a four constituent mixture where there is a high degree of overlap and only three peaks are observed. Immediately above this is the corresponding 2D separation, which shows that the addition of the second dimension separation helps resolve the mixture into four distinct peaks. The upper left chromatogram in (A) shows the ^{2}D separation of two sample constituents not resolved in the ^{1}D separation that are resolved somewhat by the second dimension, where the dashed line indicates the time at which the ^{1}D separation was sampled. In (B) the lower panel shows a reconstructed ^{1}D chromatogram that accounts for the effect of undersampling (two samples taken in the time corresponding to the 8σ width of the first dimension peak prior to sampling; the sampling intervals are indicated by the finely dashed lines), which severely broadens the ^{1}D peaks. Here, only three maxima are observed in the 2D chromatogram, and the loss of resolution due to undersampling shows up primarily in the ^{1}D contribution to the overall 2D separation. The upper left panel in (B) shows that undersampling has no effect on the width of ^{2}D peaks but that each ^{2}D separation is more crowded due to remixing of constituents that had been separated by the first dimension prior to sampling. Reprinted from Reference 38 with permission from Springer.

The relationship between t_s and $^1\sigma$ is the critical determinant of the quantitative effect of undersampling on the peak capacity of LC × LC separations. This relationship has been studied extensively by several groups [39–42], beginning with Murphy, Schure, and Foley [39], and the state of our understanding in this area was reviewed recently [9]. Equation 9.4 shows the relationship between an average ^{1}D peak broadening factor, $<\beta>$, and the sampling time (t_s) relative to the widths of ^{1}D peaks ($^1\sigma$) prior to sampling. Equations 9.5 and 9.6 then quantitatively express the impact of undersampling on the corrected ^{1}D and LC × LC peak capacities.

$$\langle\beta\rangle = \sqrt{1 + 0.21\left(\frac{t_s}{{}^1\sigma}\right)^2} \qquad (9.4)$$

$$^1 n'_c = \frac{{}^1 n_c}{\langle\beta\rangle} \qquad (9.5)$$

$$n'_{c,2D} = \frac{{}^1 n_c}{\langle\beta\rangle} \times {}^2 n_c \qquad (9.6)$$

These expressions make it clear that we should take all reasonable steps to minimize <β> by minimizing the sampling time relative to the widths of ^{1}D peaks. There are, however, significant technical and practical limitations to this notion. For example, in many cases gradient elution separations are used in the ^{2}D of LC × LC systems, and the time required to re-equilibrate the pumping system and HPLC column after a rapid solvent gradient imposes a lower boundary on the shortest sampling times that can be used with existing instrumentation. This kind of practical constraint becomes particularly problematic as ^{1}D peaks become very narrow, as is the case with short overall analysis times (<30 min) and separations executed at high temperatures (>80°C) and/or high pressures (>400 bar). Under these conditions, LC × LC can be inferior to conventional 1D separations, in spite of the attractiveness of the product rule, as discussed earlier (see Eq. 9.1). A few careful experimental comparisons of 1DLC and LC × LC have been carried out [25,26]. It is interesting to note that a similar problem exists in GC × GC, albeit for different physical reasons than in LC × LC [24].

Finally, it is important to recognize that while Eqs. 9.4–9.6 suggest that peak capacity in LC × LC can be maximized by minimizing <β>, the process of optimization is really far more complicated than simply minimizing <β>. The underlying issue is that while Eqs. 9.4–9.6 show that $n'_{c,2D}$ increases with decreasing <β>, thus decreasing t_s, it is also true that ^2n_c will increase as t_s is increased, at least under optimized conditions. Thus, these two effects directly oppose each other, and the art of optimization is primarily concerned with the factors that control the rate of change of $^1n'_c$ and ^2n_c with changes in t_s. These details are beyond the scope of this chapter; fortunately, several groups are currently engaged in excellent work in this area [26,41,43,44], and interested readers are referred to their work for further insight.

9.1.5   Orthogonality

There has been considerable attention paid to the process of choosing which separation modes (e.g., reversed-phase [RP], ion exchange [IEX], HILIC) to use as the basis for 2DLC separations since the initial emphasis by Giddings and Guiochon [21,23] on the importance of this issue to the overall effectiveness of 2D methods. A comprehensive discussion of this topic is well beyond the scope of this chapter, and the reader is referred to other sources for a more complete view [9,45]. Here we briefly introduce the concept of orthogonality and again emphasize its importance to provide a basis for evaluating the effectiveness of 2D separations involving at least one HILIC separation. While it has been argued that the term "orthogonality" is not the best word to describe the extent of correlation between retention mechanisms as originally discussed by Giddings, the term is so firmly established in the literature that we choose to continue its use here.

Figure 9.5 effectively communicates the concept and importance of orthogonality in the context of 2D separations. Panels A and B represent the best

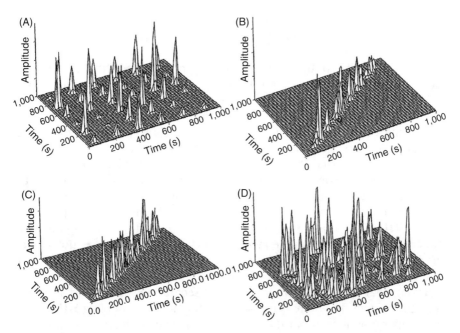

Figure 9.5. Simulated 2D chromatograms used to estimate the dimensionality of LC × LC separations. Panel A describes a perfectly orthogonal separation with a dimensionality, D, of 2.00. Panel B shows the result of a completely ordered nonorthogonal chromatogram with a dimension of 1.00 resulting from the use of the identical columns in each dimension. The 2D chromatogram shown in panel C was the result of a chromatogram simulated using random peak coordinates, but with a high correlation between the first and second dimensions, yielding a dimensionality of 0.71. Panel D shows a random distribution of peaks in both dimensions, with a dimensionality of 1.78. Reprinted from Reference 45 with permission from Elsevier.

and worst case scenarios, where the retention mechanisms of the two separation modes are either weakly correlated (i.e., highly orthogonal) and lead to spreading of peaks across the 2D separation plane, or strongly correlated, in which case, addition of the 2D separation does not add much, if any, resolving power to the existing ^{1}D separation. A great deal of effort has been dedicated to both measuring or assessing the "degree of orthogonality" of a given pair of separations, and exploring the effectiveness of different stationary phase combinations in different application contexts. The reader is referred to a recent book chapter [9] that provides a highly detailed comparison of different metrics of orthogonality that range from simple to complex, and employ concepts from far-ranging fields including animal ecology. As discussed in Sections 9.2 and 9.4, one of the most significant potential advantages of 2DLC separations involving the HILIC mode of separation is the potential for a high degree of orthogonality when combined with other commonly used modes including RPLC, and we are beginning to see this advantage realized in certain application areas.

9.2 COMPLEMENTARITY OF HILIC SELECTIVITY TO OTHER SEPARATION MODES

In a landmark study of the relative chromatographic selectivity of several different modes for peptide separations, Gilar and coworkers [46] collected and compared retention data for 196 peptides separated under different conditions including size-exclusion (SEC), reversed-phase (RP, at pH 2.6 and 10), HILIC, and strong cation exchange (SCX). Figure 9.6 shows the orthogonality of several possible pairings of separation modes. We see that there are several good combinations and that the combination of a HILIC separation with a RP separation is among the best of them. To our knowledge, this study is the most comprehensive assessment of the orthogonality of different mode combinations to date for a single set of analytes or samples. A recent pairing of RP and HILIC separations in offline LC × LC separations of the polar constituents of traditional Chinese medicines (TCM) [47] is an exceptional demonstration of the potential of this type of approach for small molecules as well. Figure 9.7 shows the excellent distribution of flavonoid and glucoside peaks, achieving nearly 100% coverage of the 2D separation space. At this point in time it is evident that there is tremendous potential for coupling HILIC with other separation modes, but that we also have much to learn with respect to this option. A recent review by Jandera [11] captured much of the activity in the previous few years and discussed applications of 2DLC involving at least one HILIC separation in a number of areas including peptides and proteins, plant extracts, biopolymers (e.g., oligonucleotides), and pharmaceutical and biomedical analyses. As our understanding of the retention mechanisms of different stationary phases suitable for HILIC separations progresses (see Chapter 1 in this volume), we will find ourselves in an increasingly improved position to choose the best HILIC phase and complementary separation mode for a particular application.

9.3 INSTRUMENTATION AND EXPERIMENTAL CONSIDERATIONS

As noted earlier, sampling techniques in 2DLC separations are divided into two main categories: comprehensive (denoted LC × LC) and heartcutting or coupled-column methodologies (denoted LC-LC). The conceptual differences between these modes in terms of their operation, in both offline and online modes, were discussed in Section 9.1.3 and are not repeated here. Each combination of modes (e.g., offline LC × LC) has distinct advantages and disadvantages, and each has been used in a variety of applications. The goal of this section is to familiarize the reader with the most important practical considerations for implementation of 2DLC separations involving a HILIC separation in at least one of the dimensions. The reader is referred to the books and reviews in References 1–8 for greater detail on these instrumental considerations.

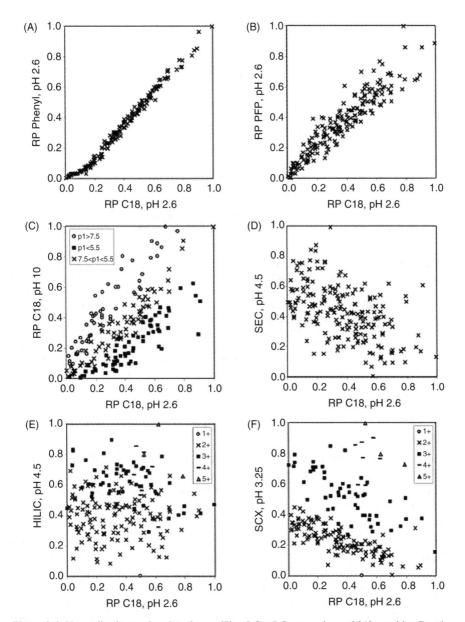

Figure 9.6. Normalized retention data from offline LC × LC separations of 243 peptides. Panels A, B, and C each show the combination of two RP columns (A: RP C18, pH 2.6 × RP Phenyl; B: RP C18, pH 2.6 × RP PFP; C: RP C18, pH 2.6 × RP C18, pH 10). Panels D, E, and F show the results of combining an RP (C18, pH 2.6) separation with SEC (pH 4.5), HILIC (pH 4.5), and SCX (pH 2.6). The RP (C18, pH 2.6) × HILIC combination was determined to be the most orthogonal combination of separation modes for this sample mixture. Reprinted with permission from *Anal. Chem.* 2005; **77**: 6426–6434. Copyright (2005) American Chemical Society.

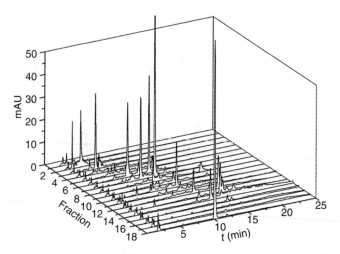

Figure 9.7. Offline HILIC × RP separation of an extract of a traditional Chinese medicine. The mobile phases used in the first dimension were water with 0.1% (v/v) formic acid (A solvent) and acetonitrile (B solvent), with a linear gradient of 5% A to 40% A in 30 min. Fractions were collected manually from 2 to 21 min at 1-min intervals, evaporated to dryness under a nitrogen stream, and reconstituted in 200 μL of methanol/water (50:50, v/v). The second dimension analysis was performed on an XUnion C18 column (150 × 2.1 mm i.d., 5 μm particle size, Sipore Co. Ltd.). The mobile phases used in the two dimensions were same. The linear gradient in the second dimension was from 10% B to 30% B in 40 min, and then reached 65% B at 50 min. UV detection was performed at 280 nm. Reprinted from Reference 47 with permission from Elsevier.

9.3.1   Online versus Offline 2DLC

While both offline and online 2DLC methods are reviewed in the applications Section 9.4, practical issues related to online methods are the primary focus of this section. The added complexity of the instrumentation required to both automate and synchronize the sample transfer process with the timescale of the ^{1}D separation is the basis for this greater depth of discussion. Offline LC × LC and LC-LC methods, while tremendously effective at generating high peak capacities in the LC × LC case and offering exceptional analysis flexibility to obtain impressive resolution of target compounds, are also more time-consuming and labor-intensive. Online methods are more user-friendly with respect to the reproducibility of replicate separations, but far more experience and expertise is required of the analyst in terms of instrument design, construction, and time required for method development.

9.3.1.1   Offline 2DLC

Offline 2DLC methods offer increased analysis flexibility over online methods because they may be executed with as little as "... a bottle located at the outlet of column where the zone(s) of interest is collected" [31]. Values for corrected peak capacity in offline LC × LC have been predicted to approach 10,000 to 15,000, but this value comes at the expense of analysis time on the order of hours to days [13,48,49]. When the increase in the peak capacity *production* (peak capacity per unit time) of offline methods is compared with online methods, the gain is only marginal. Recently, it has been reported that fully optimized online methods are capable of producing peak capacities of nearly 2500 in 1 h [30]. Traditionally the pairing of orthogonal separation modes has been limited by the sample transfer method interfacing the two dimensions of an online system. For example, normal phase (NP) and RP chromatographic modes have long been seen as a desirable combination from an orthogonality perspective, but their combination in an online system is not easily achieved because of the immiscibility of the mobile phases normally used in these systems. In offline 2DLC systems, additional sample treatment steps may be performed on each fraction prior to injection into the ^{2}D column (e.g., evaporation, reconstitution, dilution with weak ^{2}D solvent), allowing a greater degree of optimization of the ^{1}D and ^{2}D separations than can be reasonably achieved in online systems.

The key to achieving high peak capacities in offline 2DLC is optimization of both the ^{1}D and ^{2}D chromatographic conditions and the ^{1}D sampling rate. Offline 2DLC allows for the independent optimization of each of the dimensions of separation, including the key variables of solvent composition program, column temperature, column length, and mobile phase type, composition, and flow rate. Moreover, the most important perceived benefit of using offline LC × LC has been described as the lack of a "time constraint" [48]. Three significant advantages of offline 2DLC follow from this:

1. Offline separation allows for the use of a ^{2}D separation that can yield higher absolute efficiency than those typically used in online systems.
2. There is no inherent upper or lower bound on the number of fractions that must be collected from the ^{1}D separation.
3. One HPLC instrument can be used for the entire offline 2D separation rather than two that are necessary for online methods.

The first advantage is related to the optimization of each individual separation. The most significant obstacles to highly productive online 2DLC separations are solvent incompatibility between dimensions and short ^{2}D analysis times. Offline 2DLC provides simple solutions to each problem, which are highlighted above in Section 9.1.3. Second, in online separations short ^{2}D separation times are necessitated by short ^{1}D sampling times, adversely affecting the ^{2}D separation peak capacity due to its dependence on analysis time. The relationship between these two variables effectively sets an upper limit on the number of fractions of ^{1}D effluent that can be collected, which in turn limits overall 2D peak capacity. Offline 2DLC breaks this link, allowing the analyst the flexibility to have high ^{1}D sampling rates to mitigate the effects of undersampling while increasing ^{2}D analysis time as a means of producing higher ^{2}D and overall 2D peak capacity.

The added versatility and flexibility of offline methods makes them slower, less reproducible, and more labor-intensive than optimized online methods. The following example is used to illustrate the point that as the ^{1}D sampling time is made smaller in LC × LC separations, the number of fractions collected increases dramatically. If a 30-min ^{1}D separation is used with a ^{1}D sampling time of 1 min, 30 ^{1}D fractions are collected, handled by the operator, and injected into the ^{2}D system. When this sampling time is reduced to 5 s the number of manipulated samples increases from 30 to 360, resulting in a larger number of fractions that must be collected, handled, processed (e.g., solvent evaporation), reinjected, and analyzed on the ^{2}D system. Each of these processes adds to the overall analysis time and subjects samples to possible contamination. From this simple example, it becomes evident that increasing the ^{1}D sampling rate has a potentially devastating effect on the overall analysis time for offline LC × LC separations, which must be considered when establishing the goals of the analysis [48]. LC-LC separations are far less susceptible to the large amounts of time required for sample handling due to the relatively low numbers of samples handled per analysis, but are still susceptible to sample loss and contamination during the transfer process.

9.3.1.2 *Online 2DLC*

Online LC × LC and LC-LC methods are often hardware-intensive and require more judicious selection of compatible separation modes (RP, NP, IEX, SEC, HILIC, etc.) than offline methods. With this added instrumental complexity comes the benefit of increased robustness and automation. When interfacing two chromatographic instruments in an online 2DLC system, the "bridge" between the dimensions changes from the operator in an offline system to a sampling valve. In this case, the compatibility of the mobile phases used in the two dimensions is a higher priority because there is less time available to adjust the composition of the ^{1}D effluent fractions prior to injection into the ^{2}D column.

LC × LC. To achieve online LC × LC separations, fractions of ^{1}D effluent must be collected, stored, and injected into the ^{2}D column tens of times over the duration of the ^{1}D separation. Various combinations of six-, eight-, and ten-port, two-position valves have been used to execute these steps; we refer to this type of valve and its components as the "interface" between the first and second dimension separations. We discuss the basic operation of systems involving the 8- and 10-port designs in this section, and refer the reader to Cohen and Schure [1] and Mondello [4] for detailed descriptions of sampling valve designs that have been used for various applications.

The ability to make repeated, defined volume transfers of ^{1}D effluent to the ^{2}D column requires that the interface is equipped with two (or more) equal volume sample loops or trap columns. This allows one loop/trap to be filled with ^{1}D effluent while the other is emptied into the ^{2}D column. At the conclusion

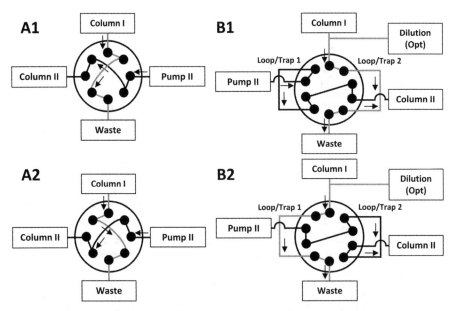

Figure 9.8. Schematics of the valve and flow configurations used in LC × LC experiments. Panels A1 and A2 show the two valve positions and corresponding flow paths for an 8-port valve that allow the effluent from Column 1 to be loaded into one loop/trap while the effluent contained by the other loop/trap is injected into Column 2. Panels B1 and B2 describe the analogous flow paths for a 10-port two-position valve. Arrows indicate the flow direction through each loop/trap. The ^{1}D effluent flow paths are shown in gray. Inspection of the flow paths through each valve shows that the flow through the black loop of A1 is opposite to that in which it is filled in A2. This back-flushing of one loop in an eight-port valve is not a problem in the 10-port configurations in B1 and B2.

of the ^{2}D separation, the sampling valve actuates to transfer the contents of the first loop/trap into the ^{2}D column while the second is loaded with ^{1}D effluent; this process is repeated for the duration of the ^{1}D separation at specified time intervals determined by the operator, referred to here as the sampling time, t_s.

In the dual sample loop setup, valve selection is limited to either an 8- or 10-port valve design, the configurations of which are described in this section and shown in Figure 9.8. In Figure 9.8 and Figure 9.9 it is implicit that two independent pumping systems are used to deliver solvent to the ^{1}D and ^{2}D columns. A single detector can be used at the outlet of the ^{2}D column, and a second detector is occasionally placed inline between the outlet of the ^{1}D column and the sampling valve. For online LC × LC the use of a 10-port, two-position valve design is most common [50]. When using a six-port valve, two valves are required to model the flow path of a single 10-port valve. It has been shown that the use of an eight-port valve does not allow for both sample loops to be loaded by the ^{1}D column and be injected into the ^{2}D column with the same direction of mobile phase flow. In the eight-port design, one of the sample

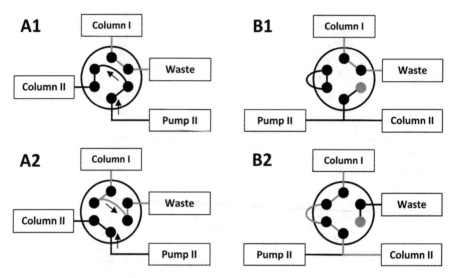

Figure 9.9. Valve diagrams used to interface the two dimensions of heartcutting 2DLC systems. Panels A1 and A2 show flow paths corresponding to the two positions of the valve as used by Apffel and coworkers. Panels B1 and B2 show the configuration used by Rogatsky and coworkers. The ^{1}D flow paths are shown in gray and arrows are used to indicate the direction of flow through the ^{2}D injection loop of A1 and A2. The gray circles of B1 and B2 indicate the use of a plug to prevent diversion of flow away from Column 2 in B1.

loops is loaded and injected into the ^{2}D column in the same direction while the second loop is loaded in one direction and emptied in the opposite direction (i.e., back flushed) into the ^{2}D column. The effect of this flow path has been shown to lead to significant retention time differences that depend on the loop from which a particular fraction is injected [51]. Back-flushing is not needed in a 10-port valve because each loop is loaded and injected in the same direction, resulting in more consistent peak shapes and ^{2}D retention times.

Sample loop volume is an important experimental parameter that needs to be sensibly determined. Appropriately sized loops for systems employing the dual-loop interface are selected by considering the volume of ^{1}D effluent that is collected during the sampling time, which is equal to the ^{2}D cycle time. In the case of isocratic elution conditions, the ^{2}D cycle time is equal to the ^{2}D analysis time, but in the case of gradient elution, the cycle time is the sum of the ^{2}D gradient time, ^2t_g, and the ^{2}D re-equilibration time. The re-equilibration time depends strongly on both the ^{2}D column dead volume and the time required to flush the fluidic components of the pumping system in preparation for a subsequent solvent gradient delivery. These issues have been described in detail elsewhere [52,53]. For example, with a ^{2}D cycle time of 30s and ^{1}D flow rate of 100μL/min, the minimum loop volume is 50μL. The key to selection of appropriate loop volumes is that they are large enough to hold the volume supplied by the ^{1}D column, but not too large as to contribute greatly to extra-column broadening of ^{2}D peaks.

As an alternative to the dual-loop interface, two trap columns packed with a stationary phase may be fitted in place of the loops. The trap configuration acts identically to the loops, where one trap is loaded by the ^{1}D column effluent while the second loop is injected into the ^{2}D column. The advantages of the trap approach are twofold: (1) there is more flexibility with respect to sampling time because in principle the trap can effectively collect and store analyte from a wide range of fraction transfer volumes, and (2) traps can potentially refocus analytes eluting from the ^{1}D column as a means of reducing the effective transfer volume and concomitant peak broadening in the second dimension.

Post-^{1}D trapping was first demonstrated in 1991 when Oda and coworkers used column switching to resolve enantiomers of the drug verapamil [54]. The advantage of this method was that the incorporation of the trap column and dilution line allowed the fraction of interest to be separated under optimal ^{1}D and ^{2}D mobile phase conditions. Use of trap columns in LC × LC separations has increased more recently; Holm and coworkers used traps in a system to quantify neuropeptides in rat brain samples [55]. In this work, an SCX ^{1}D column operated with a high fraction of acetonitrile (ACN) in the mobile phase was diluted with a 0.1% trifluroacteic acid in water after the ^{1}D separation to effectively reduce the amount of ACN in the mobile phase. The water dilution allowed refocusing of the analyte band in the two RP (C18) trap columns prior to separation on the ^{2}D RP (C18) column. In later work by the same authors, two trap columns were used in the interface between an SAX ^{1}D column with an RP ^{2}D column for the analysis of proteins in human saliva [56]. Continued work with traps was performed by Cacciola et al., who used an identical 10-port interface configured with traps on two different occasions for the RP × RP analysis of phenolic and flavone antioxidants [57,58]. They used RP (C18) traps that exhibited higher retention than the ^{1}D polyethylene glycol column allowing the trap columns to refocus the ^{1}D analyte band prior to reinjection into the ^{2}D column. Analytes were back-flushed from the traps in a small volume of mobile phase into one of two different RP (monolithic C18 or carbon clad zirconia) ^{2}D columns. Most recently, Li and coworkers utilized a HILIC ^{1}D separation with the same 10-port trap configuration and an ion-pairing RP ^{2}D separation for the analysis of mixtures of di- to deca-oligonucleotides [59]. In this case, a high aqueous "make-up flow" was introduced after the ^{1}D column but before the RP (C18) trap to dilute the organic-rich ^{1}D effluent to promote effective trapping.

The use of trap columns was extended from just two traps to 18 traps by Wilson et al., who developed an automated system for the analysis of bradykinin in rat muscle tissue dialysate and neuropeptides in rat cerebral spinal fluid dialysate dialysate [60,61]. This system was composed of 18 trap columns connected to two 10-port column selection valves that provided short-term storage of analytes eluted from the ^{1}D column, which in turn allowed for ^{2}D separation analysis times (52 min) that were much longer than the ^{1}D sampling time (3 min). As a result of the disconnect between ^{1}D sampling time and ^{2}D analysis time, this instrument configuration has been described as "an auto-mated offline system" [4]. With these features Wilson et al. were able to inter-face a ^{1}D HILIC separation with an RP ^{2}D column with RP trap columns. The high ACN concentration of the ^{1}D effluent was reduced with a water dilution line prior to analyte refocusing on the RP trap columns.

LC-LC. Previously proposed interface designs for heartcutting 2DLC are much simpler than those for their comprehensive counterparts because the number of repeated ^{1}D effluent fractions injected into the ^{2}D column is significantly lower. In the simplest sense, the sampling valve interfacing the dimensions of a LC-LC instrument acts as a gate directing the path of ^{1}D flow from waste to the ^{2}D column for a short period of time, long enough to quantitatively send the region of interest in the ^{2}D column. Then the gate closes, directing ^{1}D flow back to waste. Since its introduction by Apffel and Majors, online LC-LC has been achieved in this manner simply by directing ^{1}D effluent to the ^{2}D column using a single six-port two-position valve [62]. The disadvantage of this setup, shown in panels A1 and A2 of Figure 9.9, is that the direction of flow used to fill the sample loop is the opposite of that used to empty it. Back-flushing the loop effectively remixes components partially resolved by the ^{1}D separation.

Later, Rogatsky and Stein introduced a new flow path for a single six-port valve eliminating the back-flushing and analyte remixing of the sample loop [63]. Panels B1 and B2 of Figure 9.9 show how the two configurations of this setup direct ^{1}D column effluent either to waste or to the ^{2}D column directly via a mixing tee located after the valve interfacing the ^{1}D column with the ^{2}D pumping system. The inherent problem with this method was the increase in pressure experienced by the ^{1}D pump during the sample transfer process. This flow path puts significant instrumental pressure limits on the overall separa-tion such that the flow rate, column temperature, and particle size of each dimension must be chosen so the increase in pressure drop across both columns during the sampling period does not exceed the limit of the ^{1}D pump. In addi-tion to this practical pressure problem, changes in flow path with large differ-ences in pressure have been shown to cause flow inconsistencies and/or even backflow of mobile phases into the pumps themselves [64].

In our own LC-LC work we have borrowed sample transfer concepts from existing LC × LC methodology [37]. This sampling method used two six-port valves setup identically to the configuration used in the LC × LC work of Stoll et al. to mimic the function of a single 10-port valve (see panel B of Fig. 9.8) [65]. The twin six-port valves were equipped with two 75-μL sample loops that were alternately loaded by the ^{1}D column and emptied by the ^{2}D pumping system. The length of the ^{1}D sampling window was dictated by the number of 75 μL fractions collected and emptied into the head of the ^{2}D column operated under initial conditions of low elution strength for target compounds in the ^{2}D column. The advantage to this use of a LC × LC sampling method in LC-LC separations is that it eliminated the flow inconsistency problems characteristic of Rogatsky's approach.

9.3.2   Solvent Incompatibility

When selecting separation modes for use in a 2DLC system, the concept of orthogonality between dimensions must be a top priority if the true potential of 2D separations is to be realized [21,23,66]. Envisioning a 2DLC system composed of two highly orthogonal separation mechanisms does not pose a significant problem [9,23,46,67], but integration of these two modes into an efficient separation can be exceedingly difficult, particularly for online systems [4]. For example, consider an online LC × LC separation based on an NP ^{1}D separation followed by an RP ^{2}D separation (NP × RP). NP separations are typically operated with a nonaqueous mobile phase composed primarily of hexane or chlorinated solvents and a bare silica stationary phase, while the ^{2}D column uses a conventional C18 RP column operated with a mobile phase composed of ACN and water. Coupling these two separation modes presents two challenging problems that must be overcome prior to their incorporation into an effective online 2DLC system:

1. Solvent immiscibility: hexane and water are not miscible at concentrations relevant to 2DLC.
2. Solvent strength: hexane is an exceedingly strong RP solvent, much stronger than ACN.

Solvent immiscibility has been shown to cause detector perturbations or "bumps" in NP × RP systems using ultraviolet (UV)-visible absorbance detectors [68–70]. The second problem is due to the fact that although hexane is a "weak" solvent in NP separations, it is an extremely "strong" solvent in RP separations; thus the transfer of up to 100 μL fractions of ^{1}D effluent high in hexane concentration would have a devastating effect on peak shape in the ^{2}D separation. This effect is due to the high initial solvent strength of the hexane in the injection plug sent into the RP system resulting in minimal initial retention at the head of the ^{2}D column.

9.3.2.1   Partial Mobile Phase Evaporation

Due to these technical difficulties associated with pairing NP and RP separations, applications of online 2DLC systems involving these modes have required various solutions to the mobile phase solubility problem. One approach involves an interface utilizing partial evaporation of the solvent during the fraction transfer step by applying a vacuum to the sample loop [71]. Poor analyte recovery levels on the order of 10–50% were observed using this approach. Recently, an updated vacuum-assisted dynamic evaporation interface for an NP × RP system was evaluated, but only for high boiling point compounds [72]. The poor analyte recovery characteristic of this approach makes it impractical for routine work.

9.3.2.2   Fraction Transfer Volume Relative to the Second Dimension Column Volume

A less complicated solution to counteract the effects of solvent incompatibility between dimensions is to make the ^{2}D injection smaller than the ^{2}D column volume. This may be done either by selection of appropriately sized columns or through active post-first-dimension flow splitting. Selection of a small or capillary ^{1}D column (e.g., less than 1.0 mm i.d.) operated at flow rates of μL/min, paired with conventionally sized 2.1 or 4.6 mm i.d. ^{2}D columns minimizes the solvent incompatibility issue by effectively reducing the ^{2}D injection volume to less than 5 μL [59,68,69,73]. Injection volumes on this scale are commonly used in 1DLC, with little regard for injection solvent composition because of the small injection volume compared with the column volume. The disadvantage of this approach is that it suffers from added sample dilution and poorer detection limits by using a larger ^{2}D column i.d.[74]. Postcolumn flow splitters have been used extensively in GC × GC since their inception in 1991 but have only recently been applied to 2DLC [75]. The active flow splitter works to accurately limit the fraction of ^{1}D effluent transferred to the ^{2}D column, allowing more optimal, higher ^{1}D flow rate conditions to be used [30]. Reduction in transferred sample volume is achieved through the use of an actual pump to precisely and accurately divert flow to waste at a predetermined rate rather than relying on the more variable rates of passive flow-splitting systems. Unfortunately, this method again serves to decrease sensitivity, but the authors argue it is a small price to pay for the added resolving power of optimized separations and subsequent increase in peak capacity.

9.3.2.3   On-Column Focusing

Many elaborate and creative solutions have been proposed to solve the problem of introducing fractions of ^{1}D effluent with high solvent strengths into the ^{2}D column, all with the goal of achieving high-efficiency separations in both dimensions through the use of "on-column focusing" at the head of the ^{2}D column. Since the inception of online 2DLC in the late 1970s the benefits of on-column focusing have been known as a means to counteract the dilution of analyte bands through the course of a multidimensional separation [62,76]. The chromatographic process inherently results in dilution of the analyte band and the extent of the dilution increases with increasing analysis time. Thus, dilution and analysis time negatively affect the limits of detection for low-concentration compounds in 1DLC, and even more so for 2DLC if nothing is done to counteract this process [74]. Recently, on-column focusing has received increased attention as a means to counteract the dilution of ^{1}D analyte bands during injection into the ^{2}D column [28,43].

Achieving on-column focusing in the ^{2}D is directly related to the problem of solvent compatibility through the strength of the ^{1}D effluent relative to the ^{2}D mobile phase. The following solutions have been proposed: trapping and back-flushing [55,61,77], reverse osmosis [7], careful choice of column order [5,78], and dilution with weak solvent [33,37,55,79–81].

Reverse osmosis has been suggested as a means of reducing the ^{1}D effluent volume prior to ^{2}D injection by forcing, under pressure, the ^{1}D effluent through a semi-permeable membrane [7]. To the best of our knowledge, there have been no experimental reports utilizing this methodology.

Previously, we advocated the use of RP separations in both dimensions to effectively eliminate the solvent immiscibility problem observed in both LC-LC and LC × LC systems [5,37]. Several groups have discussed and demonstrated the effectiveness of placing the more retentive column in the ^{2}D of a RP × RP system [5]. This solution provides effective on-column focusing because compounds generally require less organic modifier for elution from the ^{1}D column than the ^{2}D one. Injection of ^{1}D effluent into the ^{2}D column under these conditions effectively concentrates analytes at the head of the ^{2}D column until the injection solvent is flushed from the ^{2}D column and the mobile phase reaches a composition that results in elution. The disadvantage of this approach is that the number of columns that both satisfy this requirement of very different retentivities, and are orthogonal, is quite limited. In an interesting variation on the use of RP columns in both dimensions, parallel ^{1}D and ^{2}D gradients have been proposed as an additional solution to the problem of achieving suitable retention in both columns because the levels of organic modifier in each separation are more closely matched over the course of the entire 2D separation [10,58,78,82].

The use of a dilution line tee-ed into the outlet of the ^{1}D column has only recently been found useful to achieve on-column focusing, although it was demonstrated two decades ago [54]. In our own work, we have relied on the effectiveness of a water (weak elution solvent) dilution of ^{1}D column effluent in a setup involving three very different RP columns applied to targeted 3DLC separations [33,37]. Water dilution has also been found effective for the separation of peptides by LC × LC [79]. More recently, an ACN dilution line was incorporated into a 2DLC system using RP in the ^{1}D and HILIC in the ^{2}D for the analysis of 11 peptides in plasma [80]. In each case, this approach efficiently lowered the elution strength of the transfer fraction solvent through the addition of a weak eluent component.

In our most recent work involving dilution of the fraction transfer solvent, we were able to show qualitatively and quantitatively that dilution of ^{1}D effluent with water decreases the peak width of ^{2}D peaks even when the transfer volume is larger than the dead volume of the column itself (e.g., 75 µL injection into a 30 mm × 2.1 mm i.d. column) [33].

9.3.3 Fast Separations

In Section 9.1.4 we discussed the problem of undersampling in multidimensional separations. Whenever the same type of separation is used in the two dimensions of a 2D system (e.g., LC and LC, GC and GC) the second dimension separation will always be too slow compared with the first, and anything that can be done to improve the speed of the second separation will mitigate undersampling and improve the overall productivity of the 2D separation. Development of ultrafast 1DLC separations has and will continue to play an important role in the development of LC × LC due to the nature of LC × LC separation methodology [83], as these separations are characterized by a relatively long, slow ^{1}D separation that is sampled tens or hundreds of times followed by the same number of ^{2}D separations. Another way of looking at this situation is to say that the ^{2}D separation is the rate determining step in the overall 2D analysis [5]. Even modest improvements in the speed of the ^{2}D separation can have a big effect on the overall 2D separation because the productivity of the ^{2}D separation affects the 2D peak capacity ($n_{c,2D}$) through both the ^{2}D peak capacity (^2n_c) and the undersampling terms ($<\beta>$) in Eq. 9.6.

9.3.3.1   General Considerations for Fast LC Separations

In a recent review article [83] we summarized developments in the theory and practice of fast LC, with a focus on the most recent decade. The fundamental driving force underlying nearly all of the most recent developments is focused on the goal of decreasing the time required for analyte transfer between the mobile and stationary phases. However, this goal manifests itself in a variety of experimental approaches. We continue to see reductions in particle size and changes in particle morphology, such that superficially porous particles in the 2- to 3-micron diameter range are quickly becoming one of the most commonly used materials, in particular for fast separations. The use of elevated column temperature (e.g., 50–200°C) improves speed by decreasing mobile phase viscosity, which both allows the use of higher eluent velocities at a given pump pressure limit and increases analyte diffusivity such that high separation efficiencies are maintained at these high velocities. In the last decade we have observed a step change in the pressure limits of pumps used for analytical LC separations, from the long-standing 6000 psi limit to the 15 to 20,000 psi range. This new capability encourages the use of smaller particles and longer columns for faster and more efficient separations.

One of the unique aspects of HILIC separations, as compared with the RP mode, is that the lower viscosities of the commonly used mobile phases (typically ACN-rich) in HILIC separations have an effect on separation speed that is similar to that of increasing the column temperature. That is, reduced eluent viscosity allows the use of higher eluent velocities and longer columns for faster and more efficient separations. The combination of HILIC and high temperature conditions, while largely unexplored at this point, should be an interesting and fruitful area of research in the years to come.

9.3.3.2   Fast HILIC Separations

Aside from the complementary selectivity of the HILIC mode to other common modes such as RP, one of the greatest advantages of incorporating

HILIC into an $LC \times LC$ system is the possibilities that it offers for fast ^2D separations. It has long been known that elevating the temperature of a mobile phase composed of ACN and water will markedly decrease its viscosity, particularly at large fractions of ACN [84]. By definition, HILIC separations are operated with a high fraction ACN in the mobile phase and therefore make the most of the decrease in viscosity. This allows the use of flow rates higher than those used with similar columns operated under RP conditions due to the smaller pressure drop across the column. The pressure drop across packed bed HILIC columns operated under fast conditions has been on par with silica monolithic columns [85]. Recently, Chauve and coworkers compared the dynamic characteristics of three commercially available HILIC phases. They observed that very flat van Deemter C-terms were observed for some of them; however, they did not actually report the fitting coefficients to allow comparison with RP phases under similar conditions [86].

9.4   APPLICATIONS

A search in SciFinder Scholar for the terms "HILIC" and "two-dimensional liquid chromatography" yields a first hit in the year 2005 [46]. Since then, interest in the topic has increased year over year with about 10 publications in each of the past three years. A comprehensive listing of articles describing applications of 2DLC involving a HILIC separation in at least one of the dimensions is given in Table 9.1. At this time the applications are quite diverse and include the analysis of foods and beverages, peptides and oligonucleotides, TCMs, plant materials, animal tissues and biofluids, and polymeric materials. Readers interested in one or more of these specific areas will find quick access to the articles of interest in Table 9.1. Here, we briefly highlight a few of these applications that specifically demonstrate the advantages of 2DLC involving HILIC separations. In these examples, the incompatibility of ACN-rich and water-rich eluents (in terms of solvent strength in HILIC vs. RPLC) is overcome either by injecting relatively small volumes of ^1D effluent into the ^2D column, by diluting the ^1D effluent with weak solvent, and/or by using trapping columns prior to the ^2D analytical column.

9.4.1   TCM

In a recent qualitative study of flavonoid and glucoside components of *Scutellaria barbata* D. Don (Barbed Skullcap), Liang and coworkers [47] carried out offline $LC \times LC$ separations using either two different HILIC phases, or one HILIC and one RP phase. Samples were prepared by first extracting the TCM with water followed by solid-phase extraction with the Oasis HLB material and elution with either 10% or 80% methanol to give "polar" and "medium-polar" fractions of the sample. Figure 9.10 shows a high degree of orthogonality for both systems, but particularly for the HILIC × RP

Table 9.1. Survey of 2DLC Separations Involving the HILIC Mode of Separation

Mode Combinations	Columns	Mobile Phases	Applications	Ref.
Comprehensive (LC × LC)				
Offline				
HILIC × RP	Bare silica × C18	Gradient acetonitrile and ammonium formate × gradient formic acid and acetonitrile	MassPREP protein tryptic digestion standards	[46]
SEC × HILIC	Superdex Peptide HR × Ultra IBD	Isocratic ammonium acetate × gradient methanol and ammonium acetate	Nickel species in plant tissue	[89]
HILIC × RP	ZIC-HILIC × C18	Gradient acetonitrile and formic acid × gradient ammonium acetate and acetonitrile	Proteomics	[90]
GPC × HILIC	Sephadex G-10 × TSKgel Amide-80	Isocratic formic acid and gradient acetonitrile and trifluoroacetic acid (TFA)	g-Glutamyl peptides in edible beans	[91]
RP × HILIC	C18 × Click β-cyclodextrin (CD)	Gradient formic acid and acetonitrile × gradient acetonitrile and formic acid	Traditional Chinese medicine	[92]
RP × HILIC	C18–Atlantis HILIC Silica	Gradient TFA and acetonitrile–gradient acetonitrile and formic acid and TFA	Peptides in milk hydrolysates	[93]
HILIC × RP	Nomura Diol-100 × C18	Gradient acetonitrile and methanol, acetic acid × gradient formic acid and acetonitrile	Oligomeric procyanidins	[94]
HILIC × HILIC	Click maltose × TSKgel Amide	Gradient acetonitrile and formic acid × gradient acetonitrile and ammonium acetate	Traditional Chinese medicine	[95]

SCX × HILIC	Luna SCX × Luna HILIC	Gradient ammonium formate and acetonitrile × gradient acetonitrile and ammonium formate	Water-soluble metabolites	[96]
RP × HILIC	C18 × β-CD	Gradient formic acid and acetonitrile × gradient acetonitrile and formic acid	Whole grass extract	[97]
HILIC × RP	Develosil Diol-100 × C18	Gradient acetonitrile and methanol, acetic acid × gradient acetonitrile and formic acid	Green-tea phenolics	[98]
HILIC × RP	ZIC-HILIC × C18	Gradient acetonitrile and ammonium acetate × gradient formic acid and acetonitrile	Urinary proteomics	[99]
RP × HILIC	C18–Click β-CD	Gradient formic acid and acetonitrile–gradient acetonitrile and formic acid	Bufadienolides	[100]
HILIC × RP	ZIC-HILIC × C18	Gradient acetonitrile, acetic acid, and ammonium acetate × gradient acetic acid and acetonitrile	Complex proteome samples	[101]
HILIC × HILIC and HILIC × RP	Atlantis HILIC Silica × XAmide; XAmide × C18	Gradient acetonitrile and ammonium formate × gradient acetonitrile and ammonium formate; gradient acetonitrile and formic acid × gradient formic acid and acetonitrile	Traditional Chinese medicine	[47]

(Continued)

293

Table 9.1. (*Continued*)

Mode Combinations	Columns	Mobile Phases	Applications	Ref.
Online				
RP × HILIC	C18 × Separon SGX NH$_2$	Gradient water and acetonitrile × isocratic dichloromethane and ethanol	Ethylene oxide-propylene oxide (co)oligomers	[73]
HILIC × RP	ZIC-HILIC × PLRP-S (polymeric)	Gradient acetonitrile and ammonium acetate × gradient formic acid and acetonitrile	Bradykinin in rat muscle tissue	[61]
HILIC × RP	ZIC-HILIC × C18	Gradient acetonitrile and ammonium acetate × gradient acetic acid and acetonitrile	Neuropeptides in rat brains	[60]
HILIC × HILIC	TSKgel Amide × PolyHydroxyethyl A	Gradient acetonitrile and formic acid × step gradient acetonitrile and ammonium acetate, acetic acid	Saponins from *Quillaja saponaria*	[102]
RP × HILIC and HILIC × RP	C18 × Luna HILIC; Luna HILIC × C18	Isocratic acetone and water × gradient water and acetone; isocratic acetone and water × isocratic acetone and water	Nonionic surfactants	[103]
HILIC × IP-RP	Ascentis Silica × C18	Gradient acetonitrile and ammonium formate × gradient triethylamine–acetate and acetonitrile	di- to deca-oligonuceotides	[59]

294

Heartcutting (LC-LC)

Mode	Stationary phase	Mobile phase	Application	Ref.
HILIC-RP	Atlantis HILIC Silica–C18	Gradient acetonitrile and ammonium formate–gradient water and acetonitrile	Hydrophilic and hydrophobic standards	[104]
RP-HILIC	C18-SeQuant ZIC-pHILIC	Gradient ammonium acetate and acetonitrile–gradient acetonitrile and ammonium acetate	Derivitized free estrogens and metabolites	[105]
HILIC-RP	Atlantis HILIC Silica–C18	Gradient acetonitrile and ammonium formate–gradient water and acetonitrile	Traditional Chinese medicine	[106]
IP-RP-HILIC	C18 × Atlantis HILIC Silica	Gradient TFA and acetonitrile × gradient acetonitrile and TFA	Peptides in rat K_3EDTA plasma	[80]
HILIC-RP	SeQuant ZIC-HILIC—C18	Gradient acetonitrile and ammonium acetate–gradient ammonium acetate and acetonitrile	Pharmaceuticals and polar metabolites	[107]

Serial Coupling

Mode	Stationary phase	Mobile phase	Application	Ref.
HILIC-RP	Atlantis HILIC Silica–C18	Isocratic acetonitrile and formic acid, ammonium formate and gradient ammonium formate, formic acid and acetonitrile	Tasponglutide in human and animal plasma	[108]
IEX × HILIC	TSKgel DEAE-TSKgel Amide/ZIC-HILIC	Gradient acetonitrile and ammonium acetate	Neutral and silayted N-glycans	[109]
RP-HILIC	C18–Ascentis Si	Gradient ammonium formate and acetonitrile–stepped acetonitrile and ammonium formate	Polar pharmaceuticals	[110]

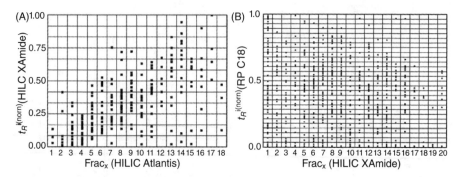

Figure 9.10. Comparison of the orthogonality of HILIC × HILIC and HILIC × RP separations of polar (A) and medium-polarity (B) fractions of an aqueous extract of traditional Chinese medicine. Reprinted from Reference 47 with permission from Elsevier.

system. The utilization of the separation space approaches 100%, which is unprecedented in 2DLC.

9.4.2 Polymers

The use of 2DLC for separations of various types of polymers has a rich history and arguably has been one of the most successful application areas for 2DLC to date [87]. Historically these separations have frequently utilized SEC and NP separation modes and the organic rich mobile phases associated with these methods [73]. As discussed previously, coupling NP and RP separations is particularly problematic because of the immiscibility of the mobile phases typically used in these modes. Recently, Wu and Marriott [88] have demonstrated the effectiveness of pairing the HILIC and RP modes in an online LC × LC separation of alkylphenol polyethoxylates (APnEOs). Figure 9.11 shows LC × LC chromatograms for the separations of four different commercial APnEO mixtures that demonstrate the complementarity of the HILIC and RP modes for this type of analysis.

9.4.3 Oligonucleotides

With increasing interest in the use of oligonucleotides and related compounds as drug products, there is a need for the development of methodologies capable of resolving these materials and process-related impurities. Recently, Li and coworkers [59] have demonstrated the combination of HILIC and ion-pairing RPLC (IP-RP) in an online LC × LC separation of a 27 component mixture of oligonucleotide homologs. Figure 9.12 shows an LC × LC chromatogram obtained with negative electrospray ionization detection for the separation of a mixture of polydeoxycytidines, polydeoxythymidines, and polydeoxyadenosines (2–10 bases in each case), which displays good orthogonality and usage of the separation space. Projection of the retention times of the analytes onto

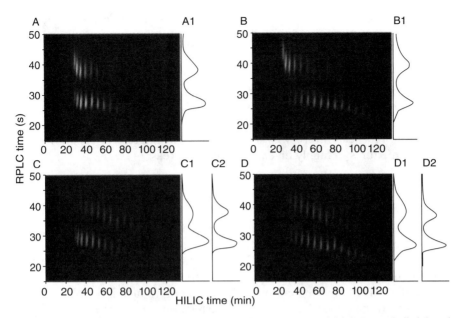

Figure 9.11. Online HILIC × RP separations of different commercial mixtures of alkylphenol polyethoxylates (APnEOs). An Ascentis Express HILIC column was used in the first dimension with an acetonitrile/water mobile phase, and an Ascentis Express C18 column was used in the second dimension with acetonitrile/water mobile phases. A-D are chromatograms for oligomer samples with varying numbers of ethoxy groups. Reprinted from Reference 88 with permission from Wiley.

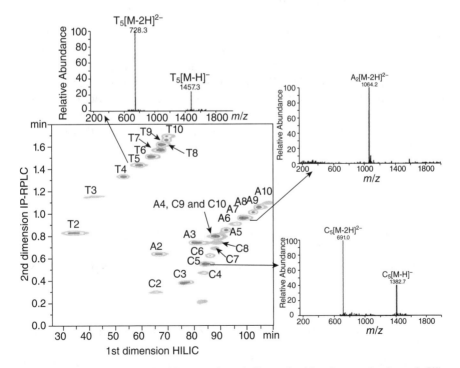

Figure 9.12. Online HILIC × IP-RP separation of oligonucleotide mixtures. An Ascentis Silica column was used in the first dimension with ammonium formate and acetonitrile mobile phases. An XBridge C18 column was used in the second dimension, with 0.1 M triethylamine acetate and acetonitrile mobile phases. The sample mixture consisted of three different homologous series of di- to deca-peptides. Reprinted from Reference 59 with permission from Elsevier.

either the HILIC or IP-RP axis shows that there is a high degree of overlap, particularly in the middle of the chromatogram, when separation of a mixture of this complexity is attempted using either of the 1D separations alone.

If these recent applications of 2DLC involving HILIC separations are any indication, the near future will continue to yield impressive separations of complex mixtures in highly diverse application areas.

9.5   THE FUTURE OF HILIC SEPARATIONS IN 2DLC

In our view, the future of 2DLC involving HILIC separations is bright. The inherent orthogonality of RP and HILIC phases for many types of separations makes coupling of these phases in 2D separations very attractive. We expect to continue to see the technique applied to more application areas with impressive results. Given the problems with the lack of reproducibility of NP separations over long periods (e.g., weeks), investigations of the reproducibility of 2DLC methods involving HILIC separations will be critical, particularly in applications such as metabolomics where reproducibility across the analysis of hundreds of samples is essential. With recent increases in the understanding of the retention mechanisms of different phases used in the HILIC mode, we are seeing more examples of the effective pairing of the HILIC mode with other HILIC separations or RP separations in 2DLC separations of analytes of high to moderate hydrophilicity. The added benefits of HILIC, including the potential for enhanced electrospray ionization compared with RP conditions, and the potential for very fast ^{2}D separations due to the low viscosities of typical HILIC eluents, only add to the attractiveness of 2DLC separations involving the HILIC mode. Although there have been a number of studies demonstrating the feasibility of using HILIC in the ^{1}D and/or ^{2}D of 2DLC systems, systematic studies are needed to establish practical guidelines for important variables including the use of trapping columns and dilution of the ^{1}D effluent prior to fraction transfer to the ^{2}D column.

REFERENCES

1. Cohen SA, Schure MR, eds. *Multidimentional Liquid Chromatography: Theory and Applications in Industrial Chemistry and the Life Sciences.* Hoboken, NJ: Wiley-Interscience; 2008.

2. Cortes H. *Multidimensional Chromatography: Techniques and Applications.* New York: Marcel Dekker; 1990.

3. Mondello L. *Multidimensional Chromatography.* West Sussex, England; New York: Wiley; 2002.

4. Mondello L, ed. *Comprehensive Chromatography in Combination with Mass Spectrometry.* Hoboken, NJ: Wiley; 2011.

5. Stoll DR, Li X, Wang X, Carr PW, Porter SEG, Rutan SC. Fast, comprehensive two-dimensional liquid chromatography. *J. Chromatogr. A* 2007; **1168**: 3–43.

6. Malerod H, Lundanes E, Greibrokk T. Recent advances in on-line multidimensional liquid chromatography. *Anal. Methods* 2010; **2**: 110–122.

7. Guiochon G, Marchetti N, Mriziq K, Shalliker R. Implementations of two-dimensional liquid chromatography. *J. Chromatogr. A* 2008; **1189**: 109–168.

8. Donato P, Cacciola F, Mondello L, Dugo P. Comprehensive chromatographic separations in proteomics. *J. Chromatogr. A* 2011; **1218**: 8777–8790.

9. Carr PW, Davis JM, Rutan SC, Stoll DR. Principles of online comprehensive multidimensional liquid chromatography. In Grushka E, Grinberg N, eds., *Advances in Chromatography*. Boca Raton, FL: CRC Press; 2012.

10. Jandera P, Hajek T, Cesla P. Effects of the gradient profile, sample volume and solvent on the separation in very fast gradients, with special attention to the second-dimension gradient in comprehensive two-dimensional liquid chromatography. *J. Chromatogr. A* 2011; **1218**: 1995–2006.

11. Jandera P. Stationary phases for hydrophilic interaction chromatography, their characterization and implementation into multidimensional chromatography concepts. *J. Sep. Sci.* 2008; **31**: 1421–1437.

12. Wang X, Barber WE, Long WJ. Applications of superficially porous particles: High speed, high efficiency or both? *J. Chromatogr. A* 2012; **1228**: 72–88.

13. Guiochon G. The limits of the separation power of unidimensional column liquid chromatography. *J. Chromatogr. A* 2006; **1126**: 6–49.

14. Dolan J, Snyder L, Djordjevic N, Hill D, Waeghe T. Reversed-phase liquid chromatographic separation of complex samples by optimizing temperature and gradient time, I. Peak capacity limitations. *J. Chromatogr. A* 1999; **857**: 1–20.

15. Neue U. Peak capacity in unidimensional chromatography. *J. Chromatogr. A* 2007; **1184**: 107–130.

16. Giddings JC. Maximum number of components resolvable by gel filtration and other elution chromatographic methods. *Anal. Chem.* 1967; **39**: 1027–1028.

17. Wang X, Stoll DR, Schellinger AP, Carr PW. Peak capacity optimization of peptide separations in reversed-phase gradient elution chromatography: Fixed column format. *Anal. Chem.* 2006; **78**: 3406–3416.

18. Wang X, Barber W, Carr P. A practical approach to maximizing peak capacity by using long columns packed with pellicular stationary phases for proteomic research. *J. Chromatogr. A* 2006; **1107**: 139–151.

19. Carr PW, Wang X, Stoll DR. Effect of pressure, particle size, and time on optimizing performance in liquid chromatography. *Anal. Chem.* 2009; **81**: 5342–5353.

20. Karger B, Snyder L, Horvath C. *An Introduction to Separation Science*. New York: Wiley-Interscience; 1973.

21. Guiochon G, Beaver LA, Gonnord MF, Siouffi AM, Zakaria M. Theoretical investigation of the potentialities of the use of a multidimensional column in chromatography. *J. Chromatogr. A* 1983; **255**: 415–437.

22. Guiochon G, Gonnord MF, Zakaria M, Beaver LA, Siouffi AM. Chromatography with a two-dimensional column. *Chromatographia* 1983; **17**: 121–124.

23. Giddings JC. Two-dimensional separations: Concept and promise. *Anal. Chem.* 1984; **56**: 1258A-70A-A-70A.

24. Blumberg L, David F, Klee M, Sandra P. Comparison of one-dimensional and comprehensive two-dimensional separations by gas chromatography. *J. Chromatogr. A* 2008; **1188**: 2–16.

25. Stoll DR, Wang X, Carr PW. Comparison of the practical resolving power of one- and two-dimensional high-performance liquid chromatography analysis of metabolomic samples. *Anal. Chem.* 2008; **80**: 268–278.

26. Huang Y, Gu H, Filgueira M, Carr PW. An experimental study of sampling time effects on the resolving power of on-line two-dimensional high performance liquid chromatography. *J. Chromatogr. A* 2011; **1218**: 2984–2994.

27. Adahchour M, Beens J, Vreuls RJJ, Brinkman UAT. Recent developments in comprehensive two-dimensional gas chromatography (GCxGC). *Trends Analyt. Chem.* 2006; **25**: 438–454.

28. Horvath K, Fairchild JN, Guiochon G. Detection issues in two-dimensional on-line chromatography. *J. Chromatogr. A* 2009; **1216**: 7785–7792.

29. Dugo P, Cacciola F, Kumm T, Dugo G, Mondello L. Comprehensive multidimensional liquid chromatography: Theory and applications. *J. Chromatogr. A* 2008; **1184**: 353–368.

30. Filgueira MR, Huang Y, Witt K, Castells C, Carr PW. Improving peak capacity in fast online comprehensive two-dimensional liquid chromatography with post-first-dimension flow splitting. *Anal. Chem.* 2011; **83**: 9531–9539.

31. Murphy R, Schure MR. Instrumentation for comprehensive multidimensional liquid chromatography. In Cohen SA, Schure MR, eds., *Multidimentional Liquid Chromatography: Theory and Applications in Industrial Chemistry and the Life Sciences.* Hoboken, NJ: Wiley-Interscience; 2008, p. 95.

32. Groskreutz SR, Swenson MM, Secor LB, Stoll DR. Selective comprehensive multi-dimensional separation for resolution enhancement in high performance liquid chromatography, Part II: Applications. *J. Chromatogr. A* 2012; **1228**: 41–50.

33. Groskreutz SR, Swenson MM, Secor LB, Stoll DR. Selective comprehensive multi-dimensional separation for resolution enhancement in high performance liquid chromatography, Part I: Principles and instrumentation. *J. Chromatogr. A* 2012; **1228**: 31–40.

34. Johnson EL, Gloor R, Majors RE. Coupled column chromatography employing exclusion and a reversed phase. A potential general approach to sequential analysis. *J. Chromatogr.* 1978; **149**: 571–585.

35. Simmons MC, Snyder LR. Two-stage gas-liquid chromatography. *Anal. Chem.* 1958; **30**: 32–35.

36. Murahashi T. Determination of mutagenic 3-nitrobenzanthrone in diesel exhaust particulate matter by three-dimensional high-performance liquid chromatography. *Analyst* 2003; **128**: 42–45.

37. Simpkins SW, Bedard JW, Groskreutz SR, Swenson MM, Liskutin TE, Stoll DR. Targeted three-dimensional liquid chromatography: A versatile tool for quantitative trace analysis in complex matrices. *J. Chromatogr. A* 2010; **1217**: 7648–7660.

38. Stoll DR. Recent progress in online, comprehensive two-dimensional high-performance liquid chromatography for non-proteomic applications. *Anal. Bioanal. Chem.* 2010; **397**: 979–986.

39. Murphy RE, Schure MR, Foley JP. Effect of sampling rate on resolution in comprehensive two-dimensional liquid chromatography. *Anal. Chem.* 1998; **70**: 1585–1594.

40. Seeley J. Theoretical study of incomplete sampling of the first dimension in comprehensive two-dimensional chromatography. *J. Chromatogr. A* 2002; **962**: 21–27.

41. Horie K, Kimura H, Ikegami T, Iwatsuka A, Saad N, Fiehn O, Tanaka N. Calculating optimal modulation periods to maximize the peak capacity in two-dimensional HPLC. *Anal. Chem.* 2007; **79**: 3764–3770.

42. Davis JM, Stoll DR, Carr PW. Effect of first-dimension undersampling on effective peak capacity in comprehensive two-dimensional separations. *Anal. Chem.* 2008; **80**: 461–473.

43. Vivo-Truyols G, van der Wal S, Schoenmakers PJ. Comprehensive study on the optimization of online two-dimensional liquid chromatographic systems considering losses in theoretical peak capacity in first- and second-dimensions: A pareto-optimality approach. *Anal. Chem.* 2010; **82**: 8525–8536.

44. Fairchild JN, Horvath K, Guiochon G. Approaches to comprehensive multidimensional liquid chromatography systems. *J. Chromatogr. A* 2009; **1216**: 1363–1371.

45. Schure MR. The dimensionality of chromatographic separations. *J. Chromatogr. A* 2011; **1218**: 293–302.

46. Gilar M, Olivova P, Daly AE, Gebler JC. Orthogonality of separation in two-dimensional liquid chromatography. *Anal. Chem.* 2005; **77**: 6426–6434.

47. Liang Z, Li K, Wang X, Ke Y, Jin Y, Liang X. Combination of off-line two-dimensional hydrophilic interaction liquid chromatography for polar fraction and two-dimensional hydrophilic interaction liquid chromatography-reversed-phase liquid chromatography for medium-polar fraction in a traditional Chinese medicine. *J. Chromatogr. A* 2012; **1224**: 61–69.

48. Horvath K, Fairchild JN, Guiochon G. Optimization strategies for off-line two-dimensional liquid chromatography. *J. Chromatogr. A* 2009; **1216**: 2511–2518.

49. Gilar M, Daly A, Kele M, Neue U, Gebler J. Implications of column peak capacity on the separation of complex peptide mixtures in single- and two-dimensional high-performance liquid chromatography. *J. Chromatogr. A* 2004; **1061**: 183–192.

50. Schoenmakers PJ, Vivo-Truyols G, Decrop W. A protocol for designing comprehensive two-dimensional liquid chromatography separation systems. *J. Chromatogr. A* 2006; **1120**: 282–290.

51. Vanderhorst A, Schoenmakers P. Comprehensive two-dimensional liquid chromatography of polymers. *J. Chromatogr. A* 2003; **1000**: 693–709.

52. Schellinger A, Stoll D, Carr P. High speed gradient elution reversed-phase liquid chromatography. *J. Chromatogr. A* 2005; **1064**: 143–156.

53. Stoll DR. Fast, comprehensive two-dimensional liquid chromatography. PhD Dissertation, Chemistry, University of Minnesota, 2007.

54. Oda Y, Asakawa N, Kajima T, Yoshida Y, Sato T. On-line determination and resolution of verapamil enantiomers by high-performance liquid chromatography with column switching. *J. Chromatogr. A* 1991; **541**: 411–418.

55. Holm A, Storbraten E, Mihailova A, Karaszewski B, Lundanes E, Greibrokk T. Combined solid-phase extraction and 2D LC-MS for characterization of the neuropeptides in rat-brain tissue. *Anal. Bioanal. Chem.* 2005; **382**: 751–759.

56. Pepaj M, Holm A, Fleckenstein B, Lundanes E, Greibrokk T. Fractionation and separation of human salivary proteins by pH-gradient ion exchange and reversed phase chromatography coupled to mass spectrometry. *J. Sep. Sci.* 2006; **29**: 519–528.

57. Cacciola F, Jandera P, Blahova E, Mondello L. Development of different comprehensive two dimensional systems for the separation of phenolic antioxidants. *J. Sep. Sci.* 2006; **29**: 2500–2513.

58. Cacciola F, Jandera P, Hajdu Z, Cesla P, Mondello L. Comprehensive two-dimensional liquid chromatography with parallel gradients for separation of phenolic and flavone antioxidants. *J. Chromatogr. A* 2007; **1149**: 73–87.

59. Li Q, Lynen F, Wang J, Li H, Xu G, Sandra P. Comprehensive hydrophilic interaction and ion-pair reversed-phase liquid chromatography for analysis of di- to deca-oligonucleotides. *J. Chromatogr. A* 2012; **1255**: 237–243. DOI: 10.1016/j.chroma.2011.11.062.

60. Mihailova A, Malerod H, Wilson SR, Karaszewski B, Hauser R, Lundanes E, Greibrokk T. Improving the resolution of neuropeptides in rat brain with on-line HILIC-RP compared to on-line SCX-RP. *J. Sep. Sci.* 2008; **31**: 459–467.

61. Wilson SR, Jankowski M, Pepaj M, Mihailova A, Boix F, Vivo Truyols G, Lundanes E, Greibrokk T. 2D LC separation and determination of bradykinin in rat muscle tissue dialysate with on-line SPE-HILIC-SPE-RP-MS. *Chromatographia* 2007; **66**: 469–474.

62. Apffel JA, Alfredson TV, Majors RE. Automated on-line multi-dimensional high-performance liquid chromatographic techniques for the clean-up and analysis of water-soluble samples. *J. Chromatogr. A* 1981; **206**: 43–57.

63. Rogatsky E, Stein DT. Two-dimensional reverse phase-reverse phase chromatography: A simple and robust platform for sensitive quantitative analysis of peptides by LC/MS. Hardware design. *J. Sep. Sci.* 2006; **29**: 538–546.

64. Rogatsky E, Braaten K, Cruikshank G, Jayatillake H, Zheng B, Stein DT. Flow inconsistency: The evil twin of column switching—Hardware aspects. *J. Chromatogr. A* 2009; **1216**: 7721–7727.

65. Stoll D, Cohen J, Carr P. Fast, comprehensive online two-dimensional high performance liquid chromatography through the use of high temperature ultra-fast gradient elution reversed-phase liquid chromatography. *J. Chromatogr. A* 2006; **1122**: 123–137.

66. Giddings JC. Concepts and comparisons in multidimensional separation. *J. High Resolut. Chromatogr.* 1987; **10**: 319–323.

67. Slonecker PJ, Li X, Ridgway TH, Dorsey JG. Informational orthogonality of two-dimensional chromatographic separations. *Anal. Chem.* 1996; **68**: 682–689.

68. Dugo P, Favoino O, Luppino R, Dugo G, Mondello L. Comprehensive two-dimensional normal-phase (adsorption)—Reversed-phase liquid chromatography. *Anal. Chem.* 2004; **76**: 2525–2530.

69. Dugo P, Skerikova V, Kumm T, Trozzi A, Jandera P, Mondello L. Elucidation of carotenoid patterns in citrus products by means of comprehensive normal-phase—reversed-phase liquid chromatography. *Anal. Chem.* 2006; **78**: 7743–7750.

70. Francaois I, de Villiers A, Sandra P. Considerations on the possibilities and limitations of comprehensive normal phase-reversed phase liquid chromatography (NPLC × RPLC). *J. Sep. Sci.* 2006; **29**: 492–498.

71. Tian H, Xu J, Xu Y, Guan Y. Multidimensional liquid chromatography system with an innovative solvent evaporation interface. *J. Chromatogr. A* 2006; **1137**: 42–48.

72. Ding K, Xu Y, Wang H, Duan C, Guan Y. A vacuum assisted dynamic evaporation interface for two-dimensional normal phase/reverse phase liquid chromatography. *J. Chromatogr. A* 2010; **1217**: 5477–5483.

73. Jandera P, Fischer J, Lahovska H, Novotna K, Cesla P, Kolarova L. Two-dimensional liquid chromatography normal-phase and reversed-phase separation of (co)oligo-mers. *J. Chromatogr. A* 2006; **1119**: 3–10.

74. Schure MR. Limit of detection, dilution factors, and technique compatibility in multidimensional separations utilizing chromatography, capillary electrophoresis, and field-flow fractionation. *Anal. Chem.* 1999; **71**: 1645–1657.

75. Liu Z, Phillips JB. Comprehensive two-dimensional gas chromatography using an on-column thermal modulator interface. *J. Chromatogr. Sci.* 1991; **29**: 227–231.

76. Erni F, Frei R. Two-dimensional column liquid chromatographic technique for resolution of complex mixtures. *J. Chromatogr. A* 1978; **149**: 561–569.

77. Trone M, Vaughn M, Cole S. Semi-automated peak trapping recycle chromatography instrument for peak purity investigations. *J. Chromatogr. A* 2006; **1133**: 104–111.

78. Jandera P, Cesla P, Hajek T, Vohralik G, Vynuchalova K, Fischer J. Optimization of separation in two-dimensional high-performance liquid chromatography by adjusting phase system selectivity and using programmed elution techniques. *J. Chromatogr. A* 2008; **1189**: 207–220.

79. Moore AW, Jorgenson JW. Comprehensive three-dimensional separation of peptides using size exclusion chromatography/reversed phase liquid chromatography/optically gated capillary zone electrophoresis. *Anal. Chem.* 1995; **67**: 3456–3463.

80. Liu A, Tweed J, Wujcik C. Investigation of an on-line two-dimensional chromatographic approach for peptide analysis in plasma by LC-MS-MS. *J. Chromatogr. B* 2009; **877**: 1873–1881.

81. Haefliger OP. Universal two-dimensional HPLC technique for the chemical analysis of complex surfactant mixtures. *Anal. Chem.* 2003; **75**: 371–378.

82. Cesla P, Hajek T, Jandera P. Optimization of two-dimensional gradient liquid chromatography separations. *J. Chromatogr. A* 2009; **1216**: 3443–3457.

83. Carr PW, Stoll DR, Wang X. Perspectives on recent advances in the speed of high-performance liquid chromatography. *Anal. Chem.* 2011; **83**: 1890–1900.

84. Li J, Carr PW. Accuracy of empirical correlations for estimating diffusion coefficients in aqueous organic mixtures. *Anal. Chem.* 1997; **69**: 2530–2536.

85. Appelblad P, Jonsson T, Jiang W, Irgum K. Fast hydrophilic interaction liquid chromatographic separations on bonded zwitterionic stationary phase. *J. Sep. Sci.* 2008; **31**: 1529–1536.

86. Chauve B, Guillarme D, Cleon P, Veuthey J-L. Evaluation of various HILIC materials for the fast separation of polar compounds. *J. Sep. Sci.* 2010; **33**: 752–764.

87. Rittig F, Pasch H. Multidimentional liquid chromatography in industrial applications. In Cohen S, Schure MR, eds., *Multidimensional Liquid Chromatography*. Hoboken, NJ: Wiley-Interscience; 2008.

88. Wu Z-Y, Marriott PJ. One- and comprehensive two-dimensional high-performance liquid chromatography analysis of alkylphenol polyethoxylates. *J. Sep. Sci.* 2011; **34**: 3322–3329.

89. Ouerdane L, Mari S, Czernic P, Lebrun M, Lobinski R. Speciation of non-covalent nickel species in plant tissue extracts by electrospray Q-TOFMS/MS after their isolation by 2D size exclusion-hydrophilic interaction LC (SEC-HILIC) monitored by ICP-MS. *J. Anal. At. Spectrom.* 2006; **21**: 676.

90. Boersema PJ, Divecha N, Heck AJR, Mohammed S. Evaluation and optimization of ZIC-HILIC-RP as an alternative MudPIT strategy. *J. Proteome Res.* 2007; **6**: 937–946.

91. Dunkel A, Koster J, Hofmann T. Molecular and sensory characterization of gamma-glutamyl peptides as key contributors to the kokumi taste of edible beans phaseolus vulgaris. *J. Agric. Food Chem.* 2007; **55**: 6712–6719.

92. Liu Y, Xue X, Guo Z, Xu Q, Zhang F, Liang X. Novel two-dimensional reversed-phase liquid chromatography/hydrophilic interaction chromatography, an excellent orthogonal system for practical analysis. *J. Chromatogr. A* 2008; **1208**: 133–140.

93. van Platerink CJ, Janssen H-GM, Haverkamp J. Application of at-line two-dimensional liquid chromatography–mass spectrometry for identification of small hydrophilic angiotensin I-inhibiting peptides in milk hydrolysates. *Anal. Bioanal. Chem.* 2008; **391**: 299–307.

94. Kalili KM, de Villiers A. Off-line comprehensive 2-dimensional hydrophilic interaction–reversed phase liquid chromatography analysis of procyanidins. *J. Chromatogr. A* 2009; **1216**: 6274–6284.

95. Liu Y, Guo Z, Feng J, Xue X, Zhang F, Xu Q, Liang X. Development of orthogonal two-dimensional hydrophilic interaction chromatography systems with the introduction of novel stationary phases. *J. Sep. Sci.* 2009; **32**: 2871–2876.

96. Fairchild JN, Horvath K, Gooding JR, Campagna SR, Guiochon G. Two-dimensional liquid chromatography/mass spectrometry/mass spectrometry separation of water-soluble metabolites. *J. Chromatogr. A* 2010; **1217**: 8161–8166.

97. Feng J-T, Guo Z-M, Shi H, Gu J-P, Jin Y, Liang X-M. Orthogonal separation on one beta-cyclodextrin column by switching reversed-phase liquid chromatography and hydrophilic interaction chromatography. *Talanta* 2010; **81**: 1870–1876.

98. Kalili KM, de Villiers A. Off-line comprehensive two-dimensional hydrophilic interaction × reversed phase liquid chromatographic analysis of green tea phenolics. *J. Sep. Sci.* 2010; **33**: 853–863.

99. Loftheim H, Nguyen TD, Malerod H, Lundanes E, Asberg A, Reubsaet L. 2-D hydrophilic interaction liquid chromatography-RP separation in urinary proteomics—Minimizing variability through improved downstream workflow compatibility. *J. Sep. Sci.* 2010; **33**: 864–872.

100. Liu Y, Feng J, Xiao Y, Guo Z, Zhang J, Xue X, Ding J, Zhang X, Liang X. Purification of active bufadienolides from toad skin by preparative reversed-phase liquid chromatography coupled with hydrophilic interaction chromatography. *J. Sep. Sci.* 2010; **33**: 1487–1494.

101. Di Palma S, Boersema PJ, Heck AJR, Mohammed S. Zwitterionic hydrophilicinteraction liquid chromatography (ZIC-HILIC and ZIC-cHILIC) provide high resolution separation and increase sensitivity in proteome analysis. *Anal. Chem.* 2011; **83**: 3440–3447.

102. Wang Y, Lu X, Xu G. Development of a comprehensive two-dimensional hydrophilic interaction chromatography/quadrupole time-of-flight mass spectrometry system and its application in separation and identification of saponins from *Quillaja saponaria. J. Chromatogr. A* 2008; **1181**: 51–59.

103. Abrar S, Trathnigg B. Separation of nonionic surfactants according to functionality by hydrophilic interaction chromatography and comprehensive two-dimensional liquid chromatography. *J. Chromatogr. A* 2010; **1217**: 8222–8229.

104. Wang Y, Lehmann R, Lu X, Zhao X, Xu G. Novel, fully automatic hydrophilic interaction/reversed-phase column-switching high-performance liquid chromatographic system for the complementary analysis of polar and apolar compounds in complex samples. *J. Chromatogr. A* 2008; **1204**: 28–34.

105. Qin F, Zhao Y, Sawyer M, Li X. Column-switching reversed phase-hydrophilic interaction liquid chromatography/tandem mass spectrometry method for determination of free estrogens and their conjugates in river water. *Anal. Chim. Acta* 2008; **627**: 91–98.

106. Wang Y, Lu X, Xu G. Simultaneous separation of hydrophilic and hydrophobic compounds by using an online HILIC-RPLC system with two detectors. *J. Sep. Sci.* 2008; **31**: 1564–1572.

107. Thomas Al, Deglon J, Steimer T, Mangin P, Daali Y, Staub C. On-line desorption of dried blood spots coupled to hydrophilic interaction/reversed-phase LC/MS/MS system for the simultaneous analysis of drugs and their polar metabolites. *J. Sep. Sci.* 2010; **33**: 873–879.

108. Heinig K, Wirz T. Determination of taspoglutide in human and animal plasma using liquid chromatography-tandem mass spectrometry with orthogonal column-switching. *Anal. Chem.* 2009; **81**: 3705–3713.

109. Deguchi K, Keira T, Yamada K, Ito H, Takegawa Y, Nakagawa H, Nishimura S. Two-dimensional hydrophilic interaction chromatography coupling anion-exchange and hydrophilic interaction columns for separation of 2-pyridylamino derivatives of neutral and sialylated *N*-glycans. *J. Chromatogr. A* 2008; **1189**: 169–174.

110. Louw S, Pereira AS, Lynen F, Hanna-Brown M, Sandra P. Serial coupling of reversed-phase and hydrophilic interaction liquid chromatography to broaden the elution window for the analysis of pharmaceutical compounds. *J. Chromatogr. A* 2008; **1208**: 90–94.

INDEX

Acetone as mobile phase solvent, 99,
148, 294
Acetylneuraminic acid, 67, 150, 196
Acetylsalicylic acid, 10, 72, 92, 102–103,
145–146
Acrylamide, 241–242
Acyclovir, 68, 140
Adenosine, 34, 63–64, 72–75, 242–243
triphosphate, 36, 61, 172
ADME (adsorption, distribution,
metabolism, excretion), 181, 220
ADP (adenosine diphosphate), 61, 172
Alanine, 207
Alkylphenol polyethoxylates, 296–297
Amanitin, 72
Amines, 4, 67, 159, 225
Amino acids, 6, 36–37, 55, 58–60, 70–71,
206–209, 222
chiral separations, 150, 158
Aminoglycoside antibiotics, 243
Aminomethyl pyridine, 94, 96
4-Amino salicylic acid, 10, 72
AMP (adenosine monophosphate), 61
Aniline(s), 35, 256
Anomerization, 4, 6, 54, 67, 150, 196–197
ANP, *see* Aqueous normal phase
Aniline, 58, 70, 243, 256
Antibiotics, 58, 70, 243, 257
Anti-oxidants, 150–152, 250
Antithrombin preparations, 150
Aqueous normal phase, 20, 61
Arginine, 123, 128, 147
Aromatic amines, 256
Arsenic compounds, 256

Ascorbic acid, 69, 145, 250
Aspartic acid, 207
Aspirin, *see* Acetylsalicylic acid
ATP (adenosine triphosphate), 36, 61,
172

Bamethane, 2
Bare silica, *see* Stationary phase, silica
Basic drugs, 5
BEH, *see* Stationary phase, silica,
ethylene bridged hybrids
Benzene, 14, 35
Benzidine, 256
Benzoic acid derivatives, 61, 70
Benzylamine, 11, 13, 18, 28–30, 34, 123
Benzyltriethylammonium, 23
Bioanalytical, 219–298
Biogenic amines, 241, 246
Biological screening, 170–174
Biomarkers, 171, 185
Biopharmaceutics, 181–183
Blood, 210, 219
Bradykinin, 69, 285, 294
Brimonidine tartrate, 184
Bromide, 122

CAD, *see* Detectors, aerosol, charged
aerosol
Caffeine, 11, 18, 21, 28, 34–35
Calcium, 123
Candidate selection, 169, 183–186
Capillary electrophoresis (CE), 118, 183,
197, 200, 221
Captisol, 146–148

Hydrophilic Interaction Chromatography: A Guide for Practitioners, First Edition.
Edited by Bernard A. Olsen and Brian W. Pack.
© 2013 John Wiley & Sons, Inc. Published 2013 by John Wiley & Sons, Inc.

CHEMICAL ANALYSIS

A SERIES OF MONOGRAPHS ON ANALYTICAL CHEMISTRY
AND ITS APPLICATIONS

Series Editor
MARK F. VITHA